T0234004

Biochar

Production, Characterization, and Applications

Urbanization, Industrialization, and the Environment

Series Editor

Ming Hung Wong

Biochar

Production, Characterization, and Applications

Edited by
Yong Sik Ok • Sophie M. Uchimiya
Scott X. Chang • Nanthi Bolan

CRC Press
Taylor & Francis Group
Boca Raton London New York

CRC Press is an imprint of the
Taylor & Francis Group, an **informa** business

CRC Press
Taylor & Francis Group
6000 Broken Sound Parkway NW, Suite 300
Boca Raton, FL 33487-2742

First issued in paperback 2020

© 2016 by Taylor & Francis Group, LLC
CRC Press is an imprint of Taylor & Francis Group, an Informa business

No claim to original U.S. Government works

ISBN-13: 978-1-4822-4229-4 (hbk)
ISBN-13: 978-0-367-65876-2 (pbk)

Library of Congress Cataloging-in-Publication Data

Biochar : production, characterization, and applications / editors, Yong Sik Ok, Sophie
    M. Uchimiya, Scott X. Chang, and Nanthi Bolan.
        pages cm
    Includes bibliographical references and index.
    ISBN 978-1-4822-4229-4 (alk. paper)
    1. Biochar. 2. Soil amendments. 3. Filters and filtration--Materials. I. Ok, Yong-sik,
1944-

    TP248.B55B5384 2015
    333.95'39--dc23   2015014707

Visit the Taylor & Francis Web site at
http://www.taylorandfrancis.com

and the CRC Press Web site at
http://www.crcpress.com

Contents

SECTION I   General Overview

Anushka Upamali Rajapaksha, Dinesh Mohan,
Avanthi Deshani Igalavithana, Sang Soo Lee, and Yong Sik Ok

Prasanna Kumarathilaka, Sonia Mayakaduwa,
Indika Herath, and Meththika Vithanage

SECTION II   Biochar Production and Characterization

Sophie M. Uchimiya

Fungai N.D. Mukome and Sanjai J. Parikh

SECTION III   Environmental Applications

Ramya Thangarajan, Nanthi Bolan, Sanchita Mandal,
Anitha Kunhikrishnan, Girish Choppala,
Rajasekar Karunanithi, and Fangjie Qi

SECTION IV Agronomic Applications and Climate Change Mitigation

SECTION V Future Prospects

Preface

Biochar is receiving worldwide interest because of its potential beneficial applications in bioenergy production, global warming mitigation, and sustainable agriculture. Unfortunately, many of the published biochar books are not suitable for use as textbooks in undergraduate and graduate courses. To help meet this need, this book provides the fundamentals of biochar, such as its concept, production technology, and characterization methods, with comprehensive examples for readers. Building on these fundamentals, this book provides examples of state-of-the-art biochar application technology in agronomy and environmental sciences, with detailed case studies. This book is designed as a textbook for graduate courses and as a handbook for practitioners and policy makers. *Biochar: Production, Characterization, and Applications* is written by internationally renowned scientists with different backgrounds.

Editors

Dr. Yong Sik Ok is a professor in the Department of Biological Environment, Kangwon National University (KNU), Chuncheon, Republic of Korea, and is currently serving as the director of the Korea Biochar Research Center in Korea. Dr. Ok has supervised more than 50 postgraduate students, post-doctoral fellows, and visiting scientists from many countries, including Korea, China, Saudi Arabia, Egypt, India, Pakistan, and Sri Lanka. In addition, he has published more than 250 peer-reviewed articles and received the KNU Excellent Research Award in recognition of his out-standing achievements. He is currently serving as an editor for *Environmental Geochemistry and Health* (coordinating editor), *Journal of Soils and Sediment* (subject editor), and *Canadian Journal of Soil Science* (associate editor). He has also served as a guest editor for *Bioresource Technology, Journal of Hazardous Materials, Chemosphere, Plant and Soil, Environmental Science and Pollution Research*, and *Journal of Soils and Sediments*. Other notable engagements include working as a visiting professor at the Department of Renewable Resources, University of Alberta, Edmonton, Canada; Faculty of Bioscience Engineering, Ghent University, Belgium; and Department of Civil and Environmental Engineering, Hong Kong Polytechnic University, Hong Kong. As a top researcher in his field, Dr. Ok has organized many international conferences, including, most recently, CLEAR 2014 (Korea) and Biological Waste 2014 (Hong Kong).

 **Dr. Sophie M. Uchimiya** earned a BS in chemistry from the University of Oregon, Eugene, Oregon, in 2000, and a PhD in environmental chemistry from Johns Hopkins University, Baltimore, Maryland, in 2005. She is currently a research chemist at the U.S. Department of Agriculture-Agricultural Research Service (USDA-ARS) Southern Regional Research Center in New Orleans, Louisiana. Dr. Uchimiya holds an adjunct professorship at the Department of Material Science and Engineering, SUNY Stony Brook, and is an associate editor of *Journal of Environmental Quality*. Dr. Uchimiya has 16 years of research experience in sorption–desorption kinetics/isotherms on biosorbents, redox and other abiotic/biotic transformation reactions in soil, thermochemical conversion of agricultural wastes, contaminant fate and transport, and environmental forensics. Dr. Uchimiya's expertise in environmental chemistry realized bottom-up development of agricultural waste–derived, value-added products by elucidating chemical speciation at surface–water interfaces that will lead to desirable functions. This mechanistic approach was necessary to gain fundamental material property–function relationships in soil and other complex media where sorption, chemical transformation, and physical transport occur concurrently. Dr. Uchimiya collaborates extensively with academia and the federal sectors to utilize high-resolution spectro- and microscopic techniques such as three-dimensional synchrotron X-ray computed microtomography (Brookhaven National Laboratory, New York) and solid-state nuclear magnetic resonance (National Institute for Agro-Environmental Sciences, Japan). During her tenure at the ARS, Dr. Uchimiya brought in several reimbursable projects funded by the USDA/National Institute of Food and Agriculture/Agriculture and Food Research Initiative, Department of Defense (DoD) Strategic Environmental Research and Development Program, and DoD Defense Threat Reduction Agency.

Dr. Scott X. Chang is a professor in forest soils and nutrient dynamics at the University of Alberta, Edmonton, Canada. He has served as the chair of the Forest, Range, and Wildland Soils Division of the Soil Science Society of America, chair of the Soil Fertility and Plant Nutrition Commission of the International Union of Soil Science, and chair of the Alberta Soil Science Workshop. His research interests include soil biogeochemistry, climate change, forest productivity, agroforestry, and land reclamation. Dr. Chang has supervised more than 80 postgraduate students, postdoctoral fellows, and visiting scientists from many countries, including Canada, the United States, New Zealand, China, the Republic of Korea, Japan, Pakistan, India, Sri Lanka, the Philippines, Nepal, Iran, Kenya, Ghana, Cameroon, Ethiopia, Libya, Brazil, and France. He has published more than 170 peer-reviewed articles and was awarded the University of Alberta Annual Killam Professorship. Dr. Chang has served as an associate editor for the *Canadian Journal of Soil Science*, as an editorial board member for *Communications in Soil Science and Plant Analysis*, and as a guest editor for the *Canadian Journal of Soil Science, Forest Ecology and Management, Journal of Soils and Sediments,* and *Journal of Environmental Quality*. He is currently serving as the special issues editor for the *Canadian Journal of Soil Science*, as an editorial board member for *Biology and Fertility of Soils* and *Pedosphere*, and as an associate editor for *Journal of Soils and Sediments*. He has reviewed grant applications for the Natural Science and Engineering Research Council of Canada, Natural Environment Research Council (UK), European Union's 6 Framework Programme for Research, the Netherlands Organization for Scientific Research, the German Federal Ministry for Education and Research, the Biotechnology and Biological Sciences Research Council (UK), National Science Foundation, Georgian National Science Foundation, Agriculture and Agri-Food Canada, and British Columbia Forest Science Program grants.

Dr. Nanthi Bolan is professor of Environmental Science at the University of Newcastle, Australia. He has served as the dean of Graduate Studies of the University of South Australia and also as the leader of the Cooperative Research Centre Contaminant Assessment and Remediation of the Environment Programme on Prevention Technologies. His teaching and research interests include agronomic value of manures, fertilizers, and soil amendments; soil acidification; nutrient cycling; pesticide and metal pollutants interactions in soils; soil remediation; and waste and wastewater management. Dr. Bolan is a fellow of the American Soil Science Society and New Zealand Soil Science Society and was awarded the Communicator of the Year award by the New Zealand Institute of Agricultural Sciences. He has supervised more than 40 postgraduate students from many countries, including Australia, New Zealand, India, Indonesia, Thailand, Malaysia, Bangladesh, Brazil, France, the Netherlands, Sri Lanka, and the Philippines, and he was awarded the Massey University Research Medal for excellence in supervision. Dr. Bolan has published more than 200 papers and was awarded the M.L. Leamy Award in recognition of the most meritorious contribution to soil science. Dr. Bolan has served on the editorial board of two international journals, *Nutrient Cycling in Agroecosystem* and *Environmental Geology and Health*, and is currently serving as an associate editor of the *Journal of Environmental Quality* and *Critical Reviews in Environmental Science and Technology*.

Contributors

Tim Anderson
Alberta Innovates Technology Futures (AITF)
Vegreville, Alberta, Canada

Anthony O. Anyia
Alberta Innovates Technology Futures (AITF)
Vegreville, Alberta, Canada

Jingzi Beiyuan
Department of Civil and Environmental Engineering
Hong Kong Polytechnic University
Hong Kong, China

Irshad Bibi
Institute of Soil and Environmental Sciences
University of Agriculture Faisalabad
Faisalabad, Pakistan

and

Southern Cross GeoScience
Southern Cross University
Lismore, New South Wales, Australia

Nanthi Bolan
Global Institute for Environmental Remediation
and
Australia and Cooperative Research Centre for
    Contamination Assessment and Remediation of
    the Environment (CRC-CARE)
University of Newcastle
Callaghan, New South Wales, Australia

Yanjiang Cai
Department of Renewable Resources
University of Alberta
Edmonton, Alberta, Canada

Scott X. Chang
Department of Renewable Resources
University of Alberta
Edmonton, Alberta, Canada

Girish Choppala
Southern Cross GeoScience
Southern Cross University
Lismore, New South Wales, Australia

Mei Deng
Department of Civil and Environmental Engineering
Hong Kong Polytechnic University
Hong Kong, China

Da Dong
Institute of Environmental Science and Technology
Zhejiang University
Hangzhou, China

Don Harfield
Alberta Innovates Technology Futures (AITF)
Vegreville, Alberta, Canada

Lizhi He
Zhejiang Provincial Key Laboratory of Carbon Cycling in
    Forest Ecosystems and Carbon Sequestration
and
School of Environmental and Resource Sciences
Zhejiang A & F University
Lin'an, China

Indika Herath
Chemical and Environmental Systems Modeling
    Research Group
National Institute of Fundamental Studies
Kandy, Sri Lanka

Huagang Huang
Yancao Production Technology Center
Bijie Yancao Company of Guizhou Province
Bijie, China

Avanthi Deshani Igalavithana
Korea Biochar Research Center
and
Department of Biological Environment
Kangwon National University
Chuncheon, Republic of Korea

Rajasekar Karunanithi
Global Institute for Environmental Remediation
and
Cooperative Research Centre for Contamination
    Assessment and Remediation of the Environment
    (CRC-CARE)
University of South Australia
Mawson Lakes, South Australia, Australia

Ataullah Khan
Alberta Innovates Technology Futures (AITF)
Vegreville, Alberta, Canada

Won-Il Kim
Chemical Safety Division
Department of Agro-Food Safety
National Academy of Agricultural Science
Suwon-si, Gyeonggi-do, Republic of Korea

Prasanna Kumarathilaka
Chemical and Environmental Systems Modeling
    Research Group
National Institute of Fundamental Studies
Kandy, Sri Lanka

Anitha Kunhikrishnan
Chemical Safety Division
Department of Agro-Food Safety
National Academy of Agricultural Science
Suwon-si, Gyeonggi-do, Republic of Korea

Sang Soo Lee
Korea Biochar Research Center
and
Department of Biological Environment
Kangwon National University
Chuncheon, Republic of Korea

Kouping Lu
School of Environmental and Resource Sciences
Zhejiang A & F University
Lin'an, China

Li Lu
School of Environmental Science and Engineering
Zhejiang Gongshang University
Hangzhou, China

Sanchita Mandal
Global Centre for Environmental Risk Assessment and
    Remediation
University of South Australia
Mawson Lakes, South Australia, Australia

Sonia Mayakaduwa
Chemical and Environmental Systems Modeling
    Research Group
National Institute of Fundamental Studies
Kandy, Sri Lanka

Kim McGrouther
Scion
Rotorua, New Zealand

Dinesh Mohan
School of Environmental Sciences
Jawaharlal Nehru University
New Delhi, India

Fungai N.D. Mukome
Department of Land, Air, and Water Resources
University of California Davis
Davis, California, USA

Nabeel Khan Niazi
Institute of Soil and Environmental Sciences
University of Agriculture Faisalabad
Faisalabad, Pakistan

and

Southern Cross GeoScience
Southern Cross University
Lismore, New South Wales, Australia

Yong Sik Ok
Korea Biochar Research Center
and
Department of Biological Environment
Kangwon National University
Chuncheon, Republic of Korea

Pranoy Pal
Ecosystems and Global Change Team
Landcare Research
Palmerston North, New Zealand

Sanjai J. Parikh
Department of Land, Air, and Water Resources
University of California Davis
Davis, California, USA

Gordon W. Price
Faculty of Agriculture
Dalhousie University
Truro, Nova Scotia, Canada

Fangjie Qi
Global Institute for Environmental Remediation
University of Newcastle
Callaghan, New South Wales, Australia

and

Cooperative Research Centre for Contamination Assessment
    and Remediation of the Environment (CRC-CARE)
University of South Australia
Mawson Lakes, South Australia, Australia

Anushka Upamali Rajapaksha
Korea Biochar Research Center
and
Department of Biological Environment
Kangwon National University
Chuncheon, Republic of Korea

Nick Savidov
Alberta Agriculture and Rural Development (ARD)
Edmonton, Alberta, Canada

Balaji Seshadri
Global Institute for Environmental Remediation
University of Newcastle
Callaghan, New South Wales, Australia

and

Cooperative Research Centre for Contamination Assessment
    and Remediation of the Environment (CRC-CARE)
Salisbury, South Australia, Australia

Ramya Thangarajan
Global Centre for Environmental Risk Assessment and
    Remediation
and
Cooperative Research Centre for Contamination Assessment
    and Remediation of the Environment (CRC-CARE)
University of South Australia
Mawson Lakes, South Australia, Australia

Daniel C.W. Tsang
Department of Civil and Environmental Engineering
Hong Kong Polytechnic University
Hong Kong, China

Sophie M. Uchimiya
U.S. Department of Agriculture-ARS Southern Regional
    Research Center
New Orleans, Louisiana, USA

Meththika Vithanage
Chemical and Environmental Systems Modeling
    Research Group
National Institute of Fundamental Studies
Kandy, Sri Lanka

R. Paul Voroney
School of Environmental Sciences
University of Guelph
Guelph, Ontario, Canada

Hailong Wang
Zhejiang Provincial Key Laboratory of Carbon Cycling in
    Forest Ecosystems and Carbon Sequestration
and
School of Environmental and Resource Sciences
Zhejiang A & F University
Lin'an, China

Weixiang Wu
Institute of Environmental Science and Technology
Zhejiang University
Hangzhou, China

Jian Yang
Alberta Innovates Technology Futures (AITF)
Vegreville, Alberta, Canada

Hongjie Zhang
School of Environmental Sciences
University of Guelph
Guelph, Ontario, Canada

John Zhang
Alberta Agriculture and Rural Development (ARD)
Edmonton, Alberta, Canada

Ming Zhang
Department of Environmental Engineering
China Jiliang University
Hangzhou, China

Xiaokai Zhang
Zhejiang Provincial Key Laboratory of Carbon Cycling in
    Forest Ecosystems and Carbon Sequestration
and
School of Environmental and Resource Sciences
Zhejiang A & F University
Lin'an, China

Ting Zhong
Institute of Environmental Science and Technology
Zhejiang University
Hangzhou, China

SECTION I

General Overview

1
Definitions and Fundamentals of Biochar

Chapter 1
Definitions and Fundamentals of Biochar

Anushka Upamali Rajapaksha, Dinesh Mohan, Avanthi Deshani Igalavithana, Sang Soo Lee, and Yong Sik Ok

Chapter Outline

1.1 Introduction

There is an increasing interest in understanding biochar as a whole, particularly its prospects and applications for environmental management. Biochar has been recently recognized as a multifunctional material related to carbon sequestration, contaminant immobilization, greenhouse gas reduction, soil fertilization, and water filtration (Ahmad et al. 2012; Awad et al. 2012; Bolan et al. 2012; Lehmann and Joseph 2009).

The complex and heterogeneous chemical and physical composition of biochars provides an excellent platform for contaminant removal (Figure 1.1). The chemical composition of biochars depends on the type of feedstock and pyrolysis conditions (e.g., residence time, temperature, heating rate, and reactor type); thus, not all biochar is the same and it is difficult to define the exact chemical composition of biochar (Lehmann and Joseph 2009). Biochars are mainly composed of carbon. The organic portion of biochar has a high carbon content, and the inorganic portion mainly contains minerals such as Ca, Mg, K, and inorganic carbonates (carbonate ion), depending on its feedstock type. The pyrolysis temperature used for biochar production does not form graphite to any significant extent (Lehmann and Joseph 2009), and aromatic rings in biochar are not arranged in perfectly stacked

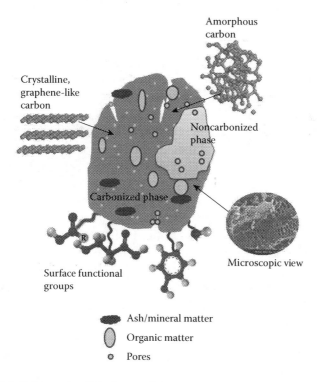

Figure 1.1 Schematic of biochar and its content.

and aligned sheets, as they are in graphite (Lehmann and Joseph 2009). More irregular arrangements of carbon are formed during biochar production, and they contain O and H (Lehmann and Joseph 2009). In some cases, mineral formation depends on the feedstock type. Generally, biochars are not fully carbonized and exhibit carbonized and noncarbonized phases.

Addition of biochar to soil results in an increase in water infiltration (Ayodele et al. 2009), soil water retention, ion exchange capacity, and nutrient retention (Laird et al. 2010; Lehmann et al. 2003) and an improvement in N use efficiency (Cheng et al. 2012). Review articles on biochar, with their objectives, are summarized in Table 1.1.

1.2 Origin and Emergence of Biochar

The origin of biochar extends back to the pre-Columbian era, when the ancient Amerindian communities in the Brazilian Amazon region first made dark earth soils (Terra Preta de Índio [black earth of the Indian]), also known as terra preta, through slash-and-char (Lehmann and Joseph 2009). These soils are characterized by high carbon content, up to 150 g C/kg soil, compared to the surrounding soils (20–30 g C/kg soil) (Glaser et al. 2002; Smith 1980). This increased carbon content is still found, even hundreds of years after these soils were abandoned, proving the persistence of the organic matter in these dark earth soils. Therefore, the total carbon accumulated in these soils appears to be higher than that in adjacent soils. The carbon persistence, soil fertility, and sustainability of biochars have attracted considerable research. Glaser et al. (2000) put forward a new idea identifying black carbon (residues of incomplete combustion—biochar) as a probable cause of and a key factor in sustainable and fertile soils. The organic matter in terra preta has shown a structural similarity to biochar, leading scientists to this explanation for the high carbon content and fertility of terra preta (Glaser et al. 2000). This concept had been proposed by Smith (1980) as well.

TABLE 1.1
Recent Review Articles on Biochars

Targeted Topic	Objective	Reference
Contaminant management	To understand the mechanistic evidence of the interaction of biochars with soil and water contaminants	Ahmad et al. (2014)
	To investigate the potential role of biochars in remediation, revegetation, and restoration of contaminated soils	Beesley et al. (2011)
	To summarize the use of slow and fast pyrolysis biochars in water and wastewater treatment	Mohan et al. (2014)
Biochar properties and composition	To review the published studies related to biochar production, characterization, and its addition to agricultural soils	Manyà (2012)
	To explain the effect of feedstock and pyrolysis temperature on biochar composition	Enders et al. (2012)
	To review biochar stability in soils and prediction of biochar properties based on O:C molar ratios	Spokas (2010)
	To discuss the biomass conversion processes into biochars and their potential applications	Libra et al. (2010)
Impacts on soil organisms	To examine the state of knowledge on soil populations of archaeans, bacteria, fungi, and fauna as well as plant root behavior as a result of biochar additions to soil	Lehmann et al. (2011)
	To explain the novel perspective for in-depth studies on biochar application to soil microbial ecology	Ding et al. (2013)
	To delineate the direct and indirect impacts of biochar on earthworms, including Enchytraeidae, and their associated soil functions	Weyers and Spokas (2011)

(Continued)

TABLE 1.1 *(Continued)*
Recent Review Articles on Biochars

Targeted Topic	Objective	Reference
Agricultural benefits and function in soil	To examine the effects of biochar addition to soils on crop productivity using meta-analysis	Jeffery et al. (2011)
	To explore various mechanisms of biochar function in soil	Sohi et al. (2010)
	To recommend further biochar research and application to soils	Verheijen et al. (2010)
Carbon management and climate mitigation strategy	To explore the implications for carbon emissions trading mechanisms in the context of a biochar soil management system from a biological perspective	Lehmann et al. (2006)
	To identify the gaps and scopes for future research activities on biochar applications for climate change mitigation and soil productivity	Sohi et al. (2009)

Lehmann and Joseph (2009) identified biochar as an excellent soil amendment that has the potential to change the concepts of environmental management.

1.3 Biochar Definitions

Researchers and other professionals have a range of definitions of the term *biochar*. Various biochar definitions based on pyrolysis conditions and its application to soil are summarized in Table 1.2 (Ahmad et al. 2014).

1.4 Biochar as a By-Product

Biochar results as a by-product or secondary product during various processes, including bio-oil and syngas production. Priority should be given to producing biochar as a by-product,

TABLE 1.2
Biochar Definitions in Literature

Definition	Special Notes	Source
A solid material obtained from the thermochemical conversion of biomass in an oxygen-limited environment	It can be applied to soil for both agricultural and environmental gains	International Biochar initiative
A carbon-rich product when biomass such as wood, manure, or leaves is heated in a closed container with little or no air	Biochar is produced by thermal decomposition of organic material under no or limited oxygen supply, at a relatively low temperature (<700°C)	Lehmann and Joseph (2009)
Solid carbon-rich residue, yield in the thermal decomposition of plant-derived biomass under partial or total absence of oxygen	It is specifically used for soil application; it cannot readily be returned to the atmosphere as carbon dioxide even under favorable environmental and biological conditions	Sohi et al. (2009)
Biomass that has been pyrolyzed in no or low-oxygen environment applied to soil at a specific site that is expected to sustainably sequester carbon and concurrently improve soil functions under current and future management, while avoiding short- and long-term detrimental effects to the wider environment as well as human and animal health		Verheijen et al. (2010)
Solid residue of incomplete combustion of biomass derived from plants		Keiluweit et al. (2010)
Biological residues combusted under low-oxygen conditions, resulting in a porous, low-density, carbon-rich material	Large surface area and cation exchange capacities of biochar have an ability to retain inorganic and organic contaminants from soil	Beesley et al. (2011)

(Continued)

TABLE 1.2 *(Continued)*
Biochar Definitions in Literature

Definition	Special Notes	Source
The porous carbonaceous solid produced by the thermochemical conversion of organic materials in an oxygen-depleted atmosphere that has physicochemical properties suitable for safe and long-term storage of carbon in the environment		Shackley et al. (2012)

rather than initially producing biochar, due to energy savings and economic viability. Pyrolysis and gasification can generate both biochar and energy. Conventional (slow) or fast pyrolysis depends upon the operating conditions, including temperature, heating rate, vapor residence time, and reactor type. Bio-oil production is favored in fast pyrolysis, whereas syngas production is favored in gasification. Bio-oil is a dark brown, free-flowing organic liquid comprised highly oxygenated compounds. Pyrolysis liquids are formed by rapidly and simultaneously depolymerizing and fragmenting cellulose, hemicellulose, and lignin with a rapid increase in temperature (Mohan et al. 2006). Syngas is generally considered to be a mixture of hydrogen gas, carbon monoxide, methane, carbon dioxide, and water and several low-molecular-weight volatile organic compounds (Ren et al. 2014).

1.4.1 Liquid Fuel Production

Many studies have highlighted the simultaneous production of bioenergy and biochar from different sources. Boateng et al. (2010) showed the possibility of the production of pyrolysis fuel intermediates with biochar as a by-product by using a biorefinery system collocated in a high-biomass soybean (*Glycine max*) cultivation system. Boateng et al. (2010) showed bio-oil production with heating values in excess of 20 MJ/kg and yields of ~70 wt%. Mineral-rich biochar (22 wt%) and combustible gases (up to 10 wt%) were obtained as by-products.

These by-products can be used to power the pyrolysis system (Boateng et al. 2010).

Fast pyrolysis with a short residence time (<2 s) is recommended to produce bio-oil from biomass, yielding ~75% bio-oil and 12% char (Mohan et al. 2006). Biochar yield depends on the pyrolysis temperature, heating rate, particle size, and lignin content of the biomass used in pyrolysis (Demirbas 2004; Laird et al. 2009; Mohan et al. 2006).

1.4.2 Syngas Production

The limiting nature of natural resources has led to the exploration of new techniques that can replace them. Biomass conversion into gases rich in carbon monoxide and hydrogen, by reacting the biomass at high temperatures (>700°C) in a limited amount of oxygen, steam, or both, is known as gasification. The resulting gas mixture is called as synthesis gas or syngas. Thus, biomass gasification gives biochar as a by-product (Mohan et al. 2006).

1.5 Related Terminologies

1.5.1 Char and Charcoal

The term *char* is used to denote any solid product resulting from the natural and synthetic organic material decomposition (Fitzer et al. 1995). Biochar and charcoal have been distinguished considering the end use (Lehmann and Joseph 2009): charcoal is used as fuel and energy, whereas biochar is directed toward environmental management and carbon sequestration.

1.5.2 Hydrochar

Hydrochar is produced from biomass hydrothermal carbonization (Libra et al. 2010). In general, dry biomass (up to 10% moisture) is used to produce carbonization, pyrolysis, or gasification biochars, whereas hydrothermal (wet) biomass carbonization

under pressure is used to produce hydrochars. Wet animal
manures, human waste, sewage sludges, municipal solid
waste, aquaculture, and algal residues can be directly used
without any drying to prepare hydrochars. Hydrothermal carbonization provides higher char yields with more water-soluble
organic compounds than pyrolysis. Biochars and hydrochars
have different chemical and physical properties; hydrochars
are characterized by higher H:C and O:C ratios than biochars (Libra et al. 2010). Due to a less aromatic structure and a
higher percentage of labile carbons, hydrochars might be readily decomposed (Libra et al. 2010), which impacts their stability
in the environment.

1.6 Biochar Science

Biochar has been receiving an increased interest as an emerging sustainable material. Due to rapid growth of biochar
research and applications, regional biochar groups, which
have connected with the International Biochar Initiative
(IBI), maintain regional activities and support local biochar
research applicable to their geographic area. The bibliographic
database Scopus provides an overview and insight into global
biochar research. To date, the United States has the highest
number of publications (601 articles), followed by China, with
571 articles (Figure 1.2). Australia, the United Kingdom, and
Germany are reported with 208, 146, and 127 biochar publications, respectively. The top 10 countries in biochar publications include the aforementioned countries along with Spain,
Canada, Italy, New Zealand, and Korea.

Progressively more research into the potential and constraints of biochars is evidenced by the number of articles published year to year (Figure 1.3). In total, 2170 journal articles
were listed in Science Citation Index up to 2015. The numbers tended to increase during 2006 and the rate of increase
was very high after 2009.

The underlying functions of biochars in soil and water systems have been investigated thoroughly (Ahmad et al. 2014;
Lehmann and Joseph 2009; Lehmann et al. 2011), but other

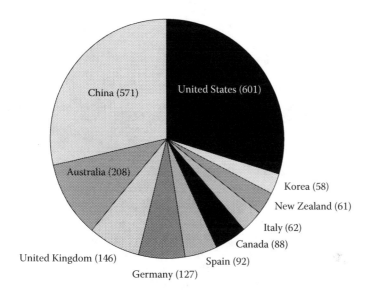

Figure 1.2 Total number of biochar publications for the top 10 publishing countries. (Based on Scopus data accessed on March 8, 2015.)

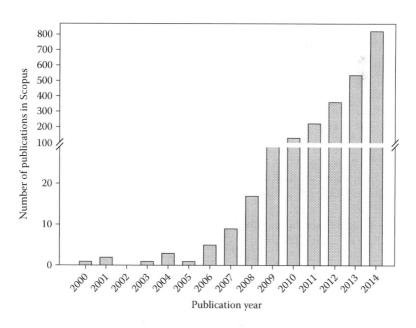

Figure 1.3 Total number of biochar articles by publication year. (Based on Scopus data accessed on March 8, 2015.)

applications remain unresolved. More information on overall chemical and physical characteristics, together with specific applications, of biochars is needed.

References

Ahmad, M., Lee, S.S., Dou, X., et al. 2012. Effects of pyrolysis temperature on soybean stover- and peanut shell-derived biochar properties and TCE adsorption in water. *Bioresource Technology* 118: 536–544.

Ahmad, M., Rajapaksha, A.U., Lim, J.E., et al. 2014. Biochar as a sorbent for contaminant management in soil and water: A review. *Chemosphere* 99: 19–33.

Awad, Y.M., Blagodatskaya, E., Ok, Y.S., et al. 2012. Effects of polyacrylamide, biopolymer, and biochar on decomposition of soil organic matter and plant residues as determined by ^{14}C and enzyme activities. *European Journal of Soil Biology* 48: 1–10.

Ayodele, A., Oguntunde, P., Joseph, A., Dias, M.d.S., Jr. 2009. Numerical analysis of the impact of charcoal production on soil hydrological behavior, runoff response and erosion susceptibility. *Revista Brasileira de Ciência do Solo* 33: 137–145.

Beesley, L., Moreno-Jiménez, E., Gomez-Eyles, J.L., et al. 2011. A review of biochars' potential role in the remediation, revegetation and restoration of contaminated soils. *Environmental Pollution* 159: 3269–3282.

Boateng, A.A., Mullen, C.A., Goldberg, N.M., et al. 2010. Sustainable production of bioenergy and biochar from the straw of high-biomass soybean lines via fast pyrolysis. *Environmental Progress & Sustainable Energy* 29: 175–183.

Bolan, N.S., Kunhikrishnan, A., Choppala, G.K., et al. 2012. Stabilization of carbon in composts and biochars in relation to carbon sequestration and soil fertility. *Science of The Total Environment* 424: 264–270.

Cheng, Y., Cai, Z.-C., Chang, S., et al. 2012. Wheat straw and its biochar have contrasting effects on inorganic N retention and N_2O production in a cultivated Black Chernozem. *Biology and Fertility of Soils* 48: 941–946.

Demirbas, A. 2004. Effects of temperature and particle size on biochar yield from pyrolysis of agricultural residues. *Journal of Analytical and Applied Pyrolysis* 72: 243–248.

Ding, Y.L., Liu, J., Wang, Y.Y. 2013. Effects of biochar on microbial ecology in agriculture soil: A review. *Chinese Journal of Applied Ecology* 24: 3311–3317.

Enders, A., Hanley, K., Whitman, T., et al. 2012. Characterization of biochars to evaluate recalcitrance and agronomic performance. *Bioresource Technology* 114: 644–653.

Fitzer, E., Kochling, K.-H., Boehm, H.P., Marsh, H. 1995. Recommended terminology for the description of carbon as a solid (IUPAC recommendations). *Pure and Applied Chemistry* 67(3): 473–506.

Glaser, B., Balashov, E., Haumaier, L., et al. 2000. Black carbon in density fractions of anthropogenic soils of the Brazilian Amazon region. *Organic Geochemistry* 31: 669–678.

Glaser, B., Lehmann, J., Zech, W. 2002. *Ameliorating physical and chemical properties of highly weathered soils in the tropics with charcoal: A review.* Springer, Berlin.

Jeffery, S., Verheijen, F.G.A., van der Velde, M., et al. 2011. A quantitative review of the effects of biochar application to soils on crop productivity using meta-analysis. *Agriculture, Ecosystems & Environment* 144: 175–187.

Keiluweit, M., Nico, P.S., Johnson, M.G., et al. 2010. Dynamic molecular structure of plant biomass-derived black carbon (Biochar). *Environmental Science & Technology* 44: 1247–1253.

Laird, D., Fleming, P., Wang, B., et al. 2010. Biochar impact on nutrient leaching from a Midwestern agricultural soil. *Geoderma* 158: 436–442.

Laird, D.A., Brown, R.C., Amonette, J.E., et al. 2009. Review of the pyrolysis platform for coproducing bio-oil and biochar. *Biofuels, Bioproducts and Biorefining* 3: 547–562.

Lehmann, J., Gaunt, J., Rondon, M. 2006. Bio-char sequestration in terrestrial ecosystems—A review. *Mitigation and Adaptation Strategies for Global Change* 11: 395–419.

Lehmann, J., Joseph, S. 2009. *Biochar for environmental management: Science and technology.* Sterling, VA: Earthscan.

Lehmann, J., Da Silva Jr, J. P., Steiner, C., et al. 2003. Nutrient availability and leaching in an archaeological Anthrosol and a Ferralsol of the Central Amazon basin: Fertilizer, manure and charcoal amendments. *Plant and Soil* 249: 343–357.

Lehmann, J., Rillig, M.C., Thies, J., et al. 2011. Biochar effects on soil biota—A review. *Soil Biology and Biochemistry* 43: 1812–1836.

Libra, J.A., Ro, K.S., Kammann, C., et al. 2010. Hydrothermal carbonization of biomass residuals: A comparative review of the chemistry, processes and applications of wet and dry pyrolysis. *Biofuels* 2: 71–106.

Manyà, J.J. 2012. Pyrolysis for biochar purposes: A review to establish current knowledge gaps and research needs. *Environmental Science & Technology* 46: 7939–7954.

Mohan, D., Pittman, C.U., Steele, P.H. 2006. Pyrolysis of wood/biomass for bio-oil: A critical review. *Energy & Fuels* 20: 848–889.

Mohan, D., Sarswat, A., Ok, Y.S., et al. 2014. Organic and inorganic contaminants removal from water with biochar, a renewable, low cost and sustainable adsorbent—A critical review. *Bioresource Technology* 160: 191–202.

Ren, S., Lei, H., Wang, L., Bu, Q., Chen, S., Wu, J. 2014. Hydrocarbon and hydrogen-rich syngas production by biomass catalytic pyrolysis and bio-oil upgrading over biochar catalysts. *RSC Advances* 4: 10731–10737.

Shackley, S., Carter, S., Knowles, T., et al. 2012. Sustainable gasification-biochar systems? A case study of rice-husk gasification in Cambodia, part 1: Context, chemical properties, environmental and health and safety issues. *Energy Policy* 42: 49–58.

Smith, N.J.H. 1980. Anthrosols and human carrying capacity in Amazonia. *Annals of the Association of American Geographers* 70: 553–566.

Sohi, S., Loez-Capel, S., Krull, E., et al. 2009. *Biochar's roles in soil and climate change: A review of research needs.* CSIRO, Clayton, Australia. CSIRO Land and Water Science Report 05/09.

Sohi, S.P., Krull, E., Lopez-Capel, E., et al. 2010. A review of biochar and its use and function in soil. In: *Advances in Agronomy*, (Ed.) D.L. Sparks. Academic Press, Burlington, pp. 47–82.

Spokas, K.A. 2010. Review of the stability of biochar in soils: Predictability of O:C molar ratios. *Carbon Management* 1: 289–303.

Verheijen, F., Jeffery, S., Bastos, A.C., et al. 2010. *Biochar application to soils. A critical scientific review on soil properties, processes and functions.* European Commission, Office for Official Publications of the European Communities, Luxembourg.

Weyers, S.L., Spokas, K.A. 2011. Impact of biochar on earthworm populations: A review. *Applied and Environmental Soil Science.* 2011: Article ID 541592, pp. 1–12.

2

Biochar

State of the Art

Chapter 2
Biochar

Prasanna Kumarathilaka, Sonia Mayakaduwa,
Indika Herath, and Meththika Vithanage

Chapter Outline

2.1 Origin of Terra Preta and Its Research

The origin of biochar is connected to the Amazon River basin where thousands of raised platforms of black and very fertile soil patches were first discovered by the explorer Herbert Smith in 1879 (Marris 2006). In relation to its dark color and origin, the soil was named Terra Preta de Índio (black earth of the Indian). Today, even addition of chemical fertilizers cannot maintain crop yields into a third consecutive growing season, yet these dark earths have retained their fertility for centuries. Hence, research has been focusing on the chemical and physical properties and cultural origin of terra preta. Field studies have provided evidence that terra preta was created through the use of slash-and-char techniques (Lehmann and Joseph 2009). Furthermore, research has confirmed and quantified that the carbonized organic matter in terra preta resulted from incomplete combustion. Similarly, it is well founded that terra preta contains charred plant material that has come to be referred to as biochar (Glaser et al. 2001). Because almost any form of natural organic material can be converted into biochar, considerable research is now being devoted to duplicating the formation of terra preta by creating biochar through various processes of heating plant debris in the absence of oxygen.

2.1.1 A Valuable Resource

Studies on terra preta have found that it consists of a large amount of carbon, with nitrogen, calcium, and phosphorus in considerably lower concentrations. The cation exchange capacity (CEC), pH, and base saturation of terra preta are significantly higher than that of surrounding soils (Glaser et al. 2000). Further investigations demonstrated that a hectare of meter-deep terra preta could hold up to 250 tons of carbon, compared with the maximum of 100 tons in adjacent soils (Glaser et al. 2001), causing the earth's dark black color. The polycyclic aromatic structure of biochar hinders biological decomposition and chemical oxidation, thereby explaining its persistence over centuries in the environment (Glaser et al. 2000). The surface of biochar has a high density of negatively charged groups of carbon and oxygen atoms called carboxylate groups. These carboxylate groups originate from superficial oxidation and this may be the reason for the increased nutrient-holding capacity of terra preta (Kim et al. 2007).

2.2 Developments in Biochar Research

During 1980–1990, research started uncover the marked positive impacts of biochar addition on various crops. Specially, short-term responses of staple grain crops to biochar application in terms of plant biomass or crop yield were assessed (Glaser et al. 2001), and biochar was shown to increase plant growth and nutrition in pot and field experiments (Lehmann et al. 2003; Steiner et al. 2007). Biochar application significantly reduced the leaching of applied N fertilizers (Lehmann et al. 2003; Steiner et al. 2007). The amendment of biochar to soil was advocated by Wim Sombroek, a soil scientist, suggesting that biochar would not only enhance soil fertility but also sequester carbon from the atmosphere–biosphere pool, as he introduced terra preta as a terrestrial carbon sink (Marris 2006). Typically, plant material is decomposed by soil microorganisms within a few years and returns carbon to the atmosphere as carbon dioxide (CO_2) as a part of the nutrient recycling role of the soil. Instead of allowing this carbon to escape from the soil as CO_2, the conversion of plant material into charcoal would provide several distinct benefits (Lehmann et al. 2003; Seifritz 1993; Steiner et al. 2007). In addition, further research findings revealed that biochar has an affinity for organic compounds and may sorb toxic by-products from the wastewater treatment process. Today, the biochar concept has a strong global context, as it is positioned strongly in areas such as climate change (carbon abatement), but intrinsically linked to renewable energy capture (biomass pyrolysis) and food production and land-use change (food and feed production), further extending to the enhancement of environmental quality (control of diffuse pollution) and management of organic wastes (stabilization and use), through management of soil nutrients. Due to the above-revealed factors, an increasing global interest in biochar research has been realized in past few years (Figure 2.1). This research is focused on dealing with environmental pollution and ensuring food security.

To date, the highest number of biochar publications is from the United States, followed by China, Australia, the United Kingdom, Germany, Canada, Spain, New Zealand, Italy, and South Korea (Figure 2.2). Most research has focused on biochar and its relationships to environmental science; agricultural and biological sciences; energy; earth and planetary sciences; and engineering (Figure 2.3).

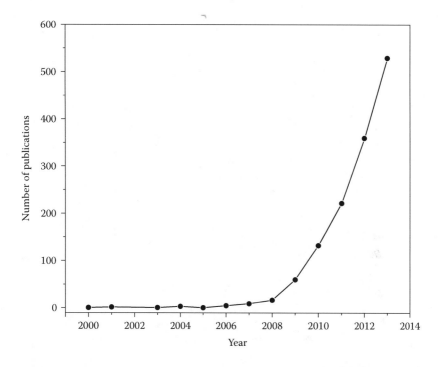

Figure 2.1 Biochar research development from 2000 to 2013. (From Scopus database.)

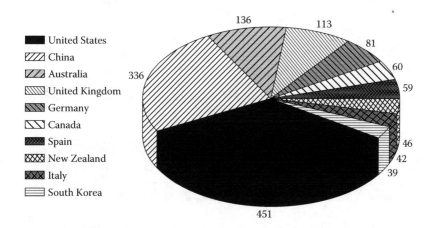

Figure 2.2 Top 10 countries contributing to biochar research. (From Scopus database.)

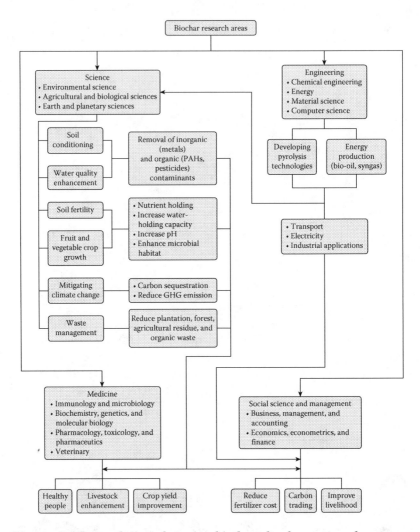

Figure 2.3 Interrelations between biochar development and economy, environment, and society.

2.3 Biochar Research for Environmental Sustainability

Many studies have been conducted on soil and water pollution using biochar. Figure 2.4 shows potential biochar applications for environmental sustainability. According to the

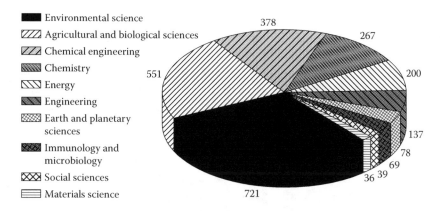

Figure 2.4 Most common subject areas in which biochar being used. (From Scopus database.)

number of studies, biochar application has shown significant environmental benefits, such as improving physiochemical and biological properties of soils and significantly increasing of plant growth and crop yield (Kookana et al. 2011; Powlson et al. 2011). Biochar is an economically viable substitute to activated carbon for removing inorganic and organic contaminants from soil and water because biochar can be made by using various biomass sources that have high volatility (Liu et al. 2011).

The economic feasibility of biochar depends on the product and the benefits to the user. In waste biomass, the benefits to the user may include increased production and reduced fertilizer requirements and the stabilization of organic carbon. In addition, production and application of biochar for electricity and waste management are economically viable. Furthermore, biochar systems contribute to the economic development of local impoverished communities. Eventually, this combination of environment, economy, and social outcomes leads to a sustainable development of the world. The following sections give a brief introduction into the broad areas to be discussed in detail in the individual chapters of this book.

2.3.1 Research on Biochar Production Technologies

Research on the different methods and parameters of biochar production, such as heating temperature and rate and residence time, has found that the properties of biochar can be changed depending on the production technology. The production process is started by the burning of fuels, by electrical heating, or by microwaves. At present, the cheap, simple, and traditional charcoal-making technologies (pits, earth mounds, and brick and metal kilns) are being switched over to modern biochar production technologies (drum-type pyrolyzers, screw-type pyrolyzers, and rotary kiln) that result in greater biochar yield. In Australia, a paddle drum pyrolyzer has been developed, whereas in many parts of Africa, screw-type pyrolyzers are being used for the production of bio-oil and biochar. In Japan, the largest commercial slow pyrolysis rotary kiln operates at 100 tons of biomass per day (Duku et al. 2011). In addition, biochar can be produced as a pyrolytic by-product during bio-oil manufacturing (Mohan et al. 2007).

According to the type of thermochemical processing (slow and fast pyrolysis, gasification, hydrothermal and flash carbonization), the products and their significant percentages will vary (Table 2.1). Bruun et al. (2011) investigated the relationship between fast pyrolysis of wheat straw at different reactor temperatures and the short-term degradability of biochar in soil. Bruun et al. (2012) also studied the effects of biochar produced through slow and fast pyrolysis on soil carbon and nitrogen dynamics after incorporation. They observed that loss of C as CO_2 for slow and fast pyrolysis conditions are significantly lower, 2.9 and 5.5%, respectively, compared to feedstock, with 53% C loss. In addition, slow pyrolysis of biochar led to net N mineralization (7%), whereas fast pyrolysis of biochar has a considerable influence on immobilized mineral N (43%), suggesting that the pyrolysis process may greatly affect the mineralization and immobilization of soil N.

Bioenergy and the biochar markets compete for the same feedstock, and biochar-based soil amendments also compete with other biochar products. According to the current technologies in use, biochar production costs vary. On average, the cost of biochar production per tonne is estimated to be between US\$51 and US\$386. Moreover, published work has mainly focused on pyrolysis technologies, whereas less information is available for gasification and hydrothermal and flash carbonization technologies to date (Meyer et al. 2011).

TABLE 2.1
Conversion Technologies and Product Distribution

Technology	Typical Temperature (°C)	Typical Residence Time	Products (%)			Carbon Content of the Solid Product
			Solid (Biochar)	Liquid (Bio-Oil)	Gas (Syngas)	
Slow pyrolysis	100–1000	Long (~2–4 h)	35	30	35	~50–95
Intermediate pyrolysis	~500	Moderate (10–20 s)	25	50	25	66–74
Fast pyrolysis	300–1000	Short (~2 s)	~12–25	~50–75	~13–25	~64–90
Gasification	~900	Moderate (10–20 s)	10	5	85	65

Sources: Ahmad M. et al., *Bioresource Technology*, 118, 536–544, 2012; Brown TR. et al., *Biofuels, Bioproducts and Biorefining*, 5, 54–68, 2011; Kim KH. et al., *Bioresource Technology*, 118, 158–162, 2012; Sohi S. et al., Biochar, climate change and soil: A review to guide future research, In *CSIRO Land and Water Science Report*, 5, 17–31, 2009; Yuan S. et al., *Bioresource Technology*, 109, 188–197, 2012.

2.3.2 State of the Art of Biochar Characteristics

Biochar is basically produced from different sources of biomass originating from plant- and animal-based waste material. Biomass such as agricultural crop residues; forestry residues; invasive plants; wood waste; and the organic portion of municipal solid waste, manures, and different types of sludge are possible feedstocks for biochar production (Duku et al. 2011). The systematic characterization of biochar is important for understanding its functions, such as contaminant sorption and soil fertilization. Furthermore, characteristics of biochar vary substantially depending on the feedstock and pyrolysis conditions (Figure 2.5).

Changes in biochar properties with different high heating temperatures can be analyzed using proximate analysis (moisture, volatile matter, fixed carbon, and ash), Brunauer–Emmett–Teller surface area, van Krevelen plot of elemental ratios, and selected elemental composition (Uchimiya et al. 2012). Noor et al. (2012) conducted the ultimate analysis (C, H, S, and O), yield variation, and the proximate analysis for cassava (*Manihot esculenta* Crantz) biochar produced at different temperatures and heating rates. Differences in pH and CEC of poultry litter, peanut hulls, and pine chips produced at 400 and 500°C have been studied before introducing the agricultural soils (Gaskin et al. 2008). In addition,

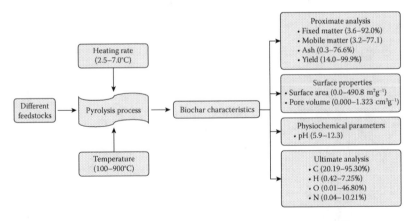

Figure 2.5 Variation in biochar characteristics. (Based on Ahmad et al. (2014b).)

Gaskin et al. (2008) analyzed total elemental concentrations of C, N, P, K, S, Ca, Mg, Al, Fe, and Na. Surface characteristics of biochar can be analyzed using Fourier transform infrared (FTIR) and Raman spectroscopies, scanning electron microscopy (SEM) with energy-dispersive x-ray spectroscopy, and x-ray photoelectron spectroscopy images (Chia et al. 2012). Liu et al. (2013) determined surface properties such as pore volume with FTIR spectroscopy and obtained SEM images for biochar samples made from oak, bamboo, and straw.

2.3.3 Research on Surface Modification

Instead of enhancing the biochar yield, low-cost modifications such as magnetization have recently been applied to study the physical and chemical activation of biochar for removing pollutants, such as phosphates, heavy metals, and organic compounds, from an aqueous medium.

Magnetic-activated biochar adsorbents can be separated via a magnetic separator (Mohan et al. 2011). For example, arsenic can be successfully removed from aqueous solution using magnetic biochar (Zhang et al. 2013). In addition, synthesized magnetic-activated carbons were capable of adsorbing pollutants such as bisphenol A and nonylphenol (Nakahira et al. 2006). Biochar/iron oxide nanopowder (γ-Fe_2O_3) increases the ability to remove contaminants; otherwise, γ-Fe_2O_3 may aggregate in aqueous solution, thereby decreasing the surface area and adsorption ability when it is applied alone.

Biochar can be modified with nitric acid, potassium permanganate ($KMnO_4$), hydrogen peroxide (H_2O_2), ammonium persulfate, air, and ozone from which oxygen-containing functional groups are increased, resulting in sorption enhancement (Uchimiya et al. 2011). Uchimiya et al. (2011) have produced steam-activated biochar (steam activation at 850°C for 1.5 h under nitrogen atmosphere with 3 ml·min⁻¹ water flow rate) by using cottonseed hull. Rajapaksha et al. (2014) have developed steam-activated tea waste biochar (biochar was subjected to 5 ml·min⁻¹ steam for an additional 45 min under peak temperature) that showed enhanced sulfamethazine removal. Uchimiya et al. (2010) were able to produce broiler litter biochar under 1600 ml·min⁻¹ nitrogen flow rate. In addition, fluid petroleum coke–activated carbon has been developed using both physical (steam and CO_2) and chemical (potassium hydroxide and phosphoric acid) activation (Rambabu et al. 2013). H_2O_2-modified biochar contains a high number of oxygen-containing functional groups for effectively removing aqueous heavy metals such as Cu, Ni, and Cd (Xue et al. 2012). Moreover, a novel biochar/MnO_x composite has been synthesized through $KMnO_4$ modification of corn straw biochar and due to the formation of inner-sphere complexes with MnO_x and O-containing groups, strong sorption capacity for Cu was reported (Song et al. 2014).

2.3.4 Climate Change Mitigation

In recent years, biochar production has been proposed as a way to mitigate climate change by sequestrating carbon in soil, and research regarding the capability of biochar to mitigate climate change is thus on the rise.

Maximum sustainable technical potential and Alpha and Beta scenarios that describe lower demands on global biomass resources can be used to estimate climate change mitigation potential of biochar produced in a sustainable manner. As a result, avoided emissions due to biochar production and biomass combustion over 100 years and on an annual basis for the above-mentioned three model scenarios have been studied. In addition, cumulative avoided greenhouse gas emissions from sustainable biochar production via rice, sugarcane, manures, biomass crops, forestry residues, and wood waste have been well documented (Woolf et al. 2010). Greenhouse gas emissions with the presence of biochars have been successfully studied in many parts of the world. Effects of fluxes of nitrous oxide (N_2O), CO_2, and methane (CH_4) with the presence of biochar (9 t·ha^{-1}) on an agricultural soil have been examined in southern Finland (Karhu et al. 2011). Liu et al. (2012) conducted a field experiment using biochar amendment (0, 20, and 40 t·ha^{-1}) for observing N_2O emission from rice fields in southern China. Mukome et al. (2013) investigated effects of walnut shell and wood feedstock biochar produced two different temperatures on CO_2 and N_2O emissions from a fertile agricultural soil amended with different types of organic and synthetic fertilizer as a short-term incubation experiment (29 days). Another 2-year field experiment with crop straw biochar (0, 10, 20, and 40 t·ha^{-1}) to examine N_2O, CO_2, and CH_4 from rice fields throughout the whole rice-growing season was carried out by Liu et al. (2012). Moreover, Van Zwieten et al. (2013) investigated N_2O emissions by adding poultry litter biochar onto the Ferrosol on which a corn (*Zea mays* L.) crop was grown. The effect of spruce chips biochar, produced at 400–450°C, on N_2O emission in mesocosms of Timothy-grass (*Phleum pretense* L.) has been studied by Saarnio et al. (2013).

2.3.5 Soil and Water Remediation

High levels of environmental contaminants are discharged into the environment from municipal, agricultural, and industrial sources at a global scale. As a result, soil and water media in an ecosystem are frequently subjected to contamination by organic and inorganic contaminants (Ahmad et al. 2014b). Biochar could be an exciting and unique solution to this contamination due to its high surface area and microporosity and thereby minimize negative environmental impacts. Many such positive research findings on organic and inorganic contaminants removal are being reported in recent years (Mohan et al. 2014). Many variables, especially pyrolysis conditions and feedstock types, determine the exact role of the biochar for inorganic and organic contaminants removal, and different interactions of the biochar with organic and inorganic contaminants (e.g., electrostatic attraction and repulsion, ion exchange, and precipitation) have been widely investigated in addition to evaluating the solution chemistry (e.g., pH and ionic strength) that also influences the sorption capacity of biochar (Ahmad et al. 2014b). In Yudhaganawa, Sri Lanka, Ni, Mn, and Cr immobilization in serpentine soil with the addition of dendro (*Gliricidia sepium* (Jacq.) Kunth ex Walp.) biochar has been investigated via incubation and pot experiments (Herath et al. 2014). Uchimiya et al. (2011) have obtained Cu sorption and desorption isotherms by adding acidic shell–derived activated carbon and broiler litter biochar for clay-rich, alkaline San Jouquin soil and eroded, acidic Norfolk sandy loam soil that have a high and low capacity, respectively, to retain heavy metals. Sorption and desorption behaviors of aminocyclopyrachlor, bentazone, and pyraclostrobin pesticides by using wood chip pellets, macadamia nut shells, and hardwood biochar produced at different conditions have been successfully investigated (Cabrera et al. 2014). In addition, differences in sorption behavior of norflurazon and fluridone herbicides to wood and grass biochars produced at different heat treatment temperatures (200–600°C) have been successfully examined (Sun et al. 2011). Sopena et al. (2012) studied desorption behavior of the herbicide isoproturon with the presence of different biochar dosages. Furthermore, biochar production using invasive plant species for removing environmental contaminants addresses two environmental issues simultaneously. In Korea, biochar produced from buffalo-weed (*Ambrosia trifida* L.), an invasive plant species, has been successfully used for removing trichloroethylene-contaminated water (Ahmad et al. 2014a). In addition, burcucumber (*Sicyos angulatus* L.) biochar is capable of removing sulfonamides, one of the most widely used antibiotic groups in the veterinary industry (Vithanage et al. 2014). Moreover, Zheng et al. (2010) investigated the sorption of pesticide atrazine and the herbicide simazine onto green waste biochar produced at 450°C under experimental parameters such as contact time, dosage, particle size, and solution pH.

Dong et al. (2011) revealed that hexavalent chromium from aqueous solution can be successfully removed by using sugar beet tailing biochar under parameters such as pH, contact time, and biochar dosage via batch experiments. The adsorptive behavior of phenol and heavy metals [Zn(II) and Cu(II)] by using switch grass, hardwood, and softwood biochar has been studied by Han et al. (2013). Marchal et al. (2013) evaluated desorption and mineralization of phenanthrene with the presence of activated carbon, biochar, and compost. Sorption, biodegradation, and leaching of the herbicide simazine with effect of biochar type and contact time have been investigated after incorporation into soil, dosage, and particle size by using commercially available biochars (Jones et al. 2011). Steam-activated tea waste biochar has been used to sorb veterinary drugs such as sulfamethazine, whereas pine needle biochar produced at different temperatures was able to sorb trichloroethylene from contaminated water (Ahmad et al. 2013; Rajapaksha et al. 2014). In addition, agricultural waste–derived biochars such as soybean stover, rice husk, rice straw, wheat residue, and sugar beet tailing have been used for sorbing organic and heavy metal pollutants (Ahmad et al. 2012b; Kloss et al. 2012; Liu and Zhang 2009; Liu et al. 2011).

2.3.6 Use in Agriculture

2.3.6.1 Improvements in Soil Quality

Although it is impossible to precisely quantify global soil degradation, 15% of land has been degraded, with the highest degradation in Europe (25%), Asia (18%), and Africa (16%) according to the 1987–1990 global assessment of soil degradation (Oldeman 1994). Soil degradation has been accelerated due to anthropogenic and natural activities in recent decades; currently, 24% of global land, of which one-fifth is cropland and 20–25% is rangeland, is degraded (Bai et al. 2008).

Yuan and Xu (2011) studied how soil quality improves with adding nine plant residues, including nonleguminous straw from canola, wheat, corn, rice, and rice hulls as well as leguminous straw from soybean, peanut, faba bean, and mung bean biochars. Soil quality parameters such as water retention capability, hydraulic conductivity, pH, total N, and organic C have been studied by adding different dosages of biochar amendments (0, 5, 10, and 20 g biochar·kg^{-1} soil) on Clarion soil (Laird et al. 2010). In addition, Laird et al. (2010) investigated the N, P, Mg, and Si leaching from the biochar amendment (0, 5, 10, and 20 g biochar·kg^{-1} soil) columns as a model experiment for temperate-region soil. Moreover, carbon sequestration on durum wheat in the Mediterranean with the presence of different biochar applications (30 and 60 t·ha^{-1}) has been investigated (Vaccari et al. 2011). Mukherjee and Zimmerman (2013) studied dissolved organic carbon (DOC), N, and P release from a variety of biochars with and without mixing with soil by using batch extraction and column leaching.

Soil enzymes play a major role in improving physiochemical properties in the soil profile. Although soil enzymatic activities with the presence of biochar applications have been studied, only limited research to date has focused on this subject. Different soil enzymes, such as urease, dehydrogenase, protease, and alkaline phosphate, have been investigated after introducing various biochar types and application rates (Oleszczuk et al. 2014; Wu et al. 2013). Biochar amendment seems to enhances the soil enzymatic activities significantly and vice versa, but the effects of biochar on soil enzymatic activities are not fully understood thus far.

2.3.6.2 Agronomic Research

The global human population is currently slightly more than 6 billion. According to the rapid rate of human population growth patterns, the world population may increase to more than 9.2 billion by 2050 (Jeffery et al. 2011). Meanwhile, malnutrition in Africa and South Asia is reported at 32 and 22%, respectively, due to lack of food security (Lehmann and Joseph 2009). Furthermore, as a result of continuous cultivation, highly weathered secondary minerals and high soil acidity, soil fertility of sub-Saharan African countries has been seriously affected (Duku et al. 2011). Therefore, improvement of crop production and achievement of food security for increasing the population pressure in a sustainable way is mandatory.

Major et al. (2010) studied the effect of biochar applications (0, 8, and 20 t·ha^{-1}) to a Colombian savanna Oxisol that was used to grow corn and soybean. They measured corn grain yield and nutrient uptake over time. Van Zwieten et al. (2010) investigated biomass and N uptake after biochar amendment into soybean, wheat, and radish in Ferrosol. The effect of biochar additions (1–5% by weight) to pepper and tomato plant development and productivity have been successfully determined (Graber et al. 2010) by observing leaf area; canopy dry weight; number of nodes; and yields of buds, flowers, and fruit.

2.3.7 Energy Production

Heat is coproduced during pyrolysis, and it can be used directly for conversion to electricity; therefore, produced heat can be used for heating buildings, houses, and schools, especially during the winter. Synthesis of fine chemicals, fertilizers, and adhesives are some of the potential benefits of liquids obtained from pyrolysis (Yaman 2004). As a result, producing fuels such as ethanol and renewable diesel and capturing energy as electricity and heat will secure the future supply of green energy in the world.

Ethanol production using straw, hulls, and distiller's dried grains in a fluidized-bed fast pyrolysis reactor has been investigated by Mullen et al. (2009). Ji-Lu (2007) showed that bio-oil and energy can be produced using rise husk pyrolyzed at temperatures between 420°C and 540°C in a fluidized bed. In addition, the importance of higher heating value for producing energy has been studied by using three different feedstocks: chicken litter, swine solids, and swine solids with rye grass (Ro et al. 2010). Quality and yield with process conditions, such as temperature, moisture content, and residence time, by using pine wood as the feedstock and characteristic properties, such as pH, water content, higher heating value, solid content, and ash of produced bio-oil, were analyzed by Thangalazhy-Gopakumar et al. (2010). The effect of alkali and alkaline earth metallic species (Na, Mg, K, and Ca) on the composition of the bio-oil produced from Mallee wood pyrolysis has been determined by Mourant et al. (2011). Moreover, economic feasibility of making bio-oil by using cellulosic biomass at different final temperatures and with different heating rates has been studied (Yoder et al. 2011). Electricity generation through slow and fast pyrolysis and to what extent it brings economic and environmental benefits have been examined in Taiwan (Kung et al. 2013).

2.3.8 Manage Waste

Both industrialized and developing countries are facing serious environmental, health, and economic issues more or less as a result of the huge amount of waste generation. But, modification of activated carbon is a complicated and time-consuming process. It is very costly (e.g., $2500/t in United States) to make activated carbon (Liu et al. 2011). Biochar production from different kinds of waste provides a sustainable way to manage in-site waste generation and thus has better economic and environmental prospects.

In Ghana, the potential of the organic portion of municipal solid wastes, livestock manures, agricultural residues, forestry residues, and wood processing wastes in the production of biochar has been evaluated (Duku et al. 2011). Also, biochar has been produced from pyrolysis of waste organic materials in the United States (Granatstein et al. 2009). Pyrolysis of livestock manures has been shown to generate nutrient-rich biochars with potential agronomic uses. Sewage sludge biochar produced at 500°C has shown effects on heavy metal (e.g., Cu, Ni, Zn, Cd, and Pb) solubility and bioavailability in a Mediterranean agricultural soil (Mendez et al. 2012).

2.4 Current Limitations and Future Challenges

Despite recent research, the environmental and economic benefits of biochar are not yet fully characterized. In short-term experiments (months to a few years), biochar generally enhances plant growth and soil nutrient status. However, the explanation for these benefits is not fully described, and neither the quantitative variability in response nor the durability of the effects is specified. Also, lack of data from long-term studies creates a problem in establishing a concrete link between biochar application and increased crop production. Due to the large variation observed in biochar properties, depending on the feedstock and various other conditions, there is an urgent need for a Manual of Procedures for its production as intended for specific applications. It may also be important to have a worldwide database of the characteristics in detail so that users are able to obtain an idea about what type of biochar to be used for a specific purpose. Furthermore, attention to long-term research and studies on untouched areas, such as microbial communities and their activities, should be encouraged. Development of predictive certainty for the longevity and durability of yield and other effects, particularly in relation to specific crops and soil types, is critical to guide selection of feedstock, production method, and application rate without long-term data.

For biochar to work as a green technology, there is a need to establish rigorously monitored supply networks and ensure that feedstocks come from sustainably managed lands and waste materials. For example, producing biochar from sewage sludge and municipal waste streams requires different assessments of risk because it is important to consider whether contaminants present in the pyrolysis feedstock are eliminated or whether they are modified to become more or less available in the biochar product. If not properly monitored, the production of biochar could lead to deforestation and processing of non-sustainable feedstocks, exacerbating the problems of decreasing biodiversity and increasing carbon emissions.

Biochar itself has been shown to potentially serve as a source of combustion-related contaminants such as polynuclear aromatic hydrocarbons and dioxins. Reported studies are less common regarding such potential negative implications of biochar application to soils, especially the impact on contaminant dynamics in a soil environment. In addition, the influence of biochar on soil chemical properties, nutrient availability, pH, and electrical conductivity as well as the environmental impact and function of biochar in water resources in terms of mechanisms has not been thoroughly evaluated. Specifically, the effect of DOC needs to be studied in detail due to the potential of contaminants to be transported together with DOC. Also, information on the physical breakdown of biochar, its long-term biological stability, and its impact on soil microorganisms, including research focused on soil enzymes, is lacking. Multicontaminant studies in soil and water are also necessary as in some cases, the change in pH may influence the release of other contaminants present in the system. These gaps in biochar research are a key challenge to be addressed as a goal toward environmental sustainability.

Acknowledgments

We thank Prof. C.B. Dissanayake, Director, IFS, and Mr. Tharanga Bandara for their support extended to this work. This study is partly supported by Indo-Sri Lanka bilateral research grant sanctioned by Ministry of Technology, Research and Atomic Energy, Sri Lanka and Department of Science and Technology (DST), Government of India

References

Ahmad M, Lee SS, Dou X, et al. (2012a). Effects of pyrolysis temperature on soybean stover- and peanut shell-derived biochar properties and TCE adsorption in water. *Bioresource Technology* **118**, 536–544.

Ahmad M, Lee SS, Rajapaksha AU, et al. (2013). Trichloroethylene adsorption by pine needle biochars produced at various pyrolysis temperatures. *Bioresource Technology* **143**, 615–622.

Ahmad M, Moon D, Lim K, et al. (2012b). An assessment of the utilization of waste resources for the immobilization of Pb and Cu in the soil from a Korean military shooting range. *Environmental Earth Sciences* **67**, 1023–1031.

Ahmad M, Moon DH, Vithanage M, et al. (2014a). Production and use of biochar from buffalo-weed (*Ambrosia trifida* L.) for trichloroethylene removal from water. *Journal of Chemical Technology & Biotechnology* **89**, 150–157.

Ahmad M, Rajapaksha AU, Lim JE, et al. (2014b). Biochar as a sorbent for contaminant management in soil and water: A review. *Chemosphere* **99**, 19–33.

Bai ZG, Dent DL, Olsson L, Schaepman ME. (2008). Proxy global assessment of land degradation. *Soil Use and Management* **24**, 223–234.

Brown TR, Wright MM, Brown RC. (2011). Estimating profitability of two biochar production scenarios: Slow pyrolysis vs fast pyrolysis. *Biofuels, Bioproducts and Biorefining* **5**, 54–68.

Bruun EW, Hauggaard-Nielsen H, Ibrahim N, et al. (2011). Influence of fast pyrolysis temperature on biochar labile fraction and short-term carbon loss in a loamy soil. *Biomass and Bioenergy* **35**, 1182–1189.

Bruun EW, Ambus P, Egsgaard H, Hauggaard-Nielsen H. (2012). Effects of slow and fast pyrolysis biochar on soil C and N turnover dynamics. *Soil Biology and Biochemistry* **46**, 73–79.

Cabrera A, Cox L, Spokas K, et al. (2014). Influence of biochar amendments on the sorption–desorption of aminocyclopyrachlor, bentazone and pyraclostrobin pesticides to an agricultural soil. *Science of the Total Environment* **470**, 438–443.

Chia CH, Gong B, Joseph SD, et al. (2012). Imaging of mineral-enriched biochar by FTIR, Raman and SEM–EDX. *Vibrational Spectroscopy* **62**, 248–257.

Dong X, Ma LQ, Li Y. (2011). Characteristics and mechanisms of hexavalent chromium removal by biochar from sugar beet tailing. *Journal of Hazardous Materials* **190**, 909–915.

Duku MH, Gu S, Hagan EB. (2011). Biochar production potential in Ghana—A review. *Renewable and Sustainable Energy Reviews* **15**, 3539–3551.

Gaskin J, Steiner C, Harris K, Das K, Bibens B. (2008). Effect of low-temperature pyrolysis conditions on biochar for agricultural use. *Transactions of the Asabe* **51**, 2061–2069.

Glaser B, Balashov E, Haumaier L, Guggenberger G, Zech W. (2000). Black carbon in density fractions of anthropogenic soils of the Brazilian Amazon region. *Organic Geochemistry* **31**, 669–678.

Glaser B, Haumaier L, Guggenberger G, Zech W. (2001). The 'Terra Preta' phenomenon: A model for sustainable agriculture in the humid tropics. *Die Naturwissenschaften* **88**, 37–41.

Graber ER, Harel YM, Kolton M, et al. (2010). Biochar impact on development and productivity of pepper and tomato grown in fertigated soilless media. *Plant and Soil* **337**, 481–496.

Granatstein D, Kruger C, Collins H, et al. (2009). *Use of biochar from the pyrolysis of waste organic material as a soil amendment.* Final project report. Center for Sustaining Agriculture and Natural Resources, Washington State University, Wenatchee, WA.

Han Y, Boateng AA, Qi PX, Lima IM, Chang J. (2013). Heavy metal and phenol adsorptive properties of biochars from pyrolyzed switchgrass and woody biomass in correlation with surface properties. *Journal of Environmental Management* **118**, 196–204.

Herath I, Kumarathilaka P, Navaratne A, Rajakaruna N, Vithanage M. (2014). Immobilization and phytotoxicity reduction of heavy metals in serpentine soil using biochar *Journal of Soils and Sediments* **15**, 126–138.

Jeffery S, Verheijen FGA, van der Velde M, Bastos AC. (2011). A quantitative review of the effects of biochar application to soils on crop productivity using meta-analysis. *Agriculture, Ecosystems & Environment* **144**, 175–187.

Ji-Lu Z. (2007). Bio-oil from fast pyrolysis of rice husk: Yields and related properties and improvement of the pyrolysis system. *Journal of Analytical and Applied Pyrolysis* **80**, 30–35.

Jones D, Edwards-Jones G, Murphy D. (2011). Biochar mediated alterations in herbicide breakdown and leaching in soil. *Soil Biology and Biochemistry* **43**, 804–813.

Karhu K, Mattila T, Bergström I, Regina K. (2011). Biochar addition to agricultural soil increased CH_4 uptake and water holding capacity—Results from a short-term pilot field study. *Agriculture, Ecosystems & Environment* **140**, 309–313.

Kim J-S, Sparovek G, Longo RM, De Melo WJ, Crowley D. (2007). Bacterial diversity of terra preta and pristine forest soil from the Western Amazon. *Soil Biology and Biochemistry* **39**, 684–690.

Kim KH, Kim J-Y, Cho T-S, Choi JW. (2012). Influence of pyrolysis temperature on physicochemical properties of biochar obtained from the fast pyrolysis of pitch pine (*Pinus rigida*). *Bioresource Technology* **118**, 158–162.

Kloss S, Zehetner F, Dellantonio A, et al. (2012). Characterization of slow pyrolysis biochars: Effects of feedstocks and pyrolysis temperature on biochar properties. *Journal of Environmental Quality* **41**, 990–1000.

Kookana R, Sarmah A, Van Zwieten L, Krull E, Singh B. (2011). Biochar application to soil: Agronomic and environmental benefits and unintended consequences. *Advances in Agronomy* **112**, 104–129.

Kung C-C, McCarl BA, Cao X. (2013). Economics of pyrolysis-based energy production and biochar utilization: A case study in Taiwan. *Energy Policy* **60**, 317–323.

Laird DA, Fleming P, Davis DD, et al. (2010). Impact of biochar amendments on the quality of a typical Midwestern agricultural soil. *Geoderma* **158**, 443–449.

Lehmann J, da Silva JP, Jr, Steiner C, et al. (2003). Nutrient availability and leaching in an archaeological Anthrosol and a Ferralsol of the Central Amazon basin: Fertilizer, manure and charcoal amendments. *Plant and Soil* **249**, 343–357.

Lehmann J, Joseph S. (2009). Biochar for environmental management: An introduction. In: (Ed.) Lehmann J, Joseph S. *Biochar for environmental management: Science and technology*, Earthscan, London, pp. 1–9.

Liu W-J, Zeng F-X, Jiang H, Zhang X-S. (2011). Preparation of high adsorption capacity bio-chars from waste biomass. *Bioresource Technology* **102**, 8247–8252.

Liu X-y, Qu J-j, Li L-q, et al. (2012). Can biochar amendment be an ecological engineering technology to depress N_2O emission in rice paddies?—A cross site field experiment from South China. *Ecological Engineering* **42**, 168–173.

Liu Z, Zhang F-S. (2009). Removal of lead from water using biochars prepared from hydrothermal liquefaction of biomass. *Journal of Hazardous Materials* **167**, 933–939.

Liu Z, Demisie W, Zhang M. (2013). Simulated degradation of biochar and its potential environmental implications. *Environmental Pollution* **179**, 146–152.

Marchal G, Smith KE, Rein A, et al. (2013). Impact of activated carbon, biochar and compost on the desorption and mineralization of phenanthrene in soil. *Environmental Pollution* **181**, 200–210.

Major J, Rondon M, Molina D, Riha SJ, Lehmann J. (2010). Maize yield and nutrition during 4 years after biochar application to a Colombian savanna oxisol. *Plant and Soil* **333**, 117–128.

Marris E. (2006). Putting the carbon back: Black is the new green. *Nature* **442**, 624–626.

Mendez A, Gomez A, Paz-Ferreiro J, Gasco G. (2012). Effects of sewage sludge biochar on plant metal availability after application to a Mediterranean soil. *Chemosphere* **89**, 1354–1359.

Meyer S, Glaser B, Quicker P. (2011). Technical, economical, and climate-related aspects of biochar production technologies: A literature review. *Environmental Science & Technology* **45**, 9473–9483.

Mohan D, Pittman CU, Jr, Bricka M, et al. (2007). Sorption of arsenic, cadmium, and lead by chars produced from fast pyrolysis of wood and bark during bio-oil production. *Journal of Colloid and Interface Science* **310**, 57–73.

Mohan D, Sarswat A, Ok YS, Pittman CU, Jr. (2014). Organic and inorganic contaminants removal from water with biochar, a renewable, low cost and sustainable adsorbent—A critical review. *Bioresource Technology* **160**, 191–202.

Mohan D, Sarswat A, Singh VK, Alexandre-Franco M, Pittman CU, Jr. (2011). Development of magnetic activated carbon from almond shells for trinitrophenol removal from water. *Chemical Engineering Journal* **172**, 1111–1125.

Mourant D, Wang Z, He M, et al. (2011). Mallee wood fast pyrolysis: Effects of alkali and alkaline earth metallic species on the yield and composition of bio-oil. *Fuel* **90**, 2915–2922.

Mukherjee A, Zimmerman AR. (2013). Organic carbon and nutrient release from a range of laboratory-produced biochars and biochar–soil mixtures. *Geoderma* **193**, 122–130.

Mukome FN, Six J, Parikh SJ. (2013). The effects of walnut shell and wood feedstock biochar amendments on greenhouse gas emissions from a fertile soil. *Geoderma* **200**, 90–98.

Mullen CA, Boateng AA, Hicks KB, Goldberg NM, Moreau RA. (2009). Analysis and comparison of bio-oil produced by fast pyrolysis from three barley biomass/byproduct streams. *Energy & fuels* **24**, 699–706.

Nakahira A, Nishida S, Fukunishi K. (2006). Synthesis of magnetic activated carbons for removal of environmental endocrine disrupter using magnetic vector. *Journal of the Ceramic Socity of Japan* **114**, 135–137.

Noor NM, Shariff A, Abdullah N. (2012). Slow Pyrolysis of Cassava Wastes for Biochar Production and Characterization. *Iranica Journal of Energy & Environment* **3**, 60–65.

Oldeman L. (1994). The global extent of soil degradation. In: (Eds.) Grrenland DJ, Szabolcs I. *Soil resilience and sustainable land use*, CAB International, Wallingford, UK, pp. 99–118.

Oleszczuk P, Josko I, Futa B, et al. (2014). Effect of pesticides on microorganisms, enzymatic activity and plant in biochar-amended soil. *Geoderma* **214–215**, 10–18.

Powlson D, Gregory P, Whalley W, et al. (2011). Soil management in relation to sustainable agriculture and ecosystem services. *Food Policy* **36**, S72–S87.

Rajapaksha AU, Vithanage M, Zhang M, et al. (2014). Pyrolysis condition affected sulfamethazine sorption by tea waste biochars. *Bioresource Technology* **166**, 303–308.

Rambabu N, Azargohar R, Dalai A, Adjaye J. (2013). Evaluation and comparison of enrichment efficiency of physical/chemical activations and functionalized activated carbons derived from fluid petroleum coke for environmental applications. *Fuel Processing Technology* **106**, 501–510.

Ro KS, Cantrell KB, Hunt PG. (2010). High-temperature pyrolysis of blended animal manures for producing renewable energy and value-added biochar. *Industrial & Engineering Chemistry Research* **49**, 10125–10131.

Saarnio S, Heimonen K, Kettunen R. (2013). Biochar addition indirectly affects N_2O emissions via soil moisture and plant N uptake. *Soil Biology and Biochemistry* **58**, 99–106.

Seifritz W. (1993). Should we store carbon in charcoal? *International Journal of Hydrogen Energy* **18**, 405–407.

Sohi S, Lopez-Capel E, Krull E, Bol R. (2009). *Biochar, climate change and soil: A review to guide future research*. In CSIRO Land and Water Science Report. **5**, 17–31.

Song Z, Lian F, Yu Z, et al. (2014). Synthesis and characterization of a novel MnO_x-loaded biochar and its adsorption properties for Cu^{2+} in aqueous solution. *Chemical Engineering Journal* **242**, 36–42.

Sopena F, Semple K, Sohi S, Bending G. (2012). Assessing the chemical and biological accessibility of the herbicide isoproturon in soil amended with biochar. *Chemosphere* **88**, 77–83.

Steiner C, Teixeira WG, Lehmann J, et al. (2007). Long term effects of manure, charcoal and mineral fertilization on crop production and fertility on a highly weathered Central Amazonian upland soil. *Plant and Soil* **291**, 275–290.

Sun K, Keiluweit M, Kleber M, Pan Z, Xing B. (2011). Sorption of fluorinated herbicides to plant biomass-derived biochars as a function of molecular structure. *Bioresource Technology* **102**, 9897–9903.

Thangalazhy-Gopakumar S, Adhikari S, Ravindran H, et al. (2010). Physiochemical properties of bio-oil produced at various temperatures from pine wood using an auger reactor. *Bioresource Technology* **101**, 8389–8395.

Uchimiya M, Bannon DI, Wartelle LH, Lima IM, Klasson KT. (2012). Lead retention by broiler litter biochars in small arms range soil: Impact of pyrolysis temperature. *Journal of Agricultural and Food Chemistry* **60**, 5035–5044.

Uchimiya M, Chang S, Klasson KT. (2011). Screening biochars for heavy metal retention in soil: Role of oxygen functional groups. *Journal of Hazardous Materials* **190**, 432–441.

Uchimiya M, Wartelle LH, Lima IM, Klasson KT. (2010). Sorption of deisopropylatrazine on broiler litter biochars. *Journal of agricultural and food chemistry* **58**, 12350–12356.

Vaccari F, Baronti S, Lugato E, et al. (2011). Biochar as a strategy to sequester carbon and increase yield in durum wheat. *European Journal of Agronomy* **34**, 231–238.

Van Zwieten L, Kimber S, Morris S, et al. (2010). Effects of biochar from slow pyrolysis of papermill waste on agronomic performance and soil fertility. *Plant and Soil* **327**, 235–246.

Van Zwieten L, Kimber S, Morris S, et al. (2013). Pyrolysing poultry litter reduces N_2O and CO_2 fluxes. *Science of The Total Environment* **465**, 279–287.

Vithanage M, Rajapaksha AU, Tang X, et al. (2014). Sorption and transport of sulfamethazine in agricultural soils amended with invasive-plant-derived biochar. *Journal of Environmental Management* **141**, 95–103.

Woolf D, Amonette JE, Street-Perrott FA, Lehmann J, Joseph S. (2010). Sustainable biochar to mitigate global climate change. *Nature Communications* **1**, 56.

Wu F, Jia Z, Wang S, Chang S, Startsev A. (2013). Contrasting effects of wheat straw and its biochar on greenhouse gas emissions and enzyme activities in a Chernozemic soil. *Biology and Fertility of Soils* **49**, 555–565.

Xue Y, Gao B, Yao Y, et al. (2012). Hydrogen peroxide modification enhances the ability of biochar (hydrochar) produced from hydrothermal carbonization of peanut hull to remove aqueous heavy metals: Batch and column tests. *Chemical Engineering Journal* **200**, 673–680.

Yaman S. (2004). Pyrolysis of biomass to produce fuels and chemical feedstocks. *Energy Conversion and Management* **45**, 651–671.

Yoder J, Galinato S, Granatstein D, Garcia-Pérez M. (2011). Economic tradeoff between biochar and bio-oil production via pyrolysis. *Biomass and Bioenergy* **35**, 1851–1862.

Yuan JH, Xu RK. (2011). The amelioration effects of low temperature biochar generated from nine crop residues on an acidic Ultisol. *Soil use and management* **27**, 110–115.

Yuan S, Dai Z-h, Zhou Z-j, et al. (2012). Rapid co-pyrolysis of rice straw and a bituminous coal in a high-frequency furnace and gasification of the residual char. *Bioresource Technology* **109**, 188–197.

Zhang M, Gao B, Varnoosfaderani S, et al. (2013). Preparation and characterization of a novel magnetic biochar for arsenic removal. *Bioresource Technology* **130**, 457–462.

Zheng W, Guo M, Chow T, Bennett DN, Rajagopala N. (2010). Sorption properties of greenwaste biochar for two triazine pesticides. *Journal of hazardous materials* **181**, 121–126.

SECTION II

Biochar Production
and Characterization

3

Biochar Production Technology

An Overview

Chapter 3

Biochar Production Technology

Sophie M. Uchimiya

Chapter Outline

3.1 Historical Perspectives

Char(coal) and the broader term *black carbon* (including soot) (Hammes et al. 2007) has long been recognized as a normal environmental (including soil) constituent resulting from fire (Shindo et al. 2004; Kaal et al. 2008; Nishimura et al. 2008) and industrial activities (Sullivan et al. 2011). Charcoal carbon can naturally make up as much as 35% of total organic carbon in U.S. agricultural soils (Skjemstad et al. 2002). The term *biochar* is used when char(coal) is deliberately added to soils. This biochar concept has received considerable attention since the widely publicized 2007 *Nature* commentary (Lehmann 2007) that proposed biochar as a tool for carbon sequestration in conjunction with bioenergy production. Biochar has since received global interest for agronomic, environmental, and industrial applications. Biochar production, i.e., pyrolysis of agricultural (Antal and Gronli 2003) and other waste (Shinogi and Kanri 2003) materials, is by no means new; it has been part of civilization for domestic and agricultural uses and as an energy source for thousands of years (Antal and Gronli 2003). Some cultures have traditionally used low-maintenance pyrolysis units such as kilns and cook stoves (Antal and Gronli 2003; Whitman et al. 2011; Sparrevik et al. 2014).

As of August 2014 (Scopus database), there are 77 (23 in January 2012) review papers, 1408 (310 in January 2012) peer-reviewed journal articles, and 24 book chapters as well as books (Lehmann and Joseph 2009) on biochar. These resources address diverse aspects of biochar that encompass agronomic benefits, including plant, microbe, and nematode interactions (Abit et al. 2012; Masiello et al. 2013; Kappler et al. 2014); economics (Woolf et al. 2010); carbon sequestration (Kuzyakov et al. 2009); pyrolysis system optimization (Antal and Gronli 2003); bioenergy production (Brewer et al. 2009); and environmental remediation (Beesley et al. 2011). In addition to the active research in both academic and federal sectors, biochar has received significant interest from policy makers and practitioners, partly because of its easily understood concept and application. Most importantly, farmers can produce biochars themselves by slow pyrolysis of agricultural wastes on-farm (Sparrevik et al. 2014). There are stakeholder organizations globally that support biochar either directly, for example, International Biochar Initiative, or indirectly via promoting green energy, for example, Southern Alliance for Clean Energy and Carbon War Room. A list of biochar manufacturers and retailers in North America is available on the U.S. Biochar Initiative website (http://biochar-us.org/manufacturers-retailers).

Field trials performed by federal laboratories, academia, and private sectors have been conducted across the United States through government-funded programs. However, biochar can give mixed results for intended applications, most importantly for crop yield. A 2012 review article concluded that only half of all biochar amendment studies showed enhanced crop yield (Spokas et al. 2012). Biochar's impact on greenhouse gas (carbon dioxide [CO_2], methane, and nitrogen oxides) fluxes to and from soil is even more confounding (Kammann et al. 2012). Laboratory findings cannot directly be extrapolated to farmland, because of a multitude of field variables (e.g., soil mixing) that impact measurable amounts of greenhouse gases. Initial rapid CO_2 release often observed in biochar-amended soils can originate from calcium carbonate and labile carbon. Therefore, ^{14}C-labeling of biochar (Kuzyakov et al. 2009) is necessary to estimate the mineralization rate of recalcitrant carbon.

This chapter provides an overview of biochar production technology, with particular emphasis on different structural components of biochar: labile and recalcitrant carbon and ash. Discussions focus on (1) feedstock heterogeneity and operational variables that pose challenges for farmers to produce biochars with a consistent quality; and (2) specific local needs of producers, for example, scale of operation, available biomass, soil type, and local climate. Recommendations are given on localized, site-specific, case-by-case biochar use based on the purpose of the biochar application appropriate for the soil properties, locally availability feedstock, and socioeconomic situations. Subsequent chapters in Section II: Biochar Production and Characterization provides detailed accounts on the chemical, physical, and surface characterization of biochar (Chapter 4); and advanced micro- and spectroscopic characterizations of biochar (Chapter 6).

3.2 Goals of Biochar Production

The recent popularity in biochar amendment partly resulted from intensively investigated traditional farming practices of the Amazon basin (Terra Preta de Índio [black earth of the Indian], or terra preta) (Lehmann et al. 2003). Terra preta is oxosol-turned-anthrosol by the intentional slash-and-char practice of pre-Columbian farmers that resulted in high soil organic carbon (SOC), cation exchange capacity (CEC), pH, and fertility that is observed in terra preta even today, although the practice ceased more than 500 years ago

(Glaser et al. 2002). These observations inspired the biochar amendment to improve soil fertility and crop yield. Today, claimed benefits of biochar amendment include improved runoff quality (reduced nutrient leaching) and the carbon negative concept arising from the recalcitrance of pyrogenic carbon and the mitigation of greenhouse gas emissions (Tilman et al. 2006). However, observations made at archeological sites of the Amazonian basin should be carefully analyzed and interpreted before implementing the same practice on a large scale elsewhere.

Pyrogenic (black) carbon is composed of short stacks of polycyclic aromatic sheets "arranged in highly disordered fashion to form a poorly interconnected microporous network" and oxidized edge functional groups (Zhu et al. 2005). Black carbon is particularly effective for sorption of planar, hydrophobic, aromatic compounds, such as polycyclic aromatic hydrocarbons and planar polychlorinated biphenyls, that are able to engage in face-to-face orientation with the graphene planes (Zhu et al. 2005). Because the smallest pores are filled first, sorption is more favorable at low surface coverage (Shih and Gschwend 2009). Consequently, nonlinear adsorption on pyrogenic carbon in soils and sediments dominates at low equilibrium solute concentration-to-solubility ratios (Cornelissen et al. 2005). Many studies have been conducted to elucidate sorption mechanisms of both natural and synthetic black carbon for polar and nonpolar solutes (Cornelissen et al. 2005).

Pyrolysis temperature impacts the degree of carbonization: the development of surface area and aromaticity in the source material (Keiluweit et al. 2010). The adsorption mechanism predominates on the carbonized fraction, whereas partitioning occurs on the labile, partially carbonized fraction (Chun et al. 2004). Polarity, aromaticity, surface area, and pore size distribution can control the sorption of nonpolar and polar solutes on 300–700°C plant biochars (Chun et al. 2004; Chen et al. 2008). Studies using natural organic matter (NOM) provided further evidence for the determining roles of sorbent surface area, polarity, and surface charge (Pignatello et al. 2006). Sorbed NOM can block micropores and increase surface acidity (Yang and Sheng 2003; Qiu et al. 2009), and prevent sorption. Isotherm becomes increasingly linear (Pignatello et al. 2006) in the presence of NOM, especially for larger solutes (from benzene to naphthalene to phenanthrene). In addition to batch isotherm experiments, advanced microscopic techniques allowed researchers to locate sorbed chemicals on porous materials at a nanometer scale (Obst et al. 2011). Scanning transmission x-ray microscopy is a useful technique to visualize surface adsorption of hydrophobic organic contaminants (Obst et al. 2011). Phenanthrene accumulated mainly on cracks and external surfaces, and within the few intraparticle pores of commercial charcoal (Obst et al. 2009) (Figure 3.1). Even on a commercial activated carbon having a less compact structure and larger surface area, micro- and mesopores were largely unoccupied by phenanthrene (Obst et al. 2011). Instead, phenanthrene adsorbed on the external surface and surfaces of cracks, and at the edges of large pores and adjacent regions of activated carbon (Obst et al. 2011).

In addition to the soil-free experiments described above, many soil amendment studies have been conducted to stabilize (i.e., decrease dissolved concentration and bioavailability) organic and inorganic contaminants (for reviews, see Kookana 2010; Beesley et al. 2011). Stabilization of heavy metals in soil strongly depends on the amount of oxygen-containing (especially carboxyl) surface functional groups to complex metal ions (Uchimiya et al. 2012). Because a lower pyrolysis temperature forms biochars with a higher amount of surface functional groups, heavy metal [e.g., Pb(II), Cu(II), Zn(II)] stabilization in a model soil system decreased as a function of pyrolysis temperature (Uchimiya et al. 2011). Biochar amendment can result in higher soluble concentration of elements such as As(V) and Sb(V) that exist predominantly in anionic forms (Wilson et al. 2010). For example, biochar increased soluble As(V) concentration by as much as 30-fold in contaminated soil, concurrently with the stabilization of Cd(II) and Zn(II) (Beesley et al. 2010). When oxoanion-forming elements such as Sb(V) and As(V) are not a potential risk driver of the target site, carboxyl-rich biochars can be used to stabilize Pb(II), Cu(II), and Zn(II) at a practical (≤5 wt%) amendment rate (Uchimiya et al. 2012).

Multilocation biochar field trials of the U.S. Department of Agriculture-Agricultural Research Service (USDA-ARS) began

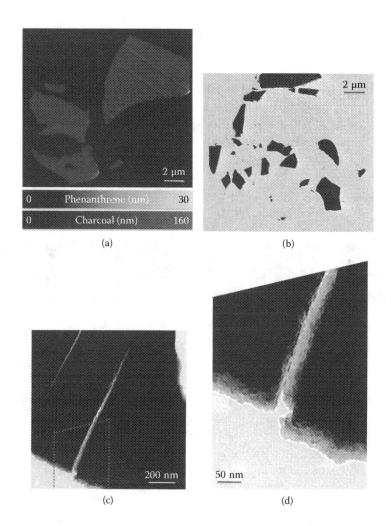

Figure 3.1 (a) Scanning transmission x-ray microscopy imaging of phenanthrene (bright regions) sorbed to charcoal (CC, gray areas). (b) Transmission electron micrscope (TEM) image of CC showing low porosity. Bright-field TEM images of (c) cracks through the CC particle. (d) and (c) at higher magnification. (Reprinted from Obst, M. et al. 2009. *Geochim. Cosmochim. Acta.* 73, 4180–4198.)

in November 2007 in Ames, Iowa, at a location where nearly 8 acres of cornfield was amended with 8 tons/acre hardwood biochar. Possibly because of the fertile Iowa soil, no significant differences in crop yield and soil quality were observed. At other ARS locations (Idaho, Kentucky, Minnesota, South Carolina, and Texas), improved water retention was observed, especially on sandy soils. This ongoing USDA-ARS multilocation biochar field plot trial, along with Greenhouse Gas Reduction and Carbon Enhancement Network (GRACEnet project) (Del Grosso et al. 2013) and ARS Biochar and Pyrolysis Initiative, currently involves 31 ARS locations across the United States, including the Southern Regional Research Center in New Orleans, Louisiana. The focus at the New Orleans location has been on the fate of inorganic (heavy metal) and organic (agrochemicals) contaminants and soil properties, as well as industrial applications (e.g., water filtration).

3.3 Thermochemical Conversion Technologies

Biomass is a general term encompassing the organic matter in living organisms and their residues originating from growing plants (including algae) and animal manure, where manure can be thought of as a processed form of plant materials (Vassilev et al. 2010). Biomass is a complex heterogeneous structure composed of cellulose, lignin, and other organic (primarily C, H, N, S, and O) and inorganic components that can exceed 40% of dry weight (Vassilev et al. 2010). Inorganic components [silicon dioxide (SO_2), aluminium oxide, iron(III) oxide, titanium oxide, calcium oxide, potassium oxide, phosphorus pentoxide, magnesium oxide, sulfur trioxide, and sodium oxide] may originally exist in biomass or be produced during thermal treatments (Song et al. 2011). An exergonic process refers to a reaction having negative standard Gibbs free energy change ($\Delta G°$); this is a measure for the maximum energy produced to bring the exothermic reaction to equilibrium (Laidler and Meiser 1999). Szargut's reference environmental model provides an estimate for the standard chemical exergies for a complete combustion of biomass to form CO_2, water, nitrogen gas (N_2), SO_2, and ash (Song et al. 2011). Specific chemical exergy for 86 biomass ranged from 11.5 to 24.2 $MJ \cdot kg^{-1}$ and increased in the following order of biomass grouping: manure < sludge < straws < grasses < hulls and shells < wood (Song et al. 2011). Exergy correlated with empirical energy content (higher heating value) by the factor of 1.047 (Song et al. 2011).

Biomass energy, or briefly bioenergy, is stored in plant and animal waste materials. Specific examples include wood from natural forests, agricultural crop residues, industrial wastes such as sludge and paper wastes, and animal waste (manure) (Demirbas 2001). Unprocessed biomass is bulky and typically has high moisture and alkaline earth metal contents, thereby complicating the direct use of biomass as a fuel (Wu et al. 2009). However, biomass can be treated thermochemically via pyrolysis to create materials with higher energy densities for subsequent bioenergy production (Wu et al. 2009). The ultimate goal of biomass waste-to-energy conversion is to produce fossil fuel substitutes for heating, electricity generation, and engine operation, with many U.S. states and other countries having legislative renewable energy targets. Currently, only ~5% of renewable biomass is used for energy production globally (Field et al. 2008).

The conversion technologies for biomass use can be separated into three categories: direct combustion, thermochemical processes, and biochemical processes. Specific examples include anaerobic digestion, (bio)gasification, and blending with fossil fuel sources (cofiring). Energy content of raw and thermochemically processed (e.g., pyrolysis and torrefaction) biomass can be used for direct combustion, cofiring with coal, or gasification (Pimchuai et al. 2010). Target end-use options remain to be (1) heat and power generation, (2) transportation fuels to replace fossil fuels, and (3) chemical feedstocks. Following cleanup and conditioning of the end products, the Fischer–Tropsch process and various other applications are available (Figure 3.2) (Butterman and Castaldi 2009). The hydrogen gas/carbon monoxide (H_2/CO) ratios close to 2 are suitable for Fischer–Tropsch synthesis of liquid hydrocarbon fuels, whereas a ratio between 4 and 6 is ideal for the operation of solid oxide fuel cells (Butterman and Castaldi 2009). However, biomass produces tars, alkali metals, water, organic acids, and dust as well as N-, S-, and Cl-containing compounds that are corrosive to processing equipment and turbines.

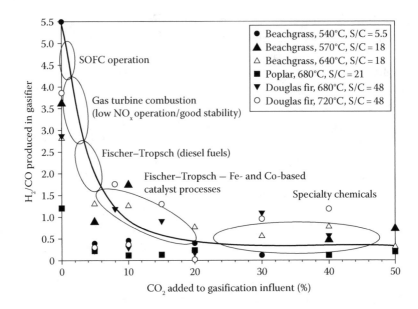

Figure 3.2 Syngas adjustment as a function of CO_2 reactant. (Reprinted from Butterman, H.C., Castaldi, M.J. 2009. *Environ. Sci. Technol.* 43, 9030–9037.)

High-quality charcoal produced from plant biomass has a calorific value of 30–33 $MJ \cdot kg^{-1}$ (within the range of high-quality coal), 21–23% volatile matter (VM) (ASTM 2009), ~70% fixed carbon, and 1–3% ash (Mok et al. 1992). These properties are ideal for cooking and heating. Theoretical yield (based on reaction stoichiometry) of carbon from cellulose is 44% on a mass fraction basis and is estimated to be 44–55% for biomass (for which exact stoichiometry is not available) (Mok et al. 1992). The highest biomass-derived, high-quality (per criteria described above) charcoal yield reported in the literature is 47% from moist wood logs in a sealed reactor (Mok et al. 1992). Char yield linearly correlated with heat of pyrolysis (measured using differential scanning calorimeter in joules per gram), and the reaction was exothermic at above 20% yield (Mok et al. 1992). Similar findings were made under atmospheric pressure (Milosavljevic et al. 1996).

Biomass pyrolysis becomes increasingly exothermic with higher system pressure, feedstock loading, and lower gas flow rate (Antal and Gronli 2003). In a sealed reactor without gas flow, higher biomass feedstock loading increased the charcoal yield (40%) as well as the exothermic heat release (-600 J·g^{-1}), and it lowered the reaction onset temperature (Mok et al. 1992). The results were attributed to high-vapor-phase concentrations of volatile products in the closed system, rather than the system pressure (Mok et al. 1992). Biomass charcoals (especially fine powders) are extremely reactive. There are reports on self-ignition (resulting from oxygen gas [O_2] chemsorption via dangling bonds of charcoal) of tightly packed charcoal fine powders with high VM content (Antal and Gronli 2003). At lower temperatures, combustion of charcoal occurs primarily by the reaction between O_2 and paraffinic carbons; at elevated temperatures, aromatic carbons become oxidized (Antal and Gronli 2003).

Biochar can be produced on-farm by slow (1–20°C·min^{-1}) pyrolysis of waste biomass (Brewer et al. 2009) typically at 350–700°C. The biochar property can be manipulated by altering pyrolysis parameters (Antal and Gronli 2003), especially the maximum heating temperature, heating rate, and feedstock. Higher temperature ($\geq$550°C) pyrolysis produces stable and highly aromatic biochars having high Brunauer–Emmett–Teller (BET) surface areas ($>$400 m^2/g) (Joseph et al. 2010; Keiluweit et al. 2010). Lower temperature ($<$550°C) pyrolysis is a less expensive option but produces less condensed biochar structure that is expected to be more biodegradable (Fuertes et al. 2010; Joseph et al. 2010; Khodadad et al. 2011). When feedstocks contain high amount of ash (e.g., grass, rice husk, and manure), a higher pyrolysis temperature results in strongly alkaline (pH as high as 10) biochar (Novak et al. 2009). Biochar-induced changes in soil pH can be attenuated by selecting appropriate feedstock as well as the pyrolysis conditions and amendment rate. For example, wood contains minimal ash and is more suitable for alkaline soils. The stability of biochar toward microbial and abiotic degradation (that results in CO_2 release) increases as a function of pyrolysis temperature (Zimmerman 2010). Lower H:C ratio and higher fixed carbon (based on thermal stability) content provide a measure for greater recalcitrance of biochar (Zimmerman 2010). Once amended, biochar undergoes physical changes in soil: coating of the biochar surface by mineral, organic, clay, and silt components of soil; root penetration (Joseph et al. 2010, 2013); sizing (Gao and Wu 2013; Spokas et al. 2014); and transport (Zhang et al. 2010).

Although other processing options (fast pyrolysis and gasification) are available for bioenergy production, slow pyrolysis produces biochars with lower mineral (compared to gasification) and higher fixed and recalcitrant carbon (compared to fast pyrolysis) components that are more suitable for carbon sequestration in soil. In addition, mobile slow pyrolysis units are commercially available for on-farm use by farmers. The purpose of fast pyrolysis is to maximize the yield of liquid-phase products (bio-oil) by a rapid heating of biomass ($>1000°C \cdot s^{-1}$ heating rate for <1 min at ~500°C) for subsequent catalytic conversion to useable liquid hydrocarbon fuels and other useful industrial chemicals. Gasification is used to maximize the syngas (often called producer gas, $CO+H_2$) yield by exposing biomass to high temperature (>800°C) under controlled amounts of O_2, steam, or both.

3.4 Overview of Biochar Characterization Methods

Biochars are postulated to improve the soil quality by the following routes (Glaser et al. 2002):

1. Increase in CEC, pH, and water retention to improve nutrient retention and bioavailability.
2. Uptake and release of inorganic and organic nutrients and pesticides in micro- and mesopores.
3. Porous environment for fungi and bacteria as a physically protected habitat from larger predators, and alteration of allelochemical signaling dynamics between plant roots and colonizing microorganisms within the rhizosphere (Warnock et al. 2007; Masiello et al. 2013).

Because the biochar research community consists of agronomists, soil chemists, and environmental and material scientists and engineers, different definitions of pore diameters can confound the interpretation of porosity. The International Union of Pure and Applied Chemistry (IUPAC) defines the pore diameter in catalysis as follows: <2 nm, micropore; <50 nm, mesopore; and >50 nm, macropore. The IUPAC defines pore diameter of macroporous polymers as 50 nm to 1 μm. In soil science and agronomy journals, a separate definition is used for pore diameter: <30 μm, micropore; water held in the pore volume is not available to plants; and >30 μm, mesopore; retained water is available for plant uptake (Yoo et al. 2006). The IUPAC definition of micro- (<2 nm) and mesopores (2–50 nm) can comprise the majority of biochar's surface area (Braida et al. 2003) measurable by N_2 (BET) and CO_2 sorption. To date, limited studies have used transmission electron microscopy to show macropores (>50 nm) (Chia et al. 2010; Joseph et al. 2010, 2013; Obst et al. 2011), and direct imaging of interactions involving micropores (<2 nm) and mesopore (2–50 nm) are scarce in the literature (Obst et al. 2011). Many reports showed scanning electron microscopy images of biochar at low magnifications (micrometer range). This size range is important for the uptake and release of water and represents "zoomed out" images away from the IUPAC definition of micro- and mesopores.

Proximate and ultimate analyses (adapted from the characterization of coal) and BET surface area (adapted from the characterization of activated carbon) measurements are the standard biochar characterization procedures (Shinogi and Kanri 2003; Wu et al. 2011; Cantrell et al. 2012). Proximate analysis is based on the American Society for Testing and Materials (ASTM) method using thermogravimetric analyzer (ASTM 2009). Moisture is determined as the weight loss after heating the char in an open crucible to 107°C and holding at this temperature until the sample weight is stabilized. VM is determined as the weight loss after heating the char in a covered crucible to 950°C and holding for 7 min. Ash is defined as the remaining mass after subsequently heating in an open crucible to 750°C and holding at this temperature until sample weight is stabilized. After the determination of moisture, ash, and VM, fixed carbon is calculated by difference. Ultimate analysis provides total CHNSO (by elemental analyzer) and mineral (by acid digestion followed by inductively coupled plasma mass spectrometry) contents of biochar. Portions of VM can desorb from biochars and can consist of >140 chemical compounds (Spokas et al. 2011), including ethylene, a known plant hormone (Spokas et al. 2010), and phenolic compounds that can stimulate microbial growth (Deenik et al. 2010).

Many economic analyses have been conducted to understand the impact of biochar on global carbon cycles (Woolf et al. 2010) and life-cycle assessments for model farms (Gaunt and Lehmann 2008; Whitman et al. 2011); some studies focused on the production variables (Meyer et al. 2011). For example, Woolf et al. (2010) concluded that biochar production is more sustainable than biomass combustion for bioenergy, except for sites where soil is fertile and coal is the primary fuel being offset by bioenergy production.

3.5 Potential Risks and Prospects

Like other carbon-rich soil amendments, for example, compost and manure, biochar is most effective for improving the quality of sandy, low CEC, and low total organic carbon soils. Additional C accumulation (after biochar amendment) must be fully investigated to understand C sequestration potential. To date, 50% of complied studies in a review article (Spokas et al. 2012) showed short-term positive yield or growth impacts, whereas 30% reported no difference, and 20% reported negative yield or growth impacts. Although biochar holds promise for reducing carbon footprint and increasing crop yield, large-scale amendment should be carefully evaluated on the basis of the local agricultural practices, soil properties, and the purposes. Producers should not assume that the oxosol-turned-anthrosol phenomena of terra preta will occur on their soils by biochar amendment, and potential risks must be examined. For example, biochar can reduce the efficacy of pesticides and other agrochemicals (Kookana 2010). Longevity of biochar in agroecosystems can compromise pesticide applications for generations. In addition, the fine dust of biochar powder can cause unacceptable air pollution problems (Laird et al. 2009).

Furthermore, biochar is flammable, and there is a potential risk of self-ignition during storage and transport (Antal and Gronli 2003; Laird et al. 2009; Gao and Wu 2011). Pelletized biochar provides a solution to some of these problems. However, depending on attrition, fine powders will form over time (Wang et al. 2012; Gao and Wu 2013). An example of promising, localized biochar use is biochar-induced increase in water-holding retention and release, and reduction of irrigation costs in arid and desert regions. For heavily contaminated (e.g., shooting range) nonagricultural soils, biochar can be used as a cost-competitive remediation strategy in conjunction with the carbon sequestration.

References

Abit, S.M., Bolster, C.H., Cai, P., Walker, S.L. 2012. Influence of feedstock and pyrolysis temperature of biochar amendments on transport of *Escherichia coli* in saturated and unsaturated soil. *Environ. Sci. Technol.* 46, 8097–8105.

Antal, M.J., Gronli, M. 2003. The art, science, and technology of charcoal production. *Ind. Eng. Chem. Res.* 42, 1619–1640.

ASTM Method D5142. 2009. *Standard Test Methods for Proximate Analysis of the Analysis Sample of Coal and Coke by Instrumental Procedures.* American Society for Testing and Materials, West Conshohocken, PA.

Beesley, L., Moreno-Jiménez, E., Gomez-Eyles, J.L. 2010. Effects of biochar and greenwaste compost amendments on mobility, bioavailability and toxicity of inorganic and organic contaminants in a multi-element polluted soil. *Environ. Pollut.* 158, 2282–2287.

Beesley, L., Moreno-Jiménez, E., Gomez-Eyles, J.L., Harris, E., Robinson, B., Sizmur, T. 2011. A review of biochars' potential role in the remediation, revegetation and restoration of contaminated soils. *Environ. Pollut.* 159, 3269–3282.

Braida, W.J., Pignatello, J.J., Lu, Y.F., Ravikovitch, P.I., Neimark, A.V., Xing, B.S. 2003. Sorption hysteresis of benzene in charcoal particles. *Environ. Sci. Technol.* 37, 409–417.

Brewer, C.E., Schmidt-Rohr, K., Satrio, J.A., Brown, R.C. 2009. Characterization of biochar from fast pyrolysis and gasification systems. *Environ. Progr. Sustain. Energ.* 28, 386–396.

Butterman, H.C., Castaldi, M.J. 2009. CO_2 as a carbon neutral fuel source via enhanced biomass gasification. *Environ. Sci. Technol.* 43, 9030–9037.

Cantrell, K.B., Hunt, P.G., Uchimiya, M., Novak, J.M., Ro, K.S. 2012. Impact of pyrolysis temperature and manure source on physicochemical characteristics of biochar. *Bioresour. Technol.* 107, 419–428.

Chen, B.L., Zhou, D.D., Zhu, L.Z. 2008. Transitional adsorption and partition of nonpolar and polar aromatic contaminants by biochars of pine needles with different pyrolytic temperatures. *Environ. Sci. Technol.* 42, 5137–5143.

Chia, C.H., Munroe, P., Joseph, S., Lin, Y. 2010. Microscopic characterisation of synthetic Terra Preta. *Aust. J. Soil Res.* 48, 593–605.

Chun, Y., Sheng, G.Y., Chiou, C.T., Xing, B.S. 2004. Compositions and sorptive properties of crop residue-derived chars. *Environ. Sci. Technol.* 38, 4649–4655.

Cornelissen, G., Gustafsson, O., Bucheli, T.D., Jonker, M.T.O., Koelmans, A.A., Van Noort, P.C.M. 2005. Extensive sorption of organic compounds to black carbon, coal, and kerogen in sediments and soils: Mechanisms and consequences for distribution, bioaccumulation, and biodegradation. *Environ. Sci. Technol.* 39, 6881–6895.

Deenik, J.L., McClellan, T., Uehara, G., Antal, M.J., Campbell, S. 2010. Charcoal volatile matter content influences plant growth and soil nitrogen transformations. *Soil Sci. Soc. Am. J.* 74, 1259–1270.

Del Grosso, S.J., White, J.W., Wilson, G., Vandenberg, B., Karlen, D.L., Follett, R.F., Johnson, J.M.F., et al. 2013. Introducing the GRACEnet/REAP data contribution, discovery, and retrieval system. *J. Environ. Qual.* 42, 1274–1280.

Demirbas, A. 2001. Biomass resource facilities and biomass conversion processing for fuels and chemicals. *Energ. Convers. Manag.* 42, 1357–1378.

Field, C.B., Campbell, J.E., Lobell, D.B. 2008. Biomass energy: The scale of the potential resource. *Trends Ecol. Evol.* 23, 65–72.

Fuertes, A.B., Arbestain, M.C., Sevilla, M., Macia-Agullo, J.A., Fiol, S., Lopez, R., Smernik, R.J., Aitkenhead, W.P., Arce, F., Macias, F. 2010. Chemical and structural properties of carbonaceous products obtained by pyrolysis and hydrothermal carbonisation of corn stover. *Aust. J. Soil Res.* 48, 618–626.

Gao, X., Wu, H. 2011. Biochar as a fuel: 4. Emission behavior and characteristics of PM 1 and PM10 from the combustion of pulverized biochar in a drop-tube furnace. *Energ. Fuel.* 25, 2702–2710.

Gao, X., Wu, H. 2013. Aerodynamic properties of biochar particles: Effect of grinding and implications. *Environ. Sci. Technol. Lett.* 1, 60–64.

Gaunt, J.L., Lehmann, J. 2008. Energy balance and emissions associated with biochar sequestration and pyrolysis bioenergy production. *Environ. Sci. Technol.* 42, 4152–4158.

Glaser, B., Lehmann, J., Zech, W. 2002. Ameliorating physical and chemical properties of highly weathered soils in the tropics with charcoal—A review. *Biol. Fertil. Soils.* 35, 219–230.

Hammes, K., Schmidt, M.W.I., Smernik, R.J., Currie, L.A., Ball, W.P., Nguyen, T.H., Louchouarn, P., et al. 2007. Comparison of quantification methods to measure fire-derived (black/elemental) carbon in soils and sediments using reference materials from soil, water, sediment and the atmosphere. *Glob. Biogeochem. Cycles.* 21, GB3016.

Joseph, S., Graber, E.R., Chia, C., Munroe, P., Donne, S., Thomas, T., Nielsen, S., et al. 2013. Shifting paradigms: Development of high-efficiency biochar fertilizers based on nano-structures and soluble components. *Carbon Manag.* 4, 323–343.

Joseph, S.D., Camps-Arbestain, M., Lin, Y., Munroe, P., Chia, C.H., Hook, J., Van Zwieten, L., et al. 2010. An investigation into the reactions of biochar in soil. *Aust. J. Soil Res.* 48, 501–515.

Kaal, J., Martínez-Cortizas, A., Buurman, P., Boado, F.C. 2008. 8000 yr of black carbon accumulation in a colluvial soil from NW Spain. *Quaternary Res.* 69, 56–61.

Kammann, C., Ratering, S., Eckhard, C., Müller, C. 2012. Biochar and hydrochar effects on greenhouse gas (carbon dioxide, nitrous oxide, and methane) fluxes from soils. *J. Environ. Qual.* 41, 1052–1066.

Kappler, A., Wuestner, M.L., Ruecker, A., Harter, J., Halama, M., Behrens, S. 2014. Biochar as an electron shuttle between bacteria and Fe(III) minerals. *Environ. Sci. Technol. Lett.* 1, 339–344.

Keiluweit, M., Nico, P.S., Johnson, M., Kleber, M. 2010. Dynamic molecular structure of plant biomass-derived black carbon (biochar). *Environ. Sci. Technol.* 44, 1247–1253.

Khodadad, C.L.M., Zimmerman, A.R., Green, S.J., Uthandi, S., Foster, J.S. 2011. Taxa-specific changes in soil microbial community composition induced by pyrogenic carbon amendments. *Soil Biol. Biochem.* 43, 385–392.

Kookana, R.S. 2010. The role of biochar in modifying the environmental fate, bioavailability, and efficacy of pesticides in soils: A review. *Aust. J. Soil Res.* 48, 627–637.

Kuzyakov, Y., Subbotina, I., Chen, H., Bogomolova, I., Xu, X. 2009. Black carbon decomposition and incorporation into soil microbial biomass estimated by 14C labeling. *Soil Biol. Biochem.* 41, 210–219.

Laidler, K.J., Meiser, J.H. 1999. *Physical Chemistry.* Houghton Mifflin Company, Boston, MA.

Laird, D.A., Brown, R.C., Amonette, J.E., Lehmann, J. 2009. Review of the pyrolysis platform for coproducing bio-oil and biochar. *Biofuels Bioprod. Biorefin.* 3, 547–562.

Lehmann, J. 2007. A handful of carbon. *Nature.* 447, 143–144.

Lehmann, J., da Silva, J.P., Steiner, C., Nehls, T., Zech, W., Glaser, B. 2003. Nutrient availability and leaching in an archaeological anthrosol and a ferralsol of the central amazon basin: Fertilizer, manure and charcoal amendments. *Plant Soil.* 249, 343–357.

Lehmann, J., Joseph, S. 2009. *Biochar for Environmental Management: Science and Technology.* Earthscan Ltd, London, UK.

Masiello, C.A., Chen, Y., Gao, X., Liu, S., Cheng, H.-Y., Bennett, M.R., Rudgers, J.A., Wagner, D.S., Zygourakis, K., Silberg, J.J. 2013. Biochar and microbial signaling: Production conditions determine effects on microbial communication. *Environ. Sci. Technol.* 47, 11496–11503.

Meyer, S., Glaser, B., Quicker, P. 2011. Technical, economical, and climate-related aspects of biochar production technologies: A literature review. *Environ. Sci. Technol.* 45, 9473–9483.

Milosavljevic, I., Oja, V., Suuberg, E.M. 1996. Thermal effects in cellulose pyrolysis: Relationship to char formation processes. *Ind. Eng. Chem. Res.* 35, 653–662.

Mok, W.S.L., Antal, M.J., Szabo, P., Varhegyi, G., Zelei, B. 1992. Formation of charcoal from biomass in a sealed reactor. *Ind. Eng. Chem. Res.* 31, 1162–1166.

Nishimura, S., Noguchi, T., Shindo, H. 2008. Distribution of charred plant fragments in particle size fractions of Japanese volcanic ash soils. *Soil Sci. Plant Nutr.* 54, 490–494.

Novak, J.M., Lima, I., Xing, B., Gaskin, J.W., Steiner, C., Das, K.C., Ahmedna, M., et al. 2009. Characterization of designer biochar produced at different temperatures and their effects on a loamy sand. *Ann. Environ. Sci.* 3, 195–206.

Obst, M., Dynes, J.J., Lawrence, J.R., Swerhone, G.D.W., Benzerara, K., Karunakaran, C., Kaznatcheev, K., Tyliszczak, T., Hitchcock, A.P. 2009. Precipitation of amorphous $CaCO_3$ (aragonite-like) by cyanobacteria: A STXM study of the influence of EPS on the nucleation process. *Geochim. Cosmochim. Acta.* 73, 4180–4198.

Obst, M., Grathwohl, P., Kappler, A., Eibl, O., Peranio, N., Gocht, T. 2011. Quantitative high-resolution mapping of phenanthrene sorption to black carbon particles. *Environ. Sci. Technol.* 45, 7314–7322.

Pignatello, J.J., Kwon, S., Lu, Y.F. 2006. Effect of natural organic substances on the surface and adsorptive properties of environmental black carbon (char): Attenuation of surface activity by humic and fulvic acids. *Environ. Sci. Technol.* 40, 7757–7763.

Pimchuai, A., Dutta, A., Basu, P. 2010. Torrefaction of agriculture residue to enhance combustible properties. *Energ. Fuel.* 24, 4638–4645.

Qiu, Y.P., Xiao, X.Y., Cheng, H.Y., Zhou, Z.L., Sheng, G.D. 2009. Influence of environmental factors on pesticide adsorption by black carbon: pH and model dissolved organic matter. *Environ. Sci. Technol.* 43, 4973–4978.

Shih, Y.H., Gschwend, P.M. 2009. Evaluating activated carbon-water sorption coefficients of organic compounds using a linear solvation energy relationship approach and sorbate chemical activities. *Environ. Sci. Technol.* 43, 851–857.

Shindo, H., Honna, T., Yamamoto, S., Honma, H. 2004. Contribution of charred plant fragments to soil organic carbon in Japanese volcanic ash soils containing black humic acids. *Org. Geochem.* 35, 235–241.

Shinogi, Y., Kanri, Y. 2003. Pyrolysis of plant, animal and human waste: Physical and chemical characterization of the pyrolytic products. *Bioresour. Technol.* 90, 241–247.

Skjemstad, J.O., Reicosky, D.C., Wilts, A.R., McGowan, J.A. 2002. Charcoal carbon in US agricultural soils. *Soil Sci. Soc. Am. J.* 66, 1249–1255.

Song, G., Shen, L., Xiao, J. 2011. Estimating specific chemical exergy of biomass from basic analysis. *Data. Ind. Eng. Chem. Res.* 50, 9758–9766.

Sparrevik, M., Lindhjem, H., Andria, V., Fet, A.M., Cornelissen, G. 2014. Environmental and socioeconomic impacts of utilizing waste for biochar in rural areas in Indonesia—A systems perspective. *Environ. Sci. Technol.* 48, 4664–4671.

Spokas, K.A., Baker, J.M., Reicosky, D.C. 2010. Ethylene: Potential key for biochar amendment impacts. *Plant Soil.* 333, 443–452.

Spokas, K.A., Cantrell, K.B., Novak, J.M., Archer, D.W., Ippolito, J.A., Collins, H.P., Boateng, A.A., et al. 2012. Biochar: A synthesis of its agronomic impact beyond carbon sequestration. *J. Environ. Qual.* 41, 973–989.

Spokas, K.A., Novak, J.M., Masiello, C.A., Johnson, M.G., Colosky, E.C., Ippolito, J.A., Trigo, C. 2014. Physical disintegration of biochar: An overlooked process. *Environ. Sci. Technol. Lett.* 1, 326–332.

Spokas, K.A., Novak, J.M., Stewart, C.E., Cantrell, K.B., Uchimiya, M., DuSaire, M.G., Ro, K.S. 2011. Qualitative analysis of volatile organic compounds on biochar. *Chemosphere.* 85, 869–882.

Sullivan, J., Bollinger, K., Caprio, A., Cantwell, M., Appleby, P., King, J., Ligouis, B., Lohmann, R. 2011. Enhanced sorption of PAHs in natural-fire-impacted sediments from Oriole Lake, California. *Environ. Sci. Technol.* 45, 2626–2633.

Tilman, D., Hill, J., Lehman, C. 2006. Carbon-negative biofuels from low-input high-diversity grassland biomass. *Science.* 314, 1598–1600.

Uchimiya, M., Bannon, D.I., Wartelle, L.H. 2012. Retention of heavy metals by carboxyl functional groups of biochars in small arms range soil. *J. Agric. Food Chem.* 60, 1798–1809.

Uchimiya, M., Wartelle, L.H., Klasson, K.T., Fortier, C.A., Lima, I.M. 2011. Influence of pyrolysis temperature on biochar property and function as a heavy metal sorbent in soil. *J. Agric. Food Chem.* 59, 2501–2510.

Vassilev, S. V., Baxter, D., Andersen, L. K., Vassileva, C. G. 2010. An overview of the chemical composition of biomass. *Fuel,* 89, 913–933.

Wang, D., Zhang, W., Hao, X., Zhou, D. 2012. Transport of biochar particles in saturated granular media: Effects of pyrolysis temperature and particle size. *Environ. Sci. Technol.* 47, 821–828.

Warnock, D.D., Lehmann, J., Kuyper, T.W., Rillig, M.C. 2007. Mycorrhizal responses to biochar in soil—Concepts and mechanisms. *Plant Soil.* 300, 9–20.

Whitman, T., Nicholson, C.F., Torres, D., Lehmann, J. 2011. Climate change impact of biochar cook stoves in western Kenyan farm households: System dynamics model analysis. *Environ. Sci. Technol.* 45, 3687–3694.

Wilson, S.C., Lockwood, P.V., Ashley, P.M., Tighe, M. 2010. The chemistry and behaviour of antimony in the soil environment with comparisons to arsenic: A critical review. *Environ. Pollut.* 158, 1169–1181.

Woolf, D., Amonette, J.E., Street-Perrott, F.A., Lehmann, J., Joseph, S. 2010. Sustainable biochar to mitigate global climate change. *Nat. Commun.* 1, pp. 56.

Wu, H., Yip, K., Kong, Z., Li, C.Z., Liu, D., Yu, Y., Gao, X. 2011. Removal and recycling of inherent inorganic nutrient species in mallee biomass and derived biochars by water leaching. *Ind. Eng. Chem. Res.* 50, 12143–12151.

Wu, H., Yip, K., Tian, F., Xie, Z., Li, C.Z. 2009. Evolution of char structure during the steam gasification of biochars produced from the pyrolysis of various mallee biomass components. *Ind. Eng. Chem. Res.* 48, 10431–10438.

Yang, Y.N., Sheng, G.Y. 2003. Pesticide adsorptivity of aged particulate matter arising from crop residue burns. *J. Agric. Food Chem.* 51, 5047–5051.

Yoo, G., Spomer, L.A., Wander, M.M. 2006. Regulation of carbon mineralization rates by soil structure and water in an agricultural field and a prairie-like soil. *Geoderma.* 135, 16–25.

Zhang, W., Niu, J., Morales, V.L., Chen, X., Hay, A.G., Lehmann, J., Steenhuis, T.S. 2010. Transport and retention of biochar particles in porous media: Effect of pH, ionic strength, and particle size. *Ecohydrology.* 3, 497–508.

Zhu, D.Q., Kwon, S., Pignatello, J.J. 2005. Adsorption of single-ring organic compounds to wood charcoals prepared under different thermochemical conditions. *Environ. Sci. Technol.* 39, 3990–3998.

Zimmerman, A.R. 2010. Abiotic and microbial oxidation of laboratory-produced black carbon (biochar). *Environ. Sci. Technol.* 44, 1295–1301.

4

Chemical, Physical, and Surface Characterization of Biochar

Chapter 4

Chemical, Physical, and Surface Characterization of Biochar

Fungai N.D. Mukome and Sanjai J. Parikh

Chapter Outline

4.1 Introduction

The term *biochar* is relatively new, although the practice of its addition to soil can be traced back thousands of years to terra preta (dark earths) in the Amazon, China, and Africa (Glaser et al. 2001; Sombroek 1966; Sombroek et al. 2003). Although the effects of these soils on plant growth and soil tilth are well documented, the characterization of the pyrolyzed biomass (biochar) believed to have been integral in the formation of these soils is somewhat limited and hampered by several issues. For example, there is no consensus on the definition of biochar based on production method (production temperature, highest treatment temperature (HTT), time, pressure, and presence or absence of oxygen). In addition, there is large variability in the physical and chemical structure of biochar due to it being part of the black carbon continuum and its production from a variety of organic biomass feedstocks (Lehmann and Joseph 2009; Schmidt et al. 2001); and also, due to the ambiguity in the classification of biochar, there are a variety of characterization protocols developed for other materials (e.g., coal, charcoal, compost, and soil) that have been modified for biochar analysis. These inconsistencies in analytical approaches have made comparisons between literature biochar data very challenging (Fidel 2012; Spokas et al. 2011).

Facilitated by the International Biochar Initiative (IBI), a diversity of biochar stakeholders came together in 2009 to address these challenges. Their work resulted in the publication of the IBI Biochar Standards that includes a working definition of biochar and a set of standardized methods for biochar analysis (IBI 2013). These methods serve as guidelines for analysis of physical and chemical properties of biochar, and according to the IBI, the methods are best practices in some cases and are not necessarily perfect, leaving room for considerable innovation and method development. Generally, biochar characteristics are determined for three main purposes: elemental composition, agronomic considerations, and potential environmental concerns. Below, the many common chemical and physical properties of biochar are presented according to these focus areas.

4.2 Elemental Composition

4.2.1 Total Carbon and Organic Carbon

One of the most important properties of biochar is its relatively high organic C (C_{org}) content that is primarily stored in recalcitrant condensed aromatic rings with some reactive functional groups (Xu et al. 2012). This composition is fundamental to the concept of biochar as a means of C sequestration. In recent years, there has been considerable support for biochar use as a soil amendment to help mitigate climate change through increased C storage in soils and reductions in net carbon dioxide (CO_2) emissions. However, not all biochar C is recalcitrant, and a labile fraction is typically released soon after incorporation into soil (Atkinson et al. 2010). Although it is possible for biochar to bind this, and other native forms of C in soils, it is the C that is "locked up" in highly condensed forms that makes biochar a proposed tool for increasing soil C stocks.

When discussing biochar C, it is important to differentiate between total carbon (TC) content (combined inorganic and C_{org}) and C_{org}. Studies of biochar have traditionally used TC, typically determined by total combustion, as the measure for carbon content. However, the use of C_{org} has been argued as a better measure of biochar C because upon pyrolysis, some biochars have high carbonate content due to high ash content. This results in overestimations in biochar C content that then have implications for elemental ratios that are used for inferring characteristics such as aromaticity (H/C), stability (O/C), and potential N immobilization (C/N). A review of the literature revealed TC and C_{org} are dependent on the biochar feedstock and HTT. The C content of biochars can range widely (36–94%, depending on feedstock), with C content increasing with higher pyrolysis temperatures (Keiluweit et al. 2010; Novak et al. 2009). Data from the UC Davis Biochar Database (http://biochar.ucdavis.edu/) reveal that for lower temperature biochars, the TC is 60–95%, for feedstocks such as wood, and 50–70%, for feedstocks such as nutshells (Mukome and Parikh 2013).

The C_{org} content of biochar is directly related to both the feedstock and the pyrolysis, or gasification, conditions (Mukome et al. 2013). Depending on the feedstock and pyrolysis conditions, up to 50% of the feedstock C_{org} can be stored in biochar (Sohi et al. 2010). Therefore, this thermal conversion process represents a scenario that increases C_{org} storage in comparison to that of biomass permitted to decompose via natural (e.g., microbial) pathways.

Importantly, although biochar C_{org} is often classified as recalcitrant, it is not a permanent mechanism for C storage. The C_{org} in biochar breaks down over time, with some biochars being less stable than others. The long-term stability of biochar is not only tied to the biochar material itself but also to how it interacts and is stabilized by soil minerals. In addition, climate and other organisms impact how long biochar C_{org} remains in the soil. The mean residence time for biochar in soil is still not well understood, with studies suggesting that half-lives could range from <50 (Bird et al. 1999) to 1400 years (Kuzyakov et al. 2009). One attempt to account for these differences has linked O/C molar ratio to biochar stability, with more stable biochars having less O-containing functional groups and thus lower O/C ratios. It has been proposed that biochar half-lives for an O/C ratio of <0.2 will be >1000 years, an O/C of 0.2–0.6 will be 100–1000 years, and an O/C of >0.6 will be <100 years (Spokas 2010).

4.2.2 C/N Ratio

The C/N ratio is often considered one of the most important chemical parameters of soil amendments as it is used to predict mineralization and N release in soils. Typically, a C/N ratio of 20:1 is acceptable for soil amendments such as manures and composts. However, because a high C/N ratio is an indicator for high N immobilization by microbes (Stevenson and Cole 1999), a range of 12:1 is often preferred to increase the plant-available N pool. The C/N for biochar varies widely and can be much higher, ranging from ~8 (algae, corn stover) to nearly 1500 (softwood) (Enders et al. 2012; Keiluweit et al. 2010; Mutanda et al. 2011), suggesting that many biochars will lead to increased N immobilization in soils. For biochar, this is true to a certain extent, but the total N immobilization that might be inferred from the C/N ratio is often restricted due to the high stability (recalcitrance) of biochar (Chan and Xu 2009).

As with most chemical properties of biochars, the values of the C/N ratio are quite variable and are highly dependent on the feedstock type. A plot of the C/N ratio versus pyrolysis temperature of various biochars shows separation based on feedstock (Figure 4.1). With this information, it is possible to select biochars based on desired agronomic effect by choosing feedstocks that result in biochars in a desired C/N ratio. For example, Figure 4.1 shows that algae-based biochars have low C/N ratios and that wood (hard and soft) feedstock biochars have the highest C/N ratios. The large variability in C/N ratios

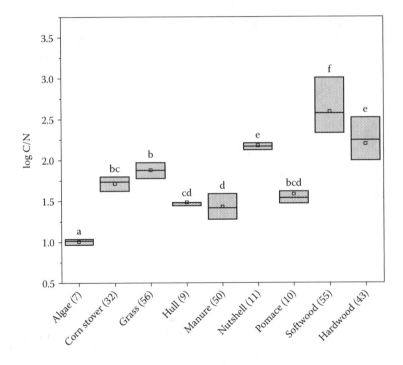

Figure 4.1 Change in the C/N ratio as a function of pyrolysis temperature of biochars ($n = 257$) derived from various feedstocks. Data from the UC Davis Biochar Database. The gray boxes show the range from first to third quartiles, with the median dividing the interquartile range into two boxes for the second and third quartiles. Letters show significant differences ($p < 0.05$) according to a one-way analysis of variance followed by Tukey's (honestly significant difference) multiple-means comparison.

of the wood feedstock biochars is attributed to the difference in the C/N ratios of hardwood and softwood, with softwood biochars having significantly higher ratios (Mukome et al. 2013). This result is in agreement with existing data on the C/N ratios of decaying woody debris showing that softwood typically has higher C/N ratios than hardwood (Saunders et al. 2011). Although feedstock is the most important variable impacting the C/N ratio of biochar, the pyrolysis temperature also influences this ratio, with increasing pyrolysis temperature generally leading to increasing C/N ratio for a given feedstock.

4.2.3 H/C versus O/C (van Krevelen Plot)

Often, it is not enough to compare between the total contributions of elements within biochars; comparisons between the relative contributions of two or more elements also can be highly informative. For example, the H/C and O/C elemental ratios can be used for interpretation of chemical data for predicting chemical structure of biochars. The H/C ratio is often used as proxy for aromaticity and as an indicator of its ability to be mineralized (Hammes et al. 2008; Krull et al. 2009), and the O/C ratio can be used to predict biochar stability in soil (Spokas 2010). Both the H/C and O/C ratios typically decrease with increased pyrolysis temperatures, indicating more stable biochars with fewer O-based functional groups (Mukome et al. 2013).

Together, these elemental ratios can be used to enhance our ability to compare between biochars through diagrams of the H/C ratio versus the O/C ratio, called van Krevelen plots. The van Krevelen plot was first developed to compare the maturity of coal products (van Krevelen 1961), but it has since been used to compare between several organic materials.

Examination of the van Krevelen plot of various biochars (Figure 4.2) reveals a strong positive relationship between the H/C atomic ratios and the O/C atomic ratios (Lehmann and Joseph 2009; Mukome et al. 2013). In this plot, the biochars are not separated based on feedstock, but instead pyrolysis temperature controls placement on the diagram. Biochars with high pyrolysis temperatures have more demethylation (low H/C) and decarboxylation (low O/C) and are located near

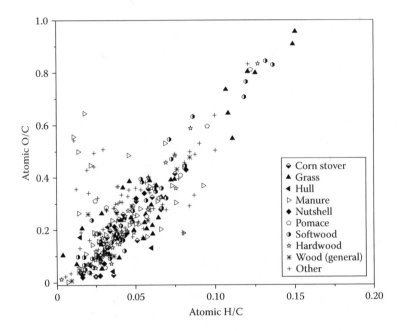

Figure 4.2 van Krevelen plot (H/C versus O/C ratios) of biochars ($n = 322$) showing variation according to feedstock. (Data from the UC Davis Biochar Database.)

the origin, representing the more recalcitrant materials. For additional information on van Krevelen plots of biochars, see Krull et al. (2009).

4.3 Agronomic Considerations

4.3.1 Mineral Ash Content

The mineral ash content is a measure of the biochar inorganic macro- and micronutrient constituents remaining after the removal of organic matter and water through combustion. Therefore, the ash content is a reflection of the inorganic content of the original feedstock. During pyrolysis and gasification processes, C and O are the main elements lost, and the ash content of biochars generally increases with increasing temperature or production method. Common inorganic components include P, S, K, Ca, Mg, Na, Fe, and Zn, with amounts varying between the biochars. These differences are mainly due to feedstocks, particularly for feedstocks that include soil and mineral materials (e.g., poultry litter, swine manure).

Biochar ash content is highly variable, with a dependence on the production method and feedstock (Enders et al. 2012; Mukome et al. 2013). As a property, ash content has been shown to be closely correlated to biochar pH, electrical conductivity (EC), and mineral composition (Lehmann et al. 2011). These correlations denote the source of ash in biochar as carbonates and oxides formed from the hydrolysis products of Ca, K, and Mg salts in the feedstocks. For example, many non-wood-derived biochars have high salt contents, such as carbonates and chlorides of potassium and calcium, and are highly basic and raise soil pH significantly (Montes-Morán et al. 2004).

Generally, for biochars made at the same HTT, wood-derived biochars have a lower ash content than non-wood-derived biochars such as waste biomass, grasses, and manures (Figure 4.3a). The high ash content from these sources can be attributed to the composition of the feedstocks in addition to the presence of silica from soil contamination. The importance of feedstock is evident in Figure 4.3b that shows statistically significant differences in the mean ash content of biochar from nine common biochar feedstocks. The data are compiled from >280 biochars in the UC Biochar Database (Mukome and Parikh 2013).

Knowledge of biochar ash content is important for several agronomic considerations. For farming activities, ash content of a biochar impacts its amenability to transportation and incorporation into soils (Blackwell et al. 2009). The fine particulate matter of high-ash-content biochars also requires special handling as these particulates may present a respiratory health hazard. Upon incorporation into the soil and depending on the application rate, ash has a liming effect (Van Zwieten et al. 2010); has been associated with increased soil hydrophobicity (Kookana et al. 2011); and may cause potential retention of hydrophobic agrochemicals, such as the herbicide group of phenylureas (Sopeña et al. 2012). Studies have also shown ash content impacts on microbial composition and activity

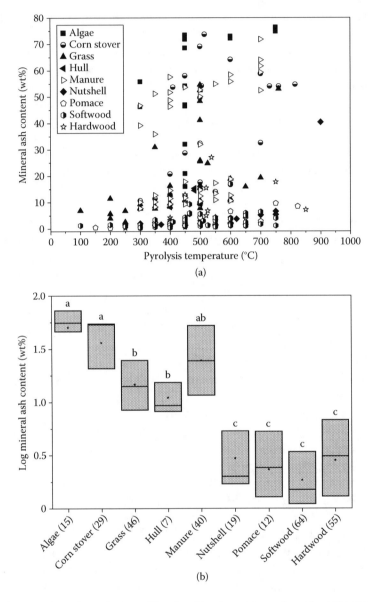

Figure 4.3 (a) Change in mineral ash content as a function of pyrolysis temperature of biochars ($n = 287$) derived from various feedstocks. Data from the UC Davis Biochar Database. (b) The gray boxes show the range from first to third quartiles, with the median dividing the interquartile range into two boxes for the second and third quartiles. Letters show significant differences ($p < 0.05$) according to a one-way analysis of variance followed by Tukey's (honestly significant difference) multiple-means comparison.

(Pietikainen et al. 2000; Van Zwieten et al. 2010) and hindrance of cell-to-cell sensing between microbes (Masiello et al. 2013).

4.3.2 Biochar pH

One of the biochar properties with key agronomic considerations is pH, or the measure of acidity or alkalinity. Most biochars are alkaline, with a pH > 7 (Figure 4.4a), and their use often raises the pH of acidic soils (liming effect) and affects mobility of cations in the soil (Lehmann et al. 2009; Van Zwieten et al. 2010). The liming effect of biochar on acid soils has been reported, with observed changes in soil pH of almost 1.5 pH units after biochar amendment (Mbagwu and Piccolo 1997). This pH change is also impacted by the biochar application rate, an area still requiring much research for optimization. Raising the soil pH results in increased base saturation (Ca and Mg concentrations) and increased soil microbial activity.

Biochar pH increases slightly with pyrolysis temperature, and this increase is due to the associated increase in ash content (Figure 4.4a). Biochar pH is also affected more by feedstock type than by pyrolysis temperature (Figure 4.4b). As evident in the Figure 4.4b, and in the literature, when produced at the same temperature, the non-wood feedstocks (e.g., manure, corn stovers, and algae) generally have higher pH than wood feedstocks (Enders et al. 2012; Mukome et al. 2013). Production process and temperature still affect biochar pH, with lower production temperatures generally resulting in more acidic biochar than higher temperatures. The basicity of the biochar arises from the presence of salts, such as carbonates and chlorides of potassium and calcium, in the ash in addition to oxygen-rich functional groups, such as γ-pyrone-type, chromene, diketone, or quinone groups (Montes-Morán et al. 2004). Although the oxygen content of feedstocks decreases with increasing pyrolysis temperature, the formation of compounds such as levoglucosan (from pyrolysis of cellulose material) and its by-products (levoglucosenone, furfural, 2,3-butanedione, and 5-methylfurfural) also results in oxygen-rich functional groups (Kawamoto et al. 2003).

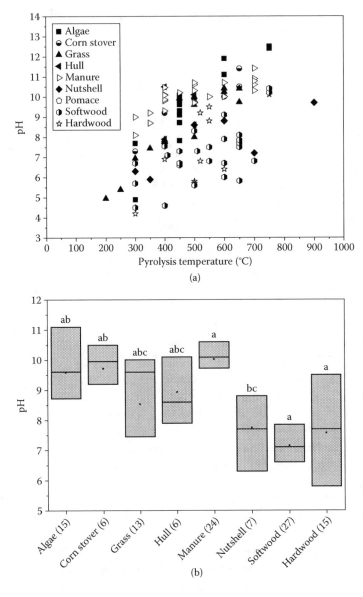

Figure 4.4 (a) Change in pH as a function of pyrolysis temperature of biochars ($n = 113$) derived from various feedstocks. Data from the UC Davis Biochar Database. (b) The gray boxes show the range from first to third quartiles, with the median dividing the interquartile range into two boxes for the second and third quartiles. Letters show significant differences ($p < 0.05$) according to a one-way analysis of variance followed by Tukey's (honestly significant difference) multiple-means comparison.

4.3.3 Cation Exchange Capacity

Cation exchange capacity (CEC) is a measure of exchangeable cations (e.g., Ca^{2+}, Mg^{2+}, K^+, and Na^+) and is an important measure of soil quality and productivity (Rhoades 1982). CEC values of biochar vary depending on the biomass feedstock and pyrolysis conditions. In a study comparing plant-based biochar (eucalyptus wood and leaves) and manure-based biochar (poultry litter and dairy manure), higher CEC and exchangeable cations in the manure-based biochar were reported (Singh et al. 2010). Fast pyrolysis production has also been shown to result in biochar with higher CEC than slow pyrolysis and gasification (Lee et al. 2010). However, temperature effects are less conclusive, with some studies showing higher CEC associated with an increasing pyrolysis temperature (Gaskin et al. 2008), but others showing a decrease in CEC with increasing temperature (Kloss et al. 2012). Figure 4.5 compiled from data from >60 biochars confirms these observations, with no observable trend effects of temperature. Different mechanisms have been proposed for the two contradicting trends. For the increasing trend, it has been proposed that the high CEC is driven by high specific surface area and the presence of surface carboxyl groups (Cheng et al. 2006; Sohi et al. 2010). In contrast, the decreasing trend is attributed to a loss of acidic functional groups with increasing pyrolysis temperature (Guo and Rockstraw 2007). It has also been suggested that biochar CEC increases as the biochar ages or weathers, with storage conditions (moisture content and temperature) significantly impacting these processes, but the rates and mechanisms have yet to be investigated (Verhejien et al. 2010).

Due to negatively charged surface of biochar, the addition of biochar to soil has been shown to increase soil CEC (i.e., greater retention of the cations through electrostatic interactions) (Liang et al. 2006; Mbagwu and Piccolo 1997). However, the duration and effectiveness of the increase from biochar soil amendments are still poorly understood.

Currently, there is poor consensus on the best approach to determining biochar CEC as the traditional methods used for soil require significant modification (Rhoades 1982). For example, biochar floats in water, and it does not readily mix

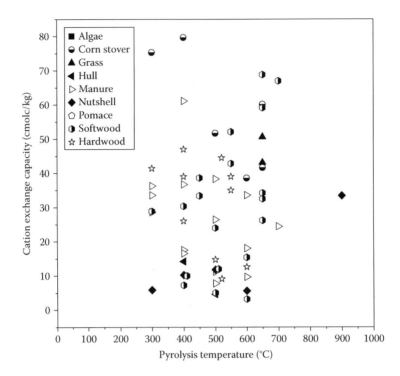

Figure 4.5 Change in cation exchange capacity as a function of pyrolysis temperature of biochars (n = 62) derived from various feedstocks. (Data from the UC Davis Biochar Database.)

or centrifuge. As a result, several modifications of soil CEC methods have been developed by different research groups (Lee et al. 2010; Mukome et al. 2013).

4.3.4 Electrical Conductivity

EC is a measure of the ability for a material to conduct an electrical current and is commonly used as a proxy for the salinity (major soluble and readily dissolvable major cations) in aqueous extracts and soil (Rhoades 1996). Biochar EC has been shown to correlate more with feedstock type than with pyrolysis temperature because it is a function of ash content and elemental composition, which are significantly determined by pyrolysis conditions and starting feedstock (Joseph et al. 2010; Rajkovich et al. 2012). As mentioned, nonmanure biochars typically have lower ash content than manure-derived biochars, and the same is true for EC values. The high EC values are associated with high amounts of soluble salts typically present in these biochar feedstocks. Pyrolysis temperature effects are not uniform, with some studies showing increases with pyrolysis temperature (Kloss et al. 2012; Singh et al. 2010) and others showing an initial increase up to 400–500°C and then a subsequent decrease (Hossain et al. 2011).

High salinity affects soil microorganisms and thus impacts some important soil processes, such as nitrification and denitrification, organic matter decomposition, and respiration. At high biochar application rates, increased salinity can also impact soil structure through flocculation. Other potential effects of increased salinity include inhibited plant growth from ion toxicity and ion imbalance (Gong et al. 2011). Standard methods of determining EC have been adopted from established soil methods; however, these approaches may underestimate the biochar EC values due to poor initial wetting of biochar, thus necessitating further investigation and refinement of current methods.

4.3.5 Density

Although potentially quite complex, the density term most commonly used to describe the density of biochar is bulk density. Biochar bulk density is defined as the mass of the biochar material as a ratio of the bulk volume of the biochar. The bulk volume comprises the volumes of the solid biochar particles, the voids within and between the biochar constituent particles (Dutta et al. 2012). Two other density terms are gaining popularity due to applicability of their measurements, namely, skeletal density and envelope density. As with bulk density, both terms consider the mass of the biochar material but differ on the volume parameter. Skeletal density considers the skeletal volume of the biochar, or the volume occupied by the solid biochar particles plus closed pores within the biochar. For the envelope density, the volume term is the external volume derived from a closely fitted imaginary envelope surrounding the biochar particles (Brewer et al. 2014; Webb 2001).

Biochar density is a function of the feedstock and the production process, with greater impact from the former (Lehmann and Joseph 2009). For example, in a study of several biochars produced through slow pyrolysis, feedstock type was determined to have a greater impact on biochar bulk density compared to production pyrolysis temperature (Rajkovich et al. 2012). Biochar from corn stover, wood, and dairy manure had comparatively lower bulk density than biochar produced from food waste, hazelnut shells, and poultry manure with sawdust. Values typically range from 1.4 to 1.7 $g \cdot cm^{-3}$, although higher values of near 2 $g \cdot cm^{-3}$ have been reported (Downie et al. 2009; Emmett 1948).

Agronomically, soil density is an important measure for predicting water infiltration, soil porosity, and aeration (Chan 2006). Research has shown that biochar soil amendments decrease soil bulk density (Glaser et al. 2002; Laird et al. 2010; Oguntunde et al. 2008) and positively impact water retention and root and microbial respiration.

4.4 Potential Environmental Concerns

Although biochar may be a beneficial agronomic strategy for some cropping systems, soil types, and climates, the potential benefits provided by biochar soil amendments do not come without some potential drawbacks. Biochar particles themselves can contain organic compounds with potential to adversely impact crop growth and soil health. During biochar production, several toxic compounds can be produced and remain as part of the biochar. If heavy metals are present in feedstock (e.g., Cr or Cu in pressure-treated lumber), the resulting biochar can be elevated in these metals. However, because most biochars are made using waste biomass from agriculture, heavy metals are not commonly of concern.

Many volatile organic compounds can be present in biochar (Spokas 2010). These compounds are formed during pyrolysis and often condense on the biochar surface as they cool. The volatiles that are of potential concern primarily include polycyclic aromatic hydrocarbons (PAHs), with trace levels of polychlorinated dibenzodioxins and polychlorinated biphenyls (Hale et al. 2012; Hilber et al. 2012). Although generally considered to be undesirable and toxic compounds, it is still not known what impact their presence in biochar will have on microbial populations and plant growth. Methods to characterize the bioavailability these compounds continue to be refined to provide more accurate information on the release of these compounds from biochar in soil.

4.5 Surface Characterization

The potential roles of biochar in C sequestration, greenhouse gas emission reduction, water retention, and sorption of contaminants are all largely dependent on the surface properties of the biochar. As a result, there have been multiple studies investigating surficial physical and chemical properties of biochar to elucidate potential mechanisms.

4.5.1 Physical Properties

4.5.1.1 Specific Surface Area

The Brunauer–Emmett–Teller (BET) surface area is area per gram of sample and is typically measured by assumed monolayer physical sorption of a gas (nitrogen gas [N_2] and CO_2) onto the biochar surface at the temperature of liquid nitrogen (77 K). Numerous studies have shown that biochar generally has a high surface area (compared to soil), and there is a positive correlation between specific surface area (SSA) and pyrolysis temperature (Keiluweit et al. 2010; Kookana et al. 2011) as shown in Figure 4.6 (data from UC Davis Biochar Database). However, for some feedstocks, this relationship is not true at higher temperatures, probably due to loss of microporous structure from plastic deformation, sintering, or fusion. This deviation is a function of the production conditions that include heating rate, pressure, and retention time (Chun et al. 2004; Downie et al. 2009).

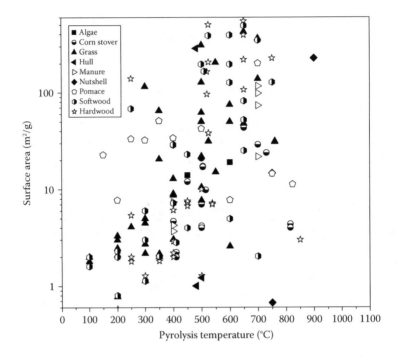

Figure 4.6 Change in specific surface area as a function of pyrolysis temperature of biochars (n = 178) derived from various feedstocks. (Data from the UC Davis Biochar Database.)

Due to the many different protocol parameters used in determining this property (e.g., sorption gas, degassing temperature, and time), compilation and comparison of literature data are challenging. No consensus has been reached on the appropriate gas to use for biochar SSA determination (CO_2 versus N_2), although it has been suggested that use of CO_2 for surface area measurements is more accurate (McLaughlin et al. 2012)

The ability to provide microbial habitats, provide soil aggregating nuclei, retain water and added nutrients, and remove contaminants is dependent on surface area. Unfortunately, the mechanisms by which biochar plays these roles are still poorly understood. In fact, there are no field study data supporting the assertion that biochar amendment results in long-term increases in soil surface area (Mukherjee and Lal 2014). Also requiring greater research is the dynamics of biochar aging and subsequent effects on soil surface area.

4.5.1.2 Porosity

Porosity involves the regions of the biochar particle that are not filled by solids (Brewer et al. 2014). Biochar porosity is derived from two main sources: macropores and micropores. The macropores (>2 nm in size) arise from the vascular structure of some biomass feedstock, and the more abundant micropores arise from gas vesicles that form in the production process due to gas release as a result of heating rates and times and the pyrolysis temperature (Downie et al. 2009). Due to the abundance of various-sized pores (range over at least five orders of magnitude; Brewer et al. 2014), no single technique can accurately characterize this surface property.

Micropores are responsible for the high sorption capacity (i.e., retention of water, gases, heavy metals, organics, and nutrients) of most biochars. They have also been shown to provide microhabitats for microorganisms, simultaneously providing protection from predation and encouraging biodiversity through the provision of different niche environments (Pietikainen et al. 2000; Warnock et al. 2007).

4.5.1.3 Morphology

Morphology is a bulk measure of the size, shape, and structure of the biochar. This characteristic is usually obtained as an image of the biochar surface and differs as a function of pyrolysis temperature and feedstock type. Images of biochar surfaces of different feedstocks show numerous micro- and macropores (Lee et al. 2010; Mukome et al. 2013; Özçimen and Ersoy-Meriçboyu 2010). Figure 4.7a and b show pine wood

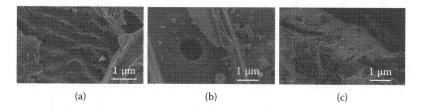

(a)   (b)   (c)

Figure 4.7 SEM images of (a) pine wood (300°C), (b) pine wood (900°C), and (c) almond shell (900°C) showing variation in surface morphology and texture.

biochar pyrolyzed at 300 and 900°C, respectively. Scanning electron microscopy images show the large differences in the sizes of the pores in the biochar as the temperature is increased. These pores are formed from gas vesicles produced during pyrolysis. This results in increased porosity and surface area on the biochar surface.

The effect of feedstock is evident in Figure 4.7a and b, showing biochar made at 900°C from pine wood and almond shells, respectively. The wood-derived biochars generally have large flat features derived from the ligno-cellulose plant tissue of the original feedstock that give rise to large pores. Non-wood-derived biochars, as the almond shell biochar, typically have smaller pores, less macrostructure, and more ash or soot particles.

4.5.2 Chemistry: Functional Group Content

Specific knowledge regarding the chemistry of biochar functional groups is paramount to predicting and understanding how a given biochar will react in the soil. For example, the addition of biochar to soil has been shown to increase sorption of a range of heavy metals (e.g., Johanson and Edgar 2006), pesticides (e.g., Spokas et al. 2009), and PAHs (e.g., Zhang et al. 2010). However, to predict which biochars are best suited for binding specific classes of compounds, a thorough characterization of the material is required. As part of this characterization, the functional group content of the biochar is critical for evaluating sorption processes.

Although elemental analysis is important, it does not provide direct information corresponding to the arrangement of the elements within the sample. To obtain information on the molecular bonds within biochar, a suite of spectroscopic analytical tools are typically used. These analytical approaches most commonly include Fourier transform infrared (FTIR), Raman, and nuclear magnetic resonance (NMR) spectroscopies. FTIR and Raman spectroscopies are both vibrational spectroscopies often used to study soil and biochar samples, and they provide complimentary spectra with peaks corresponding to the chemical vibrations of molecular bonds. NMR spectroscopy is a widely used technique that takes advantage of the magnetic properties of elements to provide information on the chemical environments of organic molecules.

Evaluation of FTIR spectroscopy data reveals that increasing pyrolysis temperature and duration of biochar production results in spectra with decreased peaks corresponding to C–O, O–H, and aliphatic C–H and with increased peaks from aromatic moieties (Keiluweit et al. 2010; Peng et al. 2011). This impact of pyrolysis temperature can be explained as the initial dehydration leads to decreased O–H bands; elevated temperatures result in increased peaks corresponding to C=C, C=O, C–O, and C–H of lignin and cellulose; and finally, at very high temperatures, there is an overall decrease in peaks as the C becomes condensed into aromatic rings structures. Parikh et al. (2014) provides a complete list of the specific FTIR band assignments for these functional groups for biochars.

Because FTIR spectroscopy only provides information on chemical bonds with dipole moments, the spectra of highly aromatic biochars to not have strong peaks. To determine the aromaticity of biochar, either Raman or NMR spectrometry is preferred. Raman spectroscopy has shown that during biochar production the Raman bands associated with the disordered carbon decrease relative to aromatic and other forms of recalcitrant carbon (Chia et al. 2012; Wu et al. 2009). Dispersive Raman spectrometry has also shown revealed how disordered carbon fraction (sp^2:sp^3 ratio) changes as a function of biochar feedstock (Jawhari et al. 1995; Mukome et al. 2013). Additional information on the Raman spectroscopy of biochar samples is available in Wu et al. (2009). NMR results also reveal increased aromaticity and condensation with increasing pyrolysis time and temperature, with differences noted between various feedstocks (Brewer et al. 2009; McBeath et al. 2014). An extended discussion on NMR spectroscopy of biochars is available in Krull et al. (2009).

The information obtained from these spectroscopic techniques is important for understanding how biochars added to soils will impact soil pH and interact with inorganic and organic compounds in soil. For example, biochars with significant contributions from carboxyl and hydroxyl groups will have higher CEC and a high sorption capacity for cations and metals. Biochars that are highly aromatic will be more hydrophobic and have high affinity for a wide range of organic compounds, particularly those with low aqueous solubility. Similarly, the functional group content dictates how biochars will interact with H^+ in soil, and thus impact the soil pH. Although biochars typically cause an increase in soil pH, biochars that have numerous oxygen-rich functional groups provide some buffering capacity, and pH increases are typically less severe. Therefore, with knowledge of the functional groups present, biochars can potentially be selected for a variety of desired agronomic or environmental impacts.

4.6 Current Shortcomings of Biochar Characterization

Although biochar physical and chemical characterization has received considerable attention, and there has been a concerted effort to standardize analytical methods, many questions related to characterization remain. For example, several studies have shown that properties such as pH, surface functional groups, and SSA change with biochar aging, and it is unclear how these changes impact soil properties and plant growth. Although some research has been directed toward this issue, there is a dearth of literature on field studies addressing biochar aging and soil property effects.

Current limitations in standard methods for biochar analysis are also numerous. For example, the currently adopted methodology for determining biochar ash content was developed for measuring ash content, moisture, and volatiles specifically for wood charcoal used for fuel, not as an agricultural amendment produced from a variety of feedstocks. As noted by Enders et al. (2012), in certain biochars, the temperature and duration of analysis used in the methodology can result in the volatilization of elements and carbonates (abundant in feedstocks such as animal manures), resulting in underestimations of ash contents and overestimations of C and O contents.

Also poorly understood are the characteristics of biochar that have human health implications. These include measurements of PAHs, dioxins, heavy metals, and other volatile pyrolysis production by-products. Although some studies have begun reporting these data, the knowledge gap in this area is still considerable. Associated with this gap is a need to investigate pyrolysis production processes and potential trends with biochar chemical and physical characteristics. Although trends in biochar properties across pyrolysis methods have been investigated, analysis of biochar characteristics related to specific production methods may elucidate potential ways to mitigate production of harmful compounds such as PAHs and dioxins.

Although considerable resources have been expended in investigating biochar and its possible applications, it is apparent that future studies need to begin with a better understanding of the material called biochar. Only through careful and appropriate physical, chemical, and surface characterization can mechanisms responsible for the effects attributed to biochar be revealed. This deliberate and mechanistic approach will improve our understanding how biochar behaves in soil, and it will be paramount in guiding the appropriate production, use, and potential innovations related to biochar.

Acknowledgments

We thank Devin Rippner for collection of the scanning electron microscopy images at the Keck Spectral Imaging Facility at UC Davis.

References

Atkinson C.J., Fitzgerald J.D., Hipps N.A. (2010). Potential mechanisms for achieving agricultural benefits from biochar application to temperate soils: A review. *Plant and Soil* 337: 1–18. DOI: 10.1007/s11104-010-0464-5.

Bird M.I., Moyo C., Veenendaal E.M., Lloyd J., Frost P. (1999). Stability of elemental carbon in a savanna soil. *Global Biogeochemical Cycles* 13: 923–932. DOI: 10.1029/1999GB900067.

Blackwell P., Riethmuller G., Collins M. (2009). Biochar application to soil. In: J. Lehmann, S. Joseph (Eds.), *Biochar for Environmental Management: Science and Technology*, Earthscan, London, UK, pp. 207–226.

Brewer C.E., Chuang V.J., Masiello C.A., Gonnermann H., Gao X., Dugan B., Driver L.E., Panzacchi P., Zygourakis K., Davies C.A. (2014). New approaches to measuring biochar density and porosity. *Biomass and Bioenergy* 66: 176–185. DOI: http://dx.doi.org/10.1016/j.biombioe.2014.03.059.

Brewer C.E., Schmidt-Rohr K., Satrio J.A., Brown R.C. (2009). Characterization of biochar from fast pyrolysis and gasification systems. *Environmental Progress & Sustainable Energy* 28: 386–396. DOI: 10.1002/ep.10378.

Chan K.Y. (2006). Bulk density. In: R. Lal (Ed.), *Encyclopedia of Soil Science*, 2n Ed. Taylor & Francis, Boca Raton, FL, pp. 191–193.

Chan K.Y., Xu Z. (2009). Biochar: Nutrient properties and their enhancement. In: J. Lehmann, S. Joseph (Eds.), *Biochar for Environmental Management: Science and Technology*, Earthscan, Sterling, VA. pp. 67–84.

Cheng C.-H., Lehmann J., Thies J.E., Burton S.D., Engelhard M.H. (2006). Oxidation of black carbon by biotic and abiotic processes. *Organic Geochemistry* 37: 1477–1488.

Chia C.H., Gong B., Joseph S.D., Marjo C.E., Munroe P., Rich A.M. (2012). Imaging of mineral-enriched biochar by FTIR, Raman and SEM–EDX. *Vibrational Spectroscopy* 62: 248–257. DOI: http://dx.doi.org/10.1016/j.vibspec.2012.06.006.

Chun Y., Sheng G., Chiou C.T., Xing B. (2004). Compositions and sorptive properties of crop residue-derived chars. *Environmental Science & Technology* 38: 4649–4655. DOI: 10.1021/es035034w.

Downie A., Crosky A., Munroe P. (2009). Physical properties of biochar. In: J. Lehmann, S. Joseph (Eds.), *Biochar for Environmental Management: Science and Technology*, Earthscan, London, UK, pp. 13–32.

Dutta B., Raghavan G.S.V., Ngadi M. (2012). Surface characterization and classification of slow and fast pyrolyzed biochar using novel methods of pycnometry and hyperspectral imaging. *Journal of Wood Chemistry and Technology* 32: 105–120. DOI: 10.1080/02773813.2011.607535.

Emmett P.H. (1948). Adsorption and pore-size measurements on charcoals and whetlerites. *Chemical Reviews* 43: 69–148. DOI: 10.1021/cr60134a003.

Enders A., Hanley K., Whitman T., Joseph S., Lehmann J. (2012). Characterization of biochars to evaluate recalcitrance and agronomic performance. *Bioresource Technology* 114: 644–653. DOI: http://dx.doi.org/10.1016/j.biortech.2012.03.022.

Fidel R.B. (2012). *Evaluation and Implementation of Methods for Quantifying Organic and Inorganic Components of Biochar Alkalinity.* Iowa State University, Ames, Iowa.

Gaskin J.W., Steiner C., Harris K., Das K.C., Bibens B. (2008). Effect of low-temperature pyrolysis conditions on biochar for agricultural use. *Transactions of the Asabe* 51: 2061–2069.

Glaser B., Haumaier L., Guggenberger G., Zech W. (2001). The 'terra preta' phenomenon: A model for sustainable agriculture in the humid tropics. *Naturwissenschaften* 88: 37–41. DOI: 10.1007/s001140000193.

Glaser B., Lehmann J., Zech W. (2002). Ameliorating physical and chemical properties of highly weathered soils in the tropics with charcoal—A review. *Biology and Fertility of Soils* 35: 219–230. DOI: 10.1007/s00374-002-0466-4.

Gong X., Chao L., Zhou M., Hong M., Luo L., Wang L., Ying W., Jingwei C., Songjie G., Fashui H. (2011). Oxidative damages of maize seedlings caused by exposure to a combination of potassium deficiency and salt stress. *Plant and Soil* 340: 443–452. DOI: 10.1007/s11104-010-0616-7.

Guo Y., Rockstraw D.A. (2007). Activated carbons prepared from rice hull by one-step phosphoric acid activation. *Microporous and Mesoporous Materials* 100: 12–19. DOI: http://dx.doi.org/10.1016/j.micromeso.2006.10.006.

Hale S.E., Lehmann J., Rutherford D., Zimmerman A.R., Bachmann R.T., Shitumbanuma V., O'Toole A., Sundqvist K.L., Arp H.P.H., Cornelissen G. (2012). Quantifying the total and bioavailable polycyclic aromatic hydrocarbons and dioxins in biochars. *Environmental Science & Technology* 46: 2830–2838. DOI: 10.1021/es203984k.

Hammes K., Smernik R.J., Skjemstad J.O., Schmidt M.W.I. (2008). Characterisation and evaluation of reference materials for black carbon analysis using elemental composition, colour, BET surface area and 13C NMR spectroscopy. *Applied Geochemistry* 23: 2113–2122.

Hilber I., Blum F., Leifeld J., Schmidt H.-P., Bucheli T.D. (2012). Quantitative determination of PAHs in biochar: A prerequisite to ensure its quality and safe application. *Journal of Agricultural and Food Chemistry* 60: 3042–3050. DOI: 10.1021/jf205278v.

Hossain M.K., Strezov V., Chan K.Y., Ziolkowski A., Nelson P.F. (2011). Influence of pyrolysis temperature on production and nutrient properties of wastewater sludge biochar. *Journal of*

Environmental Management 92: 223–228. DOI: http://dx.doi. org/10.1016/j.jenvman.2010.09.008.

IBI. (2013). *Biochar Standards*. International Biochar Initiative. http://www.biochar-international.org/sites/default/files/IBI_ Biochar_Standards_V2%200_final_2014.pdf.

Jawhari T., Roid A., Casado J. (1995). Raman spectroscopic characterization of some commercially available carbon black materials. *Carbon* 33: 1561–1565. DOI: http://dx.doi.org/10.1016/ 0008-6223(95)00117-V.

Johanson D.C., Edgar B. (2006). *From Lucy to Language: Revised, Updated, and Expanded*. Simon and Schuster, New York.

Joseph S.D., Camps-Arbestain M., Lin Y., Munroe P., Chia C.H., Hook J., van Zwieten L., et al. (2010). An investigation into the reactions of biochar in soil. *Australian Journal of Soil Research* 48: 501–515.

Kawamoto H., Murayama, M., Saka, S. (2003). Pyrolysis behavior of levoglucosan as an intermediate in cellulose pyrolysis: Polymerization into polysaccharide as a key reaction to carbonized product formation. *Journal of Wood Science* 49: 469–473.

Keiluweit M., Nico P.S., Johnson M.G., Kleber M. (2010). Dynamic molecular structure of plant biomass-derived black carbon (biochar). *Environmental Science & Technology* 44: 1247–1253. DOI: 10.1021/es9031419.

Kloss S., Zehetner F., Dellantonio A., Hamid R., Ottner F., Liedtke V., Schwanninger M., Gerzabek M.H., Soja G. (2012). Characterization of slow pyrolysis biochars: Effects of feedstocks and pyrolysis temperature on biochar properties. *Journal of Environmental Quality* 41: 990–1000. DOI: 10.2134/jeq2011.0070.

Kookana R.S., Sarmah A.K., Van Zwieten L., Krull E., Singh B. (2011). Biochar application to soil: Agronomic and environmental benefits and unintended consequences. In: L.S. Donald (Ed.), *Advances in Agronomy*, Academic Press, San Diego, CA, pp. 103–143.

Krull E.S., Baldock J.A., Skjemstad J.O., Smernik R.J. (2009). Characteristics of biochar: Organo-chemical properties. In: J. Lehmann, S. Joseph (Eds.), *Stability of Biochar in Soil. Biochar for Environmental Management: Science and Technology*, Earthscan, London, UK, pp. 53–66.

Kuzyakov Y., Subbotina I., Chen H., Bogomolova I., Xu X. (2009). Black carbon decomposition and incorporation into soil microbial biomass estimated by 14C labeling. *Soil Biology and Biochemistry* 41: 210–219.

Laird D.A., Fleming P., Davis D.D., Horton R., Wang B., Karlen D.L. (2010). Impact of biochar amendments on the quality of a typical Midwestern agricultural soil. *Geoderma* 158: 443–449.

Lee J.W., Kidder M., Evans B.R., Paik S., Buchanan Iii A.C., Garten C.T., Brown R.C. (2010). Characterization of biochars produced from cornstovers for soil amendment. *Environmental Science & Technology* 44: 7970–7974. DOI: 10.1021/es101337x.

Lehmann J., Czimczik C.I., Laird D.L., Sohi S. (2009). Stability of biochar in soil. In: L. Johannes, J. Stephen (Eds.), *Biochar for Environmental Management: Science and Technology*, Earthscan, London, UK, pp. 183–206.

Lehmann J., Joseph S. (Eds.). (2009). *Biochar for Environmental Management: Science and Technology.* Earthscan Ltd, London, UK.

Lehmann J., Rillig M.C., Thies J., Masiello C.A., Hockaday W.C., Crowley D. (2011). Biochar effects on soil biota—A review. *Soil Biology and Biochemistry* 43: 1812–1836. DOI: http://dx.doi.org/10.1016/j.soilbio.2011.04.022.

Liang B., Lehmann J., Solomon D., Kinyangi J., Grossman J., O'Neill B., Skjemstad J.O., et al. (2006). Black carbon increases cation exchange capacity in soils. *Soil Science Society of America Journal* 70: 1719–1730. DOI: 10.2136/sssaj2005.0383.

Masiello C.A., Chen Y., Gao X., Liu S., Cheng H.-Y., Bennett M.R., Rudgers J.A., Wagner D.S., Zygourakis K., Silberg J.J. (2013). Biochar and microbial signaling: Production conditions determine effects on microbial communication. *Environmental Science & Technology* 47: 11496–11503. DOI: 10.1021/es401458s.

Mbagwu J.S.C., Piccolo, A. (1997). Effects of humic substances from oxidized coal on soil chemical properties and maize yield. In: J.G. Drozd, S.S. Gonet, N. Senesi, J. Weber. (Eds.), *The Role of Humic Substances in the Ecosystems and in Environmental Protection: Proceedings of the 8th Meeting of the International Humic Substances Society, Polish Society of Humic Substances*, Polish Chapter of the International Humic Substances Society, Wroclaw, Poland, pp. 921–925.

McBeath A.V., Smernik R.J., Krull E.S., Lehmann J. (2014). The influence of feedstock and production temperature on biochar carbon chemistry: A solid-state 13C NMR study. *Biomass Bioenergy* 60: 121–129. DOI: http://dx.doi.org/10.1016/j.biombioe.2013.11.002.

McLaughlin H., Shields F., Jagiello J., Thiele G. (2012). Analytical options for biochar adsorption and surface area. In: *2012 US Biochar Conference*, July 29–August 1, 2012, Sonoma, CA, USA.

Montes-Morán M.A., Suárez D., Menéndez J.A., Fuente E. (2004). On the nature of basic sites on carbon surfaces: An overview. *Carbon* 42: 1219–1225.

Mukherjee A., Lal R. (2014). The biochar dilemma. *Soil Research* 52: 217–230. DOI: http://dx.doi.org/10.1071/SR13359.

Mukome, F.N.D., Parikh, S.J. (2013). *UC Davis Biochar Database.* University of California Davis, Davis, CA. http://ucdavis. biochar.edu

Mukome F.N.D., Zhang X., Silva L.C.R., Six J., Parikh S.J. (2013). Use of chemical and physical characteristics to investigate trends in biochar feedstocks. *Journal of Agricultural and Food Chemistry* 61: 2196–2204. DOI: 10.1021/jf3049142.

Mutanda T., Ramesh D., Karthikeyan S., Kumari S., Anandraj A., Bux F. (2011). Bioprospecting for hyper-lipid producing microalgal strains for sustainable biofuel production. *Bioresource Technology* 102: 57–70. DOI: http://dx.doi.org/10.1016/j.biortech.2010.06.077.

Novak J.M., Lima I., Xing B., Gaskin J.W., Steiner C., Das K.C., Ahmedna M., Rehrah D., Watts D.W., Busscher W.J., Schmobert H. (2009). Characterization of designer biochars produced at different temperatures and their effects on a lomay sand. *Annals of Environmental Science* 3: 195–206.

Oguntunde P.G., Abiodun B.J., Ajayi A.E., van de Giesen N. (2008). Effects of charcoal production on soil physical properties in Ghana. *Journal of Plant Nutrition and Soil Science* 171: 591–596.

Özçimen D., Ersoy-Meriçboyu A. (2010). Characterization of biochar and bio-oil samples obtained from carbonization of various biomass materials. *Renewable Energy* 35: 1319–1324.

Parikh S.J., Goyne K.W., Margenot A.J., Mukome F.N.D., Calderón F.J. (2014). Soil chemical insights provided through vibrational spectroscopy. In: L.S. Donald (Ed.), *Advances in Agronomy*, Academic Press, San Diego, CA, pp. 1–148.

Peng X., Ye L.L., Wang C.H., Zhou H., Sun B. (2011). Temperature- and duration-dependent rice straw-derived biochar: Characteristics and its effects on soil properties of an Ultisol in Southern China. *Soil and Tillage Research* 112: 159–166. DOI: 10.1016/j. still.2011.01.002.

Pietikainen J., Kiikkila O., Fritze H. (2000). Charcoal as a habitat for microbes and its effect on the microbial community of the underlying humus. *Oikos* 89: 231–242.

Rajkovich S., Enders A., Hanley K., Hyland C., Zimmerman A., Lehmann J. (2012). Corn growth and nitrogen nutrition after additions of biochars with varying properties to a temperate soil. *Biology and Fertility of Soils* 48: 271–284. DOI: 10.1007/ s00374-011-0624-7.

Rhoades J.D. (1982). Cation exchange capacity. In: A.L. Page (Ed.), *Methods of Soil Analysis. Part 2. Chemical and Microbiological*

Properties, American Society of Agronomy, Soil Science Society of America. Madison, WI, pp. 149–157.

Rhoades J.D. (1996). Salinity: Electrical conductivity and total dissolved solids. In: D.L. Sparks, et al. (Eds.), *Methods of Soil Analysis Part 3—Chemical Methods*, Soil Science Society of America, American Society of Agronomy, Madison, WI, pp. 417–435.

Saunders M.R., Fraver S., Wagner R.G. (2011). Nutrient concentration of down woody debris in mixedwood forests in central maine, USA. *Silva Fennica* 45: 197–210.

Schmidt M.W.I., Skjemstad J.O., Czimczik C.I., Glaser B., Prentice K.M., Gelinas Y., Kuhlbusch T.A.J. (2001). Comparative analysis of black carbon in soils. *Global Biogeochemical Cycles* 15: pp. 163–167

Singh B., Singh B.P., Cowie A. (2010). Characterisation and evaluation of biochars for their application as a soil amendment. *Australian Journal of Soil Research* 48: 516–525.

Sohi S.P., Krull E., Lopez-Capel E., Bol R. (2010). Chapter 2—A review of biochar and its use and function in soil. In: L.S. Donald (Ed.), *Advances in Agronomy*, Academic Press, San Diego, CA, pp. 47–82.

Sombroek W., Ruivo M.L., Fearnside P.M., Glaser B., Lehmann J. (2003). Amazonian dark earths as carbon stores and sinks. In: J. Lehmann et al. (Eds.), *Amazonian Dark Earths: Origin, Properties, Management*, Kluwer Academic Publishers, The Netherlands, pp. 125–139.

Sombroek W.G. (1966). Amazon soils a reconnaissance of the soils of the brazilian amazon region. Centre for Agricultural Publications and Documentation, Wageningen, The Netherlands. p. 292.

Sopeña F., Semple K., Sohi S., Bending G. (2012). Assessing the chemical and biological accessibility of the herbicide isoproturon in soil amended with biochar. *Chemosphere* 88: 77–83. DOI: 10.1016/j.chemosphere.2012.02.066.

Spokas K.A. (2010). Review of the stability of biochar in soils: Predictability of O:C molar ratios. *Carbon Management* 1: 289–303. DOI: 10.4155/cmt.10.32.

Spokas K.A., Koskinen W.C., Baker J.M., Reicosky D.C. (2009). Impacts of woodchip biochar additions on greenhouse gas production and sorption/degradation of two herbicides in a Minnesota soil. *Chemosphere* 77: 574–581. DOI: 10.1016/j.chemosphere.2009.06.053.

Spokas K.A., Novak J.M., Stewart C.E., Cantrell K.B., Uchimiya M., DuSaire M.G., Ro K.S. (2011). Qualitative analysis of volatile organic compounds on biochar. *Chemosphere* 85: 869–882. DOI: http://dx.doi.org/10.1016/j.chemosphere.2011.06.108.

Stevenson F.J., Cole M.A. (1999). *Cycles of Soil—Carbon, Nitrogen, Phosphorus, Sulfur, Micronutrients*. Wiley, New York.

van Krevelen D. (1961). *Coal*. Elsevier, Amsterdam.

Van Zwieten L., Kimber S., Morris S., Chan K., Downie A., Rust J., Joseph S., Cowie A. (2010). Effects of biochar from slow pyrolysis of papermill waste on agronomic performance and soil fertility. *Plant and Soil* 327: 235–246. DOI: 10.1007/s11104-009-0050-x.

Verheijen, F.G.A., Jeffery, S., Bastos, A.C., van der Velde, M., Diafas, I. *Biochar Application to Soils—A Critical Scientific Review of Effects on Soil Properties, Processes and Functions*; Office for the Official Publications of the European Communities: Luxembourg, 2009, pp. 1–149.

Warnock D.D., Lehmann J., Kuyper T.W., Rillig M.C. (2007). Mycorrhizal responses to biochar in soil – concepts and mechanisms. *Plant and Soil* 300: 9–20.

Webb P. (2001). *Volume and Density Determinations for Particle Technologists*. Micromeretics Instrument Corp., Norcross, GA.

Wu H., Yip K., Tian F., Xie Z., Li C.-Z. (2009). Evolution of char structure during the steam gasification of biochars produced from the pyrolysis of various mallee biomass components. *Industrial & Engineering Chemistry Research* 48: 10431–10438. DOI: 10.1021/ie901025d.

Xu G., Lv Y., Sun J., Shao H., Wei L. (2012). Recent advances in biochar applications in agricultural soils: Benefits and environmental implications. *Clean-Soil Air Water* 40: 1093–1098. DOI: 10.1002/clen.201100738.

Zhang H., Lin K., Wang H., Gan J. (2010). Effect of Pinus radiata derived biochars on soil sorption and desorption of phenanthrene. *Environmental Pollution* 158: 2821–2825. DOI: http://dx.doi.org/10.1016/j.envpol.2010.06.025.

SECTION III

Environmental Applications

5

Biochar for Inorganic Contaminant Management in Soil

Chapter 5

Biochar for Inorganic Contaminant Management in Soil

Ramya Thangarajan, Nanthi Bolan,
Sanchita Mandal, Anitha Kunhikrishnan,
Girish Choppala, Rajasekar Karunanithi,
and Fangjie Qi

Chapter Outline

5.1 Inorganic Contaminants in Soil

The major inorganic contaminants in soil are heavy metal(loid)s and nutrients. Some heavy metal(loid)s are biologically essential elements that they are required in low concentrations and hence are known as trace elements or micronutrients (e.g., Co, Cu, Cr, Mn, Se, and Zn) (Kesler 1994). Some nonessential heavy metal(loid)s are phytotoxic, zootoxic, or both, and hence are known as toxic elements (e.g., As, Cd, Pb, and Hg). Both groups are toxic to plants, animals, and humans at exorbitant concentrations. The two most important elemental nutrients that lead to pollution and environmental degradation are N and P (Carpenter et al. 1998).

5.1.1 Sources of Heavy Metal(loid)s

Heavy metal(loid)s reach the soil environment through both pedogenic and anthropogenic processes. Most heavy metal(loid)s occur naturally in soil parent materials, chiefly in the forms that are not readily bioavailable for plant uptake. Unlike pedogenic inputs, heavy metal(loid)s added through anthropogenic activities typically have a high bioavailability (Bolan et al. 2014).

Pedogenic sources: Most of the heavy metal(loid)s occur in nature, the major source of which is weathering of soil parent materials, including igneous and sedimentary rocks and coal. For example, the As content of igneous rocks varies widely (up to 100 mg/kg); the average content is 2–3 mg/kg, whereas in sedimentary rocks it varies from small amounts in limestone and sandstone up to 15,000 mg/kg in some Mn ores (Hindmarsh et al. 1986). Although anthropogenic As sources are increasingly becoming important, As contamination of groundwater in many countries, including Bangladesh, India, and China, is of geological origin, transported by rivers from sedimentary rocks in the Himalayas over tens of thousands of years (Mahimairaja et al. 2005).

Anthropogenic sources: Anthropogenic activities, primarily associated with industrial processes, manufacturing, and the disposal of domestic and industrial waste materials, are the major sources of metal(loid)s enrichment in soils. Biosolids is considered to be the major source of metal(loid) input in Europe and North America. Phosphate rocks (PRs) used for manufacturing P fertilizers are a major source of Cd. Regular use of P fertilizers is reported to be associated with Cd contamination in Australia and New Zealand (McLaughlin et al. 1996; Roberts et al. 1994). Although Cu is used as a trace element in agricultural and horticultural industries and as a growth promoter in piggery and poultry units (Bolan et al. 2003a, 2010a), its accumulation in agricultural soils due to continuous use of Cu-containing fungicides such as copper oxychloride and Bordeaux mixture (a combination of copper sulfate, lime, and water [H_2O]) and biosolids application has been reported in many countries (Wuana and Okieimen 2011). Chromium is used as Cr(III) in the tannery industry and as Cr(VI) in the timber treatment industry. Chromate is highly toxic and carcinogenic even when present in very low concentrations in H_2O. Large-scale use of copper chrome arsenate–treated timber as fence posts and in vineyards can also result in the release of Cu, Cr, and As to the soil environment (Robinson et al. 2006).

5.1.2 Sources of Nutrients

Although N is derived from biological N fixation and the addition of organic amendments and fertilizers, P is derived only from the latter two sources.

Biological nitrogen fixation: Nitrogen-fixing grain legumes are grown in many countries as a source of grain protein for human and animal consumption. In many countries, including Australia, New Zealand, and parts of North America and Europe, the use of legume-based pasture is the most common grazing management practice. The amount of biological N fixation depends on several factors, including legume species, soil and climatic conditions, nutrient supply, and grazing management (Ledgard and Steele 1992).

Fertilizer application: The most common N fertilizers used in agriculture include urea (46% N), ammonium sulfate (21% N), diammonium phosphate (DAP; 18% N), and organic manures such as blood and bone (8% N). The efficiency of N fertilizers varies between the fertilizer forms and is attributed to the differences in the effects of fertilizers on the rate of uptake and assimilation of N, the losses of N through ammonia (NH_3) volatilization, denitrification and leaching, the N-induced cation–anion balance in plants, and the acidifying effects of N (Bolan et al. 2004a).

Deposition of animal excreta: In grazed pastures, a substantial amount of N is recycled through the direct deposition of animal excreta. The proportion of N in urine and feces is dependent on the type of animal, the intake of dry matter, and the N concentration of the diet (Whitehead 1990; Whitehead et al. 1986). In most intensive high-producing pasture systems, where animal intake of N is high, more than half of the N is excreted as urine (Bolan et al. 2009). The N content in the diet and the volume of H_2O consumption govern the urine N concentration; urine N concentration may vary from 1 to 20 g N/L, but it is normally in the range of 8–15 g N/L (Bolan et al. 2009). The bulk of the N in feces is in organic form.

Manure application and effluent irrigation: Confined animal production (i.e., beef and dairy cattle, poultry, and swine) is the major source of manure by-products in most countries. For example, of ~900 million Mg (tonnes) of organic and inorganic agricultural recyclable by-products generated in the United States, ~72.4 million Mg is cattle, swine, and poultry manures (Thangarajan et al. 2013). These manure by-products generate annually ~7.5 and 2.3 million Mg of N and P, respectively, compared to 9 and 1.6 million Mg of N and P that is applied to agricultural land in the form of commercial fertilizers, respectively (Bolan et al. 2004b). Similarly, farm effluents contain a large reserve of plant nutrients. For example, in New Zealand, dairy and piggery effluents can supply annually N, P, and K equivalents of 17,500 tonnes of urea; 12,500 tonnes of single super phosphate; and 28,300 tonnes of potassium chloride, respectively, with a net fertilizer value worth NZ$21.1 million (Bolan et al. 2004a). The nutrients in farm effluents can meet the N and P requirements of ~50,000 ha of pasture. At a global level, dairy and piggery effluents can annually supply N and P for ~6.7 million ha of pasture (Vogeler et al. 2014).

5.2 Reactions of Inorganic Contaminants

5.2.1 Heavy Metal(loid)s

Metal(loid) ions can be retained in the soil by sorption, precipitation, and complexation reactions, and they are removed from soil through plant uptake, leaching, and volatilization. Although most metal(loid)s are not subjected to volatilization losses, some metal(loid)s, such as As, Hg, and Se, tend to form gaseous compounds (Bolan et al. 2014).

Sorption and desorption processes: Both soil properties and soil solution composition determine the dynamic equilibrium between metal(loid)s in solution and the soil solid phase. The concentration of metal(loid)s in soil solution is influenced by the nature of both organic and inorganic ligand ions and soil pH through their influence on metal(loid) sorption processes (Adriano et al. 2004). Although the specific sorption of inorganic anions increases the negative charge on soil particles, thereby increasing the sorption of metal(loid)s, it also reduces sorption of metal(loid)s by forming ion pair complexes with metal(loid)s (Bolan et al. 1997). Soil pH > 6 lowers the free metal(loid) activities in soil due to the increase in pH-dependent surface charge on oxides of Fe, Al, and Mn; chelation by organic matter; or precipitation of metal(loid) hydroxides [e.g., $Ph(OH)_3$] (Violante et al. 2010). Complexation reactions between metal(loid)s and the inorganic and organic ligand ions also contribute to metal(loid) retention by colloid particles.

Precipitation and dissolution: Precipitation is the predominant process for the immobilization of heavy metal(loid)s in high pH soils in the presence of anions such as sulfate ion, carbonate ion, hydroxide ion (OH^-), and monohydrogen phosphate ion, especially when the concentration of the heavy metal(loid) ions is high in soils (Bolan et al. 2010b). Similarly, liming typically enhances the retention of metal(loid)s in soil (Bolan et al. 2003b). Coprecipitation of metal(loid)s, especially in the presence of iron oxyhydroxides [e.g., As(V) sorption onto ferrihydrite], has also been reported as a more efficient process than sorption for metal(loid)s removal from aqueous solutions (Kwon et al. 2010).

Oxidation and reduction: Microbial oxidation and reduction of metal(loid)s influence their speciation and mobility in soil. Because As(V) is strongly retained by inorganic soil components, microbial oxidation of arsenic to AS(V) results in the immobilization of As (Bachate et al. 2012; Bolan et al. 2013). Oxidation of Cr(III) to Cr(VI), an abiotically mediated process, enhances the mobilization and bioavailability of Cr (Kim et al. 2002), whereas reduction of Cr(VI) to Cr(III) is mediated by both biotic and abiotic processes (Fendorf et al. 2000). In living systems, Se tends to be reduced rather than oxidized, and the reduction occurs under both aerobic and anaerobic conditions. Dissimilatory Se(IV) reduction to Se(0), and its subsequent precipitation, is the major biological transformation for remediation of Se oxyanions in anoxic sediments (Dungan and Frankenberger 1999). For Hg, microorganisms reduce reactive Hg(II) to nonreactive Hg(0), which may be subjected to volatilization losses.

Methylation and demethylation: Biomethylation of metal(loid)s by microorganisms in soils and sediments produces methylated derivatives that can be excreted readily from cells and that are often volatile and may be less toxic, for example, organoarsenicals (Bentley and Chasteen 2002). Organic matter also provides the methyl-donor source for methylation in soils and sediments. Methylation of Hg is controlled by low molecular weight fractions of fulvic acid in soils (Nagase et al. 1982). Methylation of Hg occurs under both aerobic and anaerobic conditions (Ravichandran 2004). In biomethylation of Hg(II) under anaerobic conditions, Hg(II) ions form either monomethyl or dimethyl Hg, which are highly toxic and more biologically mobile than the other forms (Hammerschmidt et al. 2007). Biomethylation of As forms volatile compounds (e.g., alkylarsines) that could easily be lost to the atmosphere (Bentley and Chasteen 2002). Selenium biomethylation forms less toxic dimethyl selenide, and it is a potential mechanism for removing Se from contaminated environments.

5.2.2 Nitrogen

Nitrogen added through fertilizers, manures, and composts undergoes both biotic and abiotic reactions as detailed below (Figure 5.1).

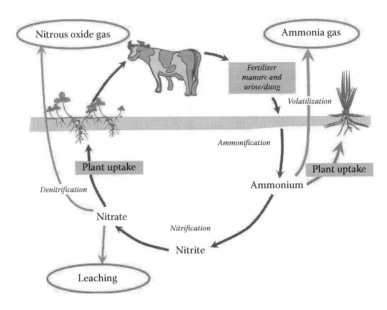

Figure 5.1 Dynamics of nitrogen transformation in soil. (From Bolan, N.S. et al. 2004a. *Advances in Agronomy*, **84**, 37–120.)

Biotic reactions: The biotic reactions include aminization, ammonification, immobilization, and denitrification. Hydrolysis of macromolecules of organic N compounds into simple N compounds is called aminization (Equation 5.1). Ammonification is the conversion of amines and amino acids into ammonium ions (NH_4^+) (Equation 5.2). Nitrification is a two-step oxidation process in which the NH_4^+ ions are converted into nitrite ions (NO_2^-) and then into nitrate ions (NO_3^-) (Equation 5.3). The nitrification process produces hydrogen ions (H^+), thereby decreasing the soil pH (Equation 5.4). During immobilization, plant-available NH_4^+ and NO_3^- ions are converted to plant-unavailable organic N by soil microorganisms. Denitrification is a four-step reduction process in which NO_3^- is reduced to NO_2^-, nitric oxide (NO), nitrous oxide (N_2O), and nitrogen gas (N_2) (Equation 5.5).

$$\text{Proteins} \xrightarrow{\text{Microorganisms}} \text{amines}(R-NH_2) + CO_2 \quad (5.1)$$

$$\text{Urea}\left[CO(NH_2)_2\right] \xrightarrow{\text{Microorganisms}} 2NH_4^+ + 2OH^- + CO_2 \quad (5.2)$$

$$2NH_4^+ + 4O_2 \rightarrow 2NO_3^- + 4H^+ + 2H_2O \qquad (5.3)$$

$$\text{Net effect } CO(NH_2)_2 \rightarrow 2NO_3 + 2H^+ \qquad (5.4)$$

$$NO_3^- \rightarrow NO_2^- \rightarrow NO \rightarrow N_2O \rightarrow N_2 \qquad (5.5)$$

Abiotic reactions: The abiotic reactions include NH_4^+ fixation, NO_3^- leaching, and NH_3 volatilization. Ammonium fixation is an adsorption process in which NH_4^+ ions (cations) are strongly retained onto clay particles or trapped inside the clay lattice, thereby reducing plant-available N. This NH_4^+ fixation process affects the N leaching and NH_3 volatilization processes. Nitrate, an anion, is weakly adsorbed onto soil particles and can be easily leached into the environment, causing groundwater pollution. Ammonium ions in an alkaline medium or in high pH soils (pH > 7.5) can dissociate into gaseous NH_3, which is subjected to volatilization losses (Equation 5.6). The initial increase in soil pH during the ammonification reaction after urea application or urine deposition is likely due to NH_3 volatilization.

$$NH_4^+ + OH^- \rightarrow NH_3 \uparrow + H_2O (pKa\ 7.6) \qquad (5.6)$$

5.2.3 Phosphorus

Phosphate compounds that are used as a fertilizer source are broadly grouped into H_2O-soluble (fast-release) and H_2O-insoluble (slow-release) fertilizers (Bolan et al. 1993; Karunanithi et al. 2014). The important H_2O-soluble P fertilizers include the following: single superphosphates (SSPs), triple superphosphates (TSPs), monoammonium phosphates (MAPs), and diammonium phosphates (DAPs). The principle P component in SSPs and TSPs is monocalcium phosphate (MCP) and in MAPs and DAPs is ammonium phosphate (AMP). The important H_2O-insoluble P fertilizers include PRs and basic slag.

Water-soluble compounds: When added to soil, MCP in SSP fertilizers dissociates to form soluble dicalcium phosphate with a release of phosphoric acid close to the fertilizer granules (Equation 5.7). Phosphoric acid subsequently dissociates into dihydrogen phosphate ion $(H_2PO_4^-)$ and H^+ (protons). The protons reduce the pH around the fertilizer granules to a very low level (pH < 2). When AMP fertilizers are added to soil, they dissociate into NH_4^+ and $H_2PO_4^-$ ions. The subsequent oxidation of NH_4^+ to NO_3^- results in the release of protons (Equation 5.8). The acidic solution around the fertilizer granules dissolves the Fe and Al compounds in the soil, resulting in the adsorption and precipitation of P and subsequent decrease of P available for plant uptake and leaching to groundwater. The acidity generated can also have important implications on the mobilization of metals.

$$Ca(H_2PO_4)_2 + H_2O \rightarrow CaHPO_4 \cdot H_2O + H_3PO_4 \qquad (5.7)$$

$$NH_4^+ + 2O_2 \rightarrow NO_3^- + 2H^+ + H_2O \qquad (5.8)$$

Water-insoluble compounds: When PRs are added to soil, they are dissolved by the acid produced or present in the soils to become plant-available forms (Equation 5.9). This is a major reason why PRs are very effective as a nutrient source in acidic soils (pH < 6.5) (Bolan et al. 1990; Chen et al. 2006) and as a metal(loid)-immobilizing agent in acid mine drainage (Evangelou and Zhang 1995; Johnson and Hallberg 2005). Once the PR is dissolved, the P released undergoes similar adsorption and precipitation reactions to those of soluble P fertilizers.

$$Ca_{10}(PO_4)_6 F_2 + 12H^+ \rightarrow 10Ca^{2+} + 6H_2PO_4^- + 2F^- \qquad (5.9)$$

5.3 Biochar: Synthesis and Characterization

5.3.1 Production Methods

Pyrolysis is the thermochemical decomposition of organic matter into noncondensable gases, condensable liquids, and a solid residual coproduct, biochar or charcoal in an inert environment (i.e., in the absence of oxygen). The end product of pyrolysis can be controlled by optimizing the pyrolysis parameters, such as temperature and residence time. For example, increasing the pyrolysis temperature (to a specific value) increases gas yield and decreases biochar production. Thus, optimized and advanced pyrolysis systems are now being used to control the biochar quality. Biochar has been extensively studied for its ability to enhance the nutrient level of soils (Schulz and Glaser 2012) and enhance plant growth (Sohi et al. 2010). Recently, biochar is widely being studied for its metal(loid) sorption efficiency and subsequent immobilization of metal(loid)s in soils.

Described below are the types of pyrolysis used to produce biochar. Some of the properties of these pyrolysis techniques are described in Table 5.1.

Slow pyrolysis is a low-temperature pyrolysis in which the biomass is less finely crushed and subjected to longer residence times and low heating rates to yield a high quantity of biochar at the expense of significant bio-oil formation (Klass 1998).

TABLE 5.1
Pyrolysis Production Techniques and the Operating Parameters

Pyrolysis System	Main Products	By-Products	Temperature (°C)	Heating Rate (°C/s)
Slow pyrolysis	Biochar, biogas	Bio-oil	300–550	0.01–2
Fast pyrolysis	Gas	Liquid and very low biochar	>600	>10^5
Flash pyrolysis	Liquid	Gas and very low biochar	400–600	10–1000
Torrefaction	Torrefied biomass or biochar	Gas	200–300	0.83
Gasification	Gas	Liquid and char	>800	—
Hydrothermal carbonization	Biocoal (biochar and water)	Gas	180–220	—

In fast pyrolysis, feedstock is finely ground to allow fast heat transfer. This finely ground feedstock is subjected to a moderate temperature with shorter residence time to produce high quality, ethylene-rich gases (syngas). Fast and flash pyrolysis differ in heating rates and hence the products derived. Due to short residence time, the production of char and tar is considerably less during these processes compared to slow pyrolysis (Bridgwater and Grassi 1991).

Torrefaction is a modified pyrolysis technique in which biomass is slowly heated to within a specified temperature range and retained there for a particular time in an inert or limited oxygen environment (Özçimen and Ersoy-Meriçboyu 2010). Gasification rearranges the molecular structure of solid or liquid biomass to produce gaseous fuel as the primary product and liquid or solid as secondary products. Unlike pyrolysis, gasification requires a gasifying medium such as steam, air, or oxygen to rearrange the molecular structure; it also adds hydrogen to the feedstock and strips hydrocarbon from the feedstock. Gasification of biomass using steam (>800°C) is called steam gasification, and gasification using oxygen enriched air or pure oxygen is called oxygen gasification (Lv et al. 2004).

Hydrothermal carbonization (HTC) is an exothermic process in which biomass is subjected to elevated temperatures in aqueous media (H_2O) in a confined system (Bridgwater and Grassi 1991). This process transforms all C from the biomass to coal (100% C efficiency) without any gaseous products. Two product streams of HTC are an insoluble, coal-like products (biocoal) and H_2O-soluble products. In addition to biocoal, HTC can also produce biochar (Özçimen and Ersoy-Meriçboyu 2010).

Microwave-assisted pyrolysis (MAP) is based on microwave heating of feedstock by molecular friction or dielectric heating in which dipolar polarization and ionic conduction are the two main mechanisms (Kappe 2004). In MAP, due to the heat being generated by the material itself in its bulk, the heat reaches the entire volume and can be much faster. Hence, when properly monitored, MAP can result in a homogeneous material, with faster production and significant reduction in energy losses compared to conventional methods (Clark and Sutton 1996). MAP has also been proven to enhance biochar yield and quality, and to a large extent, to negate undesirable secondary reactions among volatile compounds.

5.3.2 Biochar Characterization

The physical and chemical properties of biochar greatly depend on the feedstock properties and production parameters such as pyrolysis temperature and reaction rate.

Physical characterization: The magnitude of interactions between biochar and the soil environment can be explained by the surface area (SA) of the biochar. For example, Beesley and Marmiroli (2011) demonstrated that large SA of hardwood biochar and surface sorption of Cd and Zn to the biochar reduced the concentrations of these metals in leachates from a contaminated soil by 300- and 45-fold, respectively. In addition to the feedstock properties, the SA of biochar greatly depends on pyrolysis conditions such as temperature, gas flow, and the presence of air. For example, by increasing the pyrolysis temperature, the volatile matter present in the biochar pore infillings is released, producing biochars with maximum SA (Mukherjee et al. 2011).

Porosity of biochar is defined as the amount of total pore volume relative to the total bulk biochar volume. Pore sizes of <2, 2–50, and >50 nm are termed micro-, meso-, and macropores, respectively, and are often related to their respective adsorption capacities. For example, micropores have high adsorptive capacities for small molecules such as gases and common solvents (Lehmann and Joseph 2009). Thus, addition of biochar with high micropores to soil may reduce N loss as NH_3 volatilization. Similar to SA, pore structure greatly depends on pyrolysis temperature. High temperature destroys walls of the adjacent micropores, producing meso- or macropores.

Scanning electron microscopy (SEM) is a potential technique for studying the morphology (physical structure) of biochar. SEM images are very useful to obtain accurate details about the pore structure of biochars (Özçimen and Ersoy-Meriçboyu 2010). Beesley and Marmiroli (2011) used SEM for visual comparison and evaluation of the relative abundance of As, Cd, and Zn on soil and biochar surfaces before and after sorption and desorption experiments.

Chemical characterization: Biochar pH can be measured using a standard laboratory pH meter and Boehm titration. In Boehm titration, to quantify the types of acidic functional groups (AFGs) present in the biochar, the biochar is titrated with gradually increasing strengths of base. An increase in pyrolysis temperature has shown to either increase (Cao et al. 2009) or decrease (Uchimiya et al. 2010) total acidity or AFGs in biochar. Manure-based biochars have been shown to increase soil pH; because binding of metal ions by surface ligands is strongly pH dependent, it can be promoted by the biochar-induced pH increase (Uchimiya et al. 2011). For example, addition of biochar had shown to increase pH, thereby enhancing the immobilization of heavy metals [e.g., Cd(II), Cu(II), Ni(II), and Pb(I)] (Uchimiya et al. 2010). Metal(loid)s retention by biochar can also be explained by their precipitation as (hydr)oxide, carbonate, or phosphate compounds. Also, increased pH after biochar addition has been shown to contribute to higher denitrification and reduced N_2O emission from soil by Van Zwieten et al. (2010). Elemental or proximate analysis is the quantification of ash content and organic material content (carbon) in the biochar (Litaor et al. 2005). Traditionally, the thermogravimetric method has been used; but recently, elemental or CHN analysis that quantifies C, H, and N contents in the biochar is being used. In this latter method, biochar is combusted at very high temperatures with excess oxygen gas, and the carbon dioxide, H_2O, and NO produced are trapped and quantified for the C, H, and N contents, respectively (Brewer 2012). Elemental molar ratios such as H/C and O/C of biochar change with pyrolysis techniques and are used to measure the aromaticity and maturation of the biochar. Aromaticity is defined as the fraction of C in the biochar that forms aromatic bands. Biochars with dense aromatic structures are more resistant to oxidation and microbial degradation; hence, the degree of aromatic condensation in biochars is related to their recalcitrance in the environment (Spokas 2010). Low-temperature biochars are likely to have lower aromaticity than high-temperature biochars (Keiluweit et al. 2010). An easier technique for elemental analysis is x-ray fluorescence spectroscopy. In this method, when biochar is bombarded with x-rays, each element in the biochar emits fluorescence with a characteristic wavelength or set of wavelengths. The intensity of the emission is relative to the amount of that element present in the biochar.

Ash composition of a biochar sample can be quantified by digesting or leaching the sample and then measuring the concentration of the ions in the resulting solution by using atomic absorption spectroscopy or inductively coupled plasma atomic emission spectroscopy (Singh et al. 2010). Several techniques have been used to examine surface and functional chemistry of biochar (Lehmann et al. 2006): diffuse reflectance infrared Fourier transform (FTIR) spectroscopy, x-ray photoelectron spectroscopy, energy dispersive x-ray spectroscopy, and near-edge x-ray absorption fine structure spectroscopy. Of all these techniques, FTIR spectroscopy is the most frequently used technique to identify and qualitatively track changes in functional groups in biochar and soil samples. For example, the ability of the phosphorus functional group in a broiler litter biochar as measured by FTIR to bind with metal ions was examined by Uchimiya et al. (2010). They noticed a consistent trend for the affinity of orthophosphate ester ligands for the metal ions—Ni(II) < Cd(II) < Cu(II) < PB(I)—that coincided with the relative extent of removal for these four heavy metal ions from the soil. Although ^{13}C solid-state nuclear magnetic resonance spectroscopy is a complicated technique, it is informative for characterizing the C fraction of biochar (Brewer et al. 2009).

5.4 Biochar and Inorganic Contaminants Interactions

5.4.1 Heavy Metal(loid)s

The contamination of agricultural and industrially impacted urban lands with heavy metals, nutrients, and pesticides is increasing rapidly. The mobility and bioavailability of inorganic contaminants are controlled by a myriad of reactions, including redox, sorption, and complexation. Among the organic materials used for remediation of heavy metal(loid)s–contaminated sites, biochar has recently received considerable interest. For heavy metal(loid)s, biochar decreases the bioavailability through adsorption, precipitation, and redox reactions (Figure 5.2).

Biochars that are produced under anoxic conditions are rich in silicates and mineral oxides that can decrease the availability of cationic heavy metal(loid)s by increasing the soil pH and cation exchange capacity (CEC) (Joseph and Taylor 2014; Park et al. 2012). Biochar has high specific SA and the reactive functional groups in the biochar may complex and sorb metal(loid)s in contaminated soils. The sorption capacity of biochars depends on feedstock properties and also on the pyrolysis temperature. For example, sugarcane straw biochar produced at 700°C has four times higher Cd and Zn sorption capacity than that produced at 400°C (Melo et al. 2013), which may be attributed to the higher SA of the former samples. The high oxygen-containing functional groups present on the biochar surface stabilize heavy metals such as Pb(II) and Cu(II) in acidic and low organic C soils (Uchimiya et al. 2011). The sorption of Hg on biochar surface is an irreversible

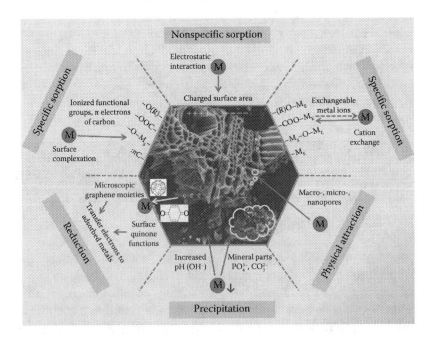

Figure 5.2 Schematic diagram of biochar sorption for metals.

reaction and sorbs via complexation with phenolic, hydroxyl, and carboxylic functional groups on low-temperature biochars (Dong et al. 2013).

At an equilibrium metal concentration of 0.06 mM, compared with control (no biochar) treatments, addition of canola straw biochar produced at 350°C increased adsorption of Cu(II), Pb(II), and Cd(II) by 26, 12, and 58.5%, respectively, through electrostatic and nonelectrostatic mechanisms; the corresponding adsorption percentage for the peanut straw biochar was 69, 64, and 157% (Xu and Zhao 2013). Cao et al. (2009) concluded that sorption of Pb by biochar is controlled by sorption–precipitation reactions and that Pb is precipitated as phosphate and carbonate compounds on the biochar surface. A close relationship between the nonelectrostatic adsorption of Pb to biochar and contents of free Fe oxides in soils amended with biochar was noted by Jiang et al. (2012). High Fe(III) oxide content in biochar may have increased the anionic contaminants sorption via precipitation. However, the adsorption capacity of anionic contaminants by biochars is not very strong as a significant amount of adsorbed metals was released during desorption experiments (Agrafioti et al. 2014). The reaction between oxyanions and biochar depends on the type of oxyanions, soil properties, and soil pH. Kappler et al. (2014) suggested that biochar can influence soil biogeochemistry not only by changing soil structure but also by directly mediating the electron transfer process by functioning as an electron shuttle. Choppala et al. (2012) concluded that reduction Cr(VI) to Cr(III) by the oxygen-containing functional groups on the biochar had decreased Cr(VI) concentration in soils.

5.4.2 Nutrients

Biochar application to soil has been shown to decrease the loss of N through leaching and gaseous emissions, thereby increasing the N use efficiency by plants (Figure 5.3). Similarly, biochar application enhances the uptake of P by plants.

Ammonia volatilization: The physical (e.g., SA) and chemical (e.g., CEC) characteristics of biochars influence their effectiveness in controlling NH_3 volatilization. For example,

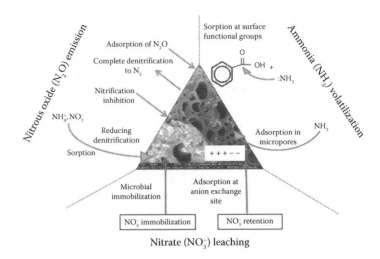

Figure 5.3 Schematic diagram of mechanisms for biochar-induced mitigation of ammonia volatilization, nitrous oxide emission, and nitrate leaching.

Spokas et al. (2012) indicated that biochar can act as soil amendment through capturing NH_3 from the soil. Research has been conducted to understand the decreased NH_3 volatilization resulting from biochar application to soil (Table 5.2). Biochar addition to soil can decrease soil pH, thereby reducing NH_3 volatilization. Biochars with high SA that contains several surface functional groups can capture NH_3. With NH_3 being an alkaline gas, the acidic surface groups on biochar with their low pH can protonate the NH_3 to form NH_4^+, thereby promoting their adsorption onto cation exchange sites (Bandosz 2006) (Equation 5.10). The NH_3 sorbed by biochar can subsequently become available for plants (Taghizadeh-Toosi et al. 2011).

$$NH_3 + H^+ \rightarrow NH_4^+ \qquad (5.10)$$

N_2O emission: Biochar addition has often been shown to decrease total N_2O emission from soils treated with N sources such as manures, urea, and composts (Bruun et al. 2012; Singh et al. 2010; Spokas et al. 2009; Van Zwieten et al. 2009) (Table 5.2). Denitrification is the biological process leading to increased N_2O emission from soil. Elevated mineral N in soils, the substrate for N_2O production, is generally associated with

TABLE 5.2
**Effect of Biochar Addition on Ammonia (NH$_3$) Volatilization
and Nitrous Oxide (N$_2$O) Emission from Soils**

Biochar	N Source (Amount Added)	Soil Type	% NH$_3$ or N$_2$O Reduction	Reference
Monterey pine biochar	Ruminant urine patches	Templeton silt loam soil	N$_2$O: >50	Taghizadeh-Toosi et al. (2011)
Monterey pine wood chips	Urine (509 kg N/ha)	Silt loam	NH$_3$: 45	Taghizadeh-Toosi et al. (2011)
Wood biochar, poultry manure biochar	—	Alfisol Vertisol	N$_2$O Alfisol: 14–73 Vertisol: 23–52	Singh et al. (2010)
Wheat straw biochar	No urea Urea (300 kg N/ha)	Hydroagric Stagnic Anthrosol	N$_2$O No urea: 40–51 Urea: 21–28	Zhang et al. (2010)
Peanut hull biochar, maize biochar, wood chip biochar, beech charcoal biochar	—	Sandy loam	N$_2$O: ~60	Kammann et al. (2012)
Wood chip biochar	—	Waukegan silt loam	N$_2$O: 20, 40, and 60 (w/w) biochar → 57, 57, and 74 reduction, respectively	Spokas et al. (2009)
Municipal biowaste charcoal	—	Typic Hapludand (loam to clay loam)	N$_2$O: 89	Yanai et al. (2007)

higher N$_2$O emissions (Vaughan and Lenton 2011). Research suggests that possible explanations for reduced N$_2$O loss from biochar-treated soils include reduced denitrification, complete denitrification, or a combination. A decrease in denitrification is likely to occur due to sorption of mineral N compounds

$\left(NH_4^+, NO_3^-\right)$ to biochar surface and lattice, thus reducing the substrate for denitrification (Taghizadeh-Toosi et al. 2011). Complete denitrification, leading to N_2 emission due to biochar addition, was explained by enhanced aeration (Taghizadeh-Toosi et al. 2012), the presence of labile C in biochar, and elevated soil pH and enhanced microbial activities (Anderson et al. 2011). Lehmann et al. (2006) also hypothesized that biochar can reduce N_2O emissions by inducing microbial immobilization of the mineral N is the soil.

NO_3^- leaching: NO_3^- leaching is the major loss of N from soils, and it can be minimized by using char materials such as biochar. Biochar has inherent chemical and physical properties (e.g., porous structure, high SA, surface functional groups [carboxyl and phenolic groups]) and anion exchange sites that could influence soil physical, chemical, and biological properties (e.g., pH, CEC, bulk density, H_2O-holding capacity) (Glaser et al. 2002; Singh et al. 2010). These properties of biochar facilitate the sorption of ions from the soil solution by a combination of electrostatic and capillary forces on their surface, thereby reducing nutrient leaching from soils. For example, it has been shown that biochar produced at high temperatures reduced NO_3^- leaching more significantly that biochar produced at low temperatures, which may be attributed to greater sorption by former samples (Clough et al. 2013; Knowles et al. 2011; Singh et al. 2010). Similarly, Yao et al. (2012) found that application of biochar reduced NO_3^- leaching by 34% through adsorption. However, it has also been observed that the adsorption, retention, and sorption capacity of NO_3^- by biochar varied with biochar type (feedstock material, low or high temperature, slow or fast pyrolysis) and characteristics. Therefore, soil amended with biochar can adsorb NO_3^- through its anion exchange sites, thereby reducing its loss and increasing its retention in the soil.

Phosphorus leaching: Excessive application of P fertilizers has caused leaching of P from agricultural fields to aquatic systems (Karunanithi et al. 2014; Loganathan et al. 2014). Nutrient leaching not only poses a threat to environmental health but also may deplete soil fertility, accelerate soil acidification, increase fertilizer costs, and reduce crop yield (Laird et al. 2010). Biochar has proven to alter P availability in soils by reducing P leaching through sorption and adsorption characteristics (Figure 5.4). Laird et al. (2010) reported that the addition of hardwood biochar produced using slow pyrolysis to typical midwestern agricultural soils significantly reduced total P leaching by 69%. Similarly, in a column study, biochar produced from Brazilian pepperwood at 600°C reduced the total amount of phosphate in the leachates by ~20.6% (Yao et al. 2012). Addition of biochar has also been shown to increase P availability in soil solution (Atkinson et al. 2010; Hossain et al. 2010). For example, in the study by Doydora et al. (2011), application of peanut hull biochar increased the amount of phosphate in the soil solution by 39%. The possible mechanisms suggested for the influence of biochar on P availability are a change in soil pH and the subsequent influence on the interaction of P with other cations, or enhanced retention

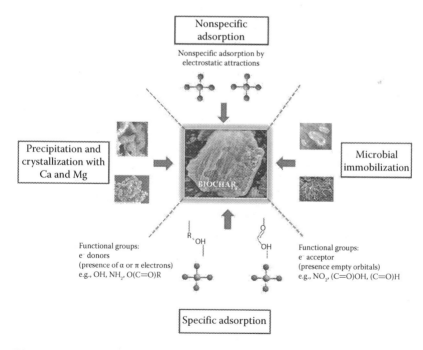

Figure 5.4 Schematic diagram of mechanisms of biochar-induced immobilization of phosphorus.

through anion exchange (Atkinson et al. 2010). In the natural environment, P is strongly adsorbed onto the surface of Fe(III) (hydr)oxides in soils (Jaisi et al. 2010). Addition of biochars reduced the amount and rate of P sorbed onto ferrihydrite (the most effective Fe oxide for P) (Cui et al. 2011). Chen et al. (2011) demonstrated that biochars with magnetized Fe^{3+}/Fe^{2+} had enhanced phosphate sorption compared to nonmagnetic biochars.

5.5 Case Studies Involving Biochar for the Remediation of Contaminated Sites

5.5.1 Heavy Metal(loid)s

Biochar has been proposed as a low-cost remediation strategy to increase soil pH, reduce leaching of toxic elements, and improve plant establishment. In a laboratory column study, Kelly et al. (2014) assessed biochar made from pine wood as a soil amendment by mixing soil material from two mine sites (Bonner Mine and Joe John mine) collected near Silverton, Colorado. The mineralization event resulted from sulfur-rich hydrothermal fluids that produced mineral assemblages that generally include abundant pyrite. Mining resulted in mine-waste rock that has been pulverized to remove sulfide ore, characterized by high acidity and toxic element concentrations and very little organic material. The mine site is highly mineralized and has been mined for Au, Ag, Cu, Pb, and Zn for >100 years. The Bonner mine tailing had a high S content (22,850 mg S/kg), indicative of pyrite-containing materials, and was characterized by an initial soil pH and acidity content of 2.99 and 1.11 mEq/g soil, respectively. The Joe John mine tailing had an initial pH of 4.41 and an acidity content of 0.26 mEq/g soil. Total contents of Al, As, Cd, Cu, Fe, Pb, S, and Zn are above levels described as phytotoxic. The previous remediation effort had low effectiveness, and the site remains largely unvegetated. The Bonner tailing was considered more phytotoxic than the Joe John material and potentially more resistant to remediation.

Pine wood biochar was applied at four application rates: 0, 10, 20, and 30 (v/v). Columns were leached seven times over 65 days, and leachate pH and concentration of toxic elements and base cations were measured at each leaching. Soil pH increased in both tailings with biochar additions, but significant differences were only detected in the Bonner material at the highest biochar addition level, increasing from 3.33 in the pure tailing to 3.63 with 30% biochar ($p = 0.0044$). In the Joe John tailing, a significant increase in pH was noted with just 10% biochar addition and ranged from 4.07 in the pure material to 4.77 with 30% biochar ($p < 0.0001$).

In leachate solution, biochar increased base cations from both materials and reduced the concentrations of Cd, Cu, Pb, and Zn in leachate solution from Joe John material (Figure 5.5). However, in the material with greater

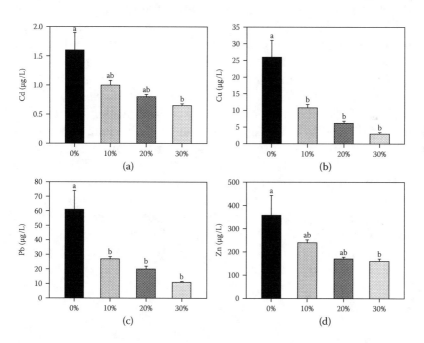

Figure 5.5 Concentration of dissolved metals [(a) Cd, (b) Cu, (c) Pb, and (d) Zn] in leachate from the Joe John material as affected by biochar treatments ($n = 3$). Means followed by different letters are considered different due to biochar application according to Tukey's honestly significant test (at $\alpha = 0.05$). Error bars show SEM. (From Kelly, C.N. et al. 2014. *Applied Geochemistry*, **43**, 35–48.)

toxic elements content (Bonner), biochar did not reduce concentrations of any of the measured dissolved toxic elements in leachate and resulted in a potentially detrimental release of Cd and Zn into solution at concentrations above that of the pure mine material.

Several studies have demonstrated the effectiveness of some biochars to improve metal toxicity in mine soils. For example, after incorporation of an orchard waste–derived biochar on a Pb and Zn mine tailing of pH 8.2, the pH, nutrient retention, CEC, and H_2O holding capacity increased and concentrations of bioavailable Cd, Pb, Tl, and Zn decreased (Fellet et al. 2011). In another Cu mine tailing of pH 5.4, hardwood-derived biochar additions decreased concentrations of Cu in pore H_2O and decreased Cu and Pb concentrations in the shoots of ryegrass grown in the soils (Karami et al. 2011). Beesley and Marmiroli (2011) noted reduced concentrations of Cd and Zn from leachate of a multi-element-polluted soil of pH 6.2 when a hardwood-derived biochar was added. Other studies have reported element-specific immobilization by biochar, where some elements are preferentially immobilized and others are mobilized (Namgay et al. 2010; Uchimiya et al. 2011). The effectiveness of biochar as a soil amendment on the two mine tailings differed depending on the initial characteristics of the tailings. The pine wood biochar was more effective at increasing soil pH and increasing cation content in the tailing, with relatively higher initial pH and less effective in the tailing with lower pH. The authors of these studies recommended that biochar may be an effective strategy for reclamation of soils if the application rate is great enough to provide sufficient sorption capacity and sufficiently raise the pH to mitigate potential toxicity.

5.5.2 Nutrients

Biochars are well demonstrated for reducing nutrient losses (gaseous and leaching) because of their nutrient retention and sorption capacities. Singh et al. (2010) assessed the influence of four biochars on soil N transformation processes from two different soils. Biochars were produced at two temperatures (400 and 550°C) from poultry manure mixed with rice hull (PM) and chipped stemwood of *Eucalyptus saligna* (W). To develop a highly porous structure, toward the end of the pyrolysis process, the high-temperature biochars (PM550 and W550) were activated using steam. Soil used in this study were Alfisol and Vertisol with a pH of 6.13 and 8.80, respectively. The pH of biochar was 9.20, 10.56, 6.93, and 9.49 for PM400, PM550, W400, and W550, respectively.

Soil–biochars and soil–sand (control) mixtures were set up in a specially designed column with an outlet for leachate collection $(NH_4^+\text{-N}, NO_3^-\text{-N}$ analysis$)$ and an airtight screw cap for gas sampling (N_2O analysis). Columns were subjected to three wetting–drying (W-D) cycles to achieve a range of H_2O-filled pore spaces. Leachate and gas samples were collected during every W-D cycle. To restock the inorganic N and C levels in the soils, a nutrient solution containing N and glucose was added during the first W-D cycle.

During the first two W-D cycles, PM400 showed higher N_2O emission than control treatment, which was attributed to the presence of higher labile intrinsic N content in PM400 compared to the other biochars. N_2O emission from other treatments (PM550, W400, and W550) were inconsistent during the first two W-D cycles. A possible explanation for N_2O emission during these cycles was attributed to the denitrification activity triggered due to high moisture content along with the application of nutrient solution and glucose (Bolan et al. 2004a). N_2O emission declined after the initial peak in all the treatments and was significantly lower in biochar treatments than that in the control treatment, which was attributed to the biochar-induced increased soil pH and occurrence of complete denitrification (N_2O to N_2) (Van Zwieten et al. 2009). During the first leaching event, PM400 increased NO_3^--N in the leachate, indicating the inability of biochars to reduce N leaching losses in short-term studies. Limited anion exchange capacity (AEC) of biochars was one the reasons for the low retention of NO_3^--N on to the biochars. However, significantly lower NO_3^--N in the leachate from biochar treatments compared to the control treatment was observed during the latter leaching, which was attributed to NH_4^+-N fixation in the interlayer, thereby decreasing the substrate for nitrification.

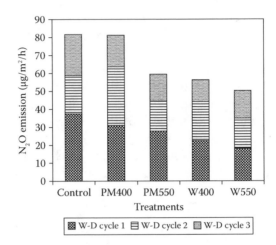

Figure 5.6 Mean nitrous oxide emissions ($\mu g/m^2/h$) from control soils (Alfisol and Vertisol) and biochar treatments during the three wetting–drying (W-D) cycles. PM400 and PM550, poultry manure and rice hull mixture biochar produced at 400 and 550°C, respectively; W400 and W550, chipped stemwood of *Eucalyptus saligna* biochar produced at 400 and 500°C, respectively. (From Singh, B.P. et al. 2010. *Journal of Environmental Quality*, **39**(4), 1224–1235.)

The authors concluded that the biochars decreased N loss as N_2O by up to 73% compared to the control treatment (Figure 5.6). They also demonstrated that oxidation of biochar surfaces over the study period had potentially increased the sorption capacity of the biochars. They recommended further study on biochar–mineral interactions to enhance the biochar-induced denitrification (N_2O to N_2) in soil.

Acknowledgments

The Postdoctoral fellowship program PJ010923 at the National Academy of Agricultural Science, Rural Development Administration, Republic of Korea, supported Dr. Anitha Kunhikrishnan's contribution.

Dr. Girish Choppala would like to thank Southern Cross GeoScience (http://scu.edu.au/geoscience) for providing resources to write this book chapter.

Appendix

1. Biochar rich in K can provide a source of K input to soil. A soil analysis has indicated that the soil contains 0.5 cmol·K/kg soil of exchangeable K. The exchangeable K level in the soil needs to be increased to 2.0 cmol · K/kg soil to overcome the K deficiency in soils. From the following information, calculate the amount of biochar (kg/ha) required to increase the K level to a depth of 5 cm. Assume all the K added through biochar is available as exchangeable K in soils.

K content of biochar	= 45 g/kg
Exchangeable K level in the soil	= 0.5 cmol/kg
Level of exchangeable K required	= 2.0 cmol/kg
Atomic weight of K	= 39
Depth of soil	= 5 cm
Bulk density of the soil	= 0.89 Mg/m^3
1 ha	= 10,000 m^2

2. From the following data, calculate the maximum depth in which all the added Cd is retained from an accidental spill of 2000 L of solution containing cadmium.

Volume of spill	= 2000 L
Concentration of Cd	= 337 mg/L
Area of spill	= 15 m^2
CEC of soil	= 7 cmol$_{(+)}$/kg
Bulk density of the soil	= 0.92 Mg/m^3
Atomic weight of Cd	= 112

   Cadmium is retained as Cd^{2+} and is assumed to occupy only 85% of the exchange sites.

3. Biochar enhances the adsorption of heavy metal cations by providing cation exchange sites and also by increasing the soil pH. Three thousand liters of an industrial effluent with 1500 mg Cd/L was discharged over a 50-m^2 area. The site has a variable charge soil. From the following information, calculate the amount of biochar (kg) required to raise the cation exchange

sites of the soil to retain all the Cd within 15-mm depth. Assume biochar increases the CEC of soil only by increasing the soil pH.

pH buffering capacity of the soil   = 9 cmol H^+ or OH^-/kg soil/pH

Average increase in cation exchange sites per unit pH   = 3 $cmol_{(+)}$/kg soil

Atomic weight of Cd   = 112

Liming (calcium carbonate [$CaCO_3$]) value of biochar   = 22%

Bulk density of the soil   = 1.11 Mg/m^3

Initial CEC of the soil   = 5 $cmol_{(+)}$/kg soil

1 mole of $CaCO_3$ = 100 g

1 mole of $CaCO_3$ produces   = 2 moles of OH^-

Cadmium is retained as Cd^{2+} ion and is assumed to occupy only 85% of the exchange sites.

4. An animal rendering plant has been planning to discharge the effluent through land application. The concentration of nitrogen (as ammonium cation) and phosphorus (as phosphate anion) in the efflu- ent is 25 mg N/L and 8 mg P/L, respectively. The soil at the disposal site is sandy and is low in CEC [4.4 $cmol_{(+)}$/kg soil], AEC [0.32 $cmol_{(-)}$/kg], organic matter content, and amorphous iron and aluminium hydrous oxides.

   a. Calculate the maximum amount of ammonium and phosphate retained (kg/ha) by the soil to a depth of 10 cm. Assume ammonium is retained as ammonium cations (NH_4^+) onto the cation exchange sites and phosphate as phosphate $(H_2PO_4^-)$ to the anion exchange sites. The bulk density of the soil is 1020 kg/m^3.

   b. From (a), calculate the volume of effluent (m^3/ha) required to reach the maximum ammonium reten- tion capacity of 1 ha of the soil to depth of 10 cm.

   c. From (b), if the rate of effluent discharge is 30,000 m^3/day over 10 ha, calculate the number of days required to saturate the ammonium retention capacity of the 1 ha of soil to a depth of 10 cm.

    d. If the rate of effluent discharge is 30,000 m³/day over 10 ha, calculate the total amount of biochar required (kg) to retain all ammonium in the effluent [CEC of biochar is 45 cmol$_{(+)}$/kg]

Answers

1.

Weight of soil/ha $= 10{,}000 \text{ m}^2/\text{ha} \times 0.89 \text{ Mg/m}^3 \times 0.05$
$= 445 \text{ Mg/ha}$

Calculations to find the extra potassium needed to increase the CEC of the soil:

Extra K needed $=$ (level of exchangeable K needed
$-$ exchangeable K level in the soil)
$= 2 - 0.5 = 1.5 \text{ cmol/kg} = 0.015 \text{ mol/kg}$
$= 0.015 \text{ mol/kg} \times 39 \text{ g/mol} = 0.585 \text{ g/kg}$
$= 0.585 \text{ kg/Mg}$

Extra K needed/ha $= 445 \text{ Mg/ha} \times 0.585 \text{ kg/Mg} = 260 \text{ kg/ha}$

Calculations to find the amount of biochar to increase the extra potassium (260 kg/ha) needed to increase the CEC of the soil:

K content of biochar $= 45 \text{ g/kg} = 45 \text{ kg/Mg}$

Biochar required $= \dfrac{1 \times 260}{45} = 5.777 \text{ Mg biochar/ha}$
$= 5777 \text{ kg biochar/ha}$

2.

Quantity of Cd spilled $=$ Volume of spill $\times$ Cd concentration in the spill
$= 2000 \text{ L} \times 337 \text{ mg/L} = 674{,}000 \text{ mg}$
$= 674 \text{ g}$

Moles of Cd in the spill $= \dfrac{\text{Amount of Cd spilled}}{\text{Atomic weight of Cd}}$

$= \dfrac{674}{112} = 6.02 \text{ moles of Cd}^{2+}$

Cation sites required $= 6.02 \times 2 = 12.04 \text{ moles}$

CEC of the soil $\qquad$ = 7 $cmol_{(+)}/kg$ = 0.07 $moles_{(+)}/kg$ soil

If the spilled Cd occupies only 85% of the soil exchange sites,

Quantity of soil required to retain the spilled Cd

$$= \frac{12.04}{0.07} \times 100/85 = 202 \text{ kg soil}$$

Calculations to find the depth of the soil to which spilled Cd was retained in the soil:

Weight of soil = spillage area × depth × soil bulk density

202 kg $\qquad$ = 15 m^2 × m × 920 kg/m^3

Depth $\qquad = \dfrac{202}{15 \times 920} = 0.0146 \text{ m} = 1.463 \text{ cm}$

3.

Quantity of Cd spilled = volume of spill × Cd concentration in the spill

$$= 3000 \text{ L} \times 1500 \text{ mg/L} = 4,500,000 \text{ mg}$$
$$= 4500 \text{ g}$$

Moles of Cd in the spill $= \dfrac{\text{Amount of Cd spilled}}{\text{Atomic weight of Cd}}$

$$= \frac{4500}{112} = 40.18 \text{ moles of } Cd^{2+}$$

Cation sites required = 40.18 × 2 = 80.36 moles

85% occupied = 80.36/0.85 = 94.54 moles of charge required

Weight of soil = spillage area × depth × soil bulk density

= 50 m^2 × 0.015 m × 1.11 Mg/m^3

= 0.832 Mg = 832 kg

Initial CEC of the soil = 5 $cmol_{(+)}/kg$ soil = 0.05 $mol_{(+)}/kg$

CEC of the 832 kg soil = 832 × 0.05 = 41.6 $mol_{(+)}$ charge

Extra charge required = 94.54 – 41.6 = 52.94 $moles_{(+)}$ charge

Average increase in cation exchange sites/unit pH

= 3 $cmol_{(+)}/kg$ soil = 0.03 $mol_{(+)}/kg$ soil

Increase in CEC/pH/832 kg soil = 0.03 × 832 = 24.96 mol/pH

pH raise required for 52.94 moles $= \dfrac{52.94}{24.96} = 2.12 \text{ pH}$

pH buffering capacity of the soil = 9 cmol H^+ or OH^-/kg soil/
$$pH = 0.09 \text{ mmol/kg/pH}$$
pH buffering capacity of the 832 kg soil
$$= 0.09 \times 832 = 74.88 \text{ mol/832 kg soil/pH}$$
$$= 158.74 \text{ mol/832 kg soil/2.12 pH}$$
1 mol $CaCO_3$ = 2 mol OH
Amount of $CaCO_3$ required = 158.74/2 = 79.37 = 79.37 mol
$$CaCO_3/832 \text{ kg soil/2.52 pH}$$
Calculation to find the quantity of biochar required to raise the cation exchange sites of the soil to retain the Cd within 15-mm depth:
Quantity of biochar required = amount of $CaCO_3$ required/ liming value of biochar

$$= \frac{7.937}{22} \times 100 = 36.08 \text{ kg biochar}$$

4.

    a. Calculation to find the maximum amount of ammonium and phosphate retained (kg/ha) by the soil to a depth of 10 cm:
Amount of soil = area × soil bulk density × depth
$$= 10,000 \text{ m}^2\text{/ha} \times 1020 \text{ kg/m}^3 \times 0.10 \text{ m}$$
$$= 1,020,000 \text{ kg soil/ha}$$

Ammonium:
Amount of NH_4^+ retained (cmol/kg)
$$= \text{CEC (cmol}_{(+)}\text{/kg)} = 4.4 \text{ cmol NH}_4\text{/kg}$$
$$= 0.044 \text{ mol NH}_4\text{/kg}$$
Amount of NH_4^+ retained
$$= \text{molecular weight of NH}_4^+ \times \text{CEC}$$
$$= 18 \text{ g/mol} \times 0.044 \text{ mol} = 0.79 \text{ g/kg soil}$$
Amount of ammonium retained (kg/ha) by 1,020,000 kg soil
$$= 0.79 \text{ g NH}_4^+ \times 1,020,000 \text{ kg soil}$$
$$= 805.8 \text{ kg NH}_4^+ \text{ or} = 805.8 \times 14/18$$
$$= 626.7 \text{ kg N as NH}_4^+ - \text{N/ha}$$

Phosphate:
Amount of phosphate retained (cmol/kg)
$$= \text{AEC (cmol}_{(-)}\text{/kg)} = 0.32 \text{ cmol phosphate/kg}$$
$$= 0.0032 \text{ mol phosphate/kg}$$

Amount of phosphate retained
$$= \text{AEC} \times \text{molecular weight of } H_2PO_4^-$$
$$= 0.0032 \times 94.97 = 0.303 \text{ g } H_2PO_4^-$$
$$= 0.0989 \text{ P as } H_2PO_4^-$$

Amount of phosphate retained (kg/ha) by 1,020,000 kg soil
$$= 0.0989 \text{ g P} \times 1,020,000 \text{ kg soil}$$
$$= 100.88 \text{ kg P/ha}$$

b. Calculation to find the volume of effluent (m³/ha) required to reach the maximum ammonium retention capacity of 1 ha of the soil to depth of 10 cm:
Volume of effluent

$$= \frac{1 \text{ L effluent}}{25 \text{ mg N as } NH_4^+} \times 626.7 \text{ kg N as } NH_4^+/\text{ha}$$
$$= 25068000 \text{ L or } 25068 \text{ m}^3/\text{ha}$$

c. Calculation to find the number of days required to saturate the ammonium retention capacity of the 1 ha of soil to a depth of 10 cm:
Discharge in 1 day = 30,000 m³/10 ha = 3000 m³/ha
Number of days to discharge 25,068 m³ of effluent

$$= \frac{1 \text{ day}}{300 \text{ m}^3} \times 25,068 \text{ m}^3 = 8.356 \left(\sim 8 \text{ days} \right)$$

d. Calculation to find the amount of biochar required to retain all ammonium in the effluent:
Amount of $NH_4^+ - N$ in 30,000 m³ of effluent
$$= 30,000 \text{ m}^3 \times 25 \text{ mg/L} = 750 \text{ kg } NH_4^+/\text{day}$$
Amount of NH_4^+ retained by 1 kg biochar
$$= 0.45 \text{ mol/kg} \times 18 \text{ g/mol} = 8.1 \text{ g } NH_4^+/\text{kg}$$
$$= 8.1 \times 14/18 = 6.3 \text{ g } NH_4^+ - N/\text{kg}$$
Amount of biochar required to retain all an NH_4^+ in the effluent of 30,000 m³

$$= \frac{1 \text{ kg biochar}}{6.3 \text{ g } NH_4 - N} \times 750 \text{ kg } NH_4 - N$$
$$= 119,047 \text{ kg biochar}$$
$$= 119 \text{ Mg biochar}$$

References

Adriano, D., Wenzel, W., Vangronsveld, J., Bolan, N.S. 2004. Role of assisted natural remediation in environmental cleanup. *Geoderma*, **122**(2), 121–142.

Agrafioti, E., Kalderis, D., Diamadopoulos, E. 2014. Arsenic and chromium removal from water using biochars derived from rice husk, organic solid wastes and sewage sludge. *Journal of Environmental Management*, **133**, 309–314.

Anderson, C.R., Condron, L.M., Clough, T.J., Fiers, M., Stewart, A., Hill, R.A., Sherlock, R.R. 2011. Biochar induced soil microbial community change: Implications for biogeochemical cycling of carbon, nitrogen and phosphorus. *Pedobiologia*, **54**(5), 309–320.

Atkinson, C.J., Fitzgerald, J.D., Hipps, N.A. 2010. Potential mechanisms for achieving agricultural benefits from biochar application to temperate soils: A review. *Plant and Soil*, **337**(1–2), 1–18.

Bachate, S.P., Khapare, R.M., Kodam, K.M. 2012. Oxidation of arsenite by two β-proteobacteria isolated from soil. *Applied Microbiology and Biotechnology*, **93**(5), 2135–2145.

Bandosz, T.J. 2006. Desulfurization on activated carbons. In: *Interface Science and Technology*, Bandosz, T.J. (Ed). Elsevier, UK, **7**, 231–292.

Beesley, L., Marmiroli, M. 2011. The immobilisation and retention of soluble arsenic, cadmium and zinc by biochar. *Environmental Pollution*, **159**(2), 474–480.

Bentley, R., Chasteen, T.G. 2002. Microbial methylation of metalloids: Arsenic, antimony, and bismuth. *Microbiology and Molecular Biology Reviews*, **66**(2), 250–271.

Bolan, N., Adriano, D., Mahimairaja, S. 2004b. Distribution and bioavailability of trace elements in livestock and poultry manure by-products. *Critical Reviews in Environmental Science and Technology*, **34**(3), 291–338.

Bolan, N., Kunhikrishnan, A., Thangarajan, R., Kumpiene, J., Park, J., Makino, T., Kirkham, M.B., Scheckel, K. 2014. Remediation of heavy metal(loid)s contaminated soils—To mobilize or to immobilize? *Journal of Hazardous Materials*, **266**, 141–166.

Bolan, N., Naidu, R., Choppala, G., Park, J., Mora, M.L., Budianta, D., Panneerselvam, P. 2010b. Solute interactions in soils in relation to the bioavailability and environmental remediation of heavy metals and metalloids. *Pedologist*, **53**, 1–18.

Bolan, N.S., Adriano, D.C., Curtin, D. 2003b. Soil acidification and liming interactions with nutrient and heavy metal transformation and bioavailability. *Advances in Agronomy*, **78**, 215–272.

Bolan, N.S., Choppala, G., Kunhikrishnan, A., Park, J., Naidu, R. 2013. Microbial transformation of trace elements in soils in relation to bioavailability and remediation. *Reviews of Environmental Contamination and Toxicology*, **225**, 1–56.

Bolan, N.S., Elliott, J., Gregg, P.E.H., Weil, S. 1997. Enhanced dissolution of phosphate rocks in the rhizosphere. *Biology and Fertility of Soils*, **24**(2), 169–174.

Bolan, N.S., Hedley, M.J., Harrison, R., Braithwaite, A.C. 1990. Influence of manufacturing variables on characteristics and the agronomic value of partially acidulated phosphate fertilizers. *Fertilizer Research*, **26**(1–3), 119–138.

Bolan, N.S., Hedley, M.J., Loganathan, P. 1993. Preparation, forms and properties of controlled-release phosphate fertilizers. *Fertilizer Research*, **35**(1–2), 13–24.

Bolan, N.S., Khan, M., Donaldson, J., Adriano, D., Matthew, C. 2003a. Distribution and bioavailability of copper in farm effluent. *Science of the Total Environment*, **309**(1), 225–236.

Bolan, N.S., Laurenson, S., Luo, J., Sukias, J. 2009. Integrated treatment of farm effluents in New Zealand's dairy operations. *Bioresource Technology*, **100**(22), 5490–5497.

Bolan, N.S., Saggar, S., Luo, J., Bhandral, R., Singh, J. 2004a. Gaseous emissions of nitrogen from grazed pastures: Processes, measurements and modelling, environmental implications, and mitigation. *Advances in Agronomy*, **84**, 37–120.

Bolan, N.S., Szogi, A., Chuasavathi, T., Seshadri, B., Rothrock, M., Panneerselvam, P. 2010a. Uses and management of poultry litter. *World's Poultry Science Journal*, **66**(4), 673–698.

Brewer, C.E. 2012. Biochar characterization and engineering. *PhD Dissertation*, Iowa State University.

Brewer, C.E., Schmidt-Rohr, K., Satrio, J.A., Brown, R.C. 2009. Characterization of biochar from fast pyrolysis and gasification systems. *Environmental Progress & Sustainable Energy*, **28**(3), 386–396.

Bridgwater, A.V., Grassi, G. 1991. *Biomass pyrolysis liquids: Upgrading and utilisation*. Springer, Netherlands.

Bruun, E.W., Ambus, P., Egsgaard, H., Hauggaard-Nielsen, H. 2012. Effects of slow and fast pyrolysis biochar on soil C and N turnover dynamics. *Soil Biology and Biochemistry*, **46**, 73–79.

Cao, X., Ma, L., Gao, B., Harris, W. 2009. Dairy-manure derived biochar effectively sorbs lead and atrazine. *Environmental Science & Technology*, **43**(9), 3285–3291.

Carpenter, S.R., Caraco, N.F., Correll, D.L., Howarth, R.W., Sharpley, A.N., Smith, V.H. 1998. Nonpoint pollution of surface waters with phosphorus and nitrogen. *Ecological Applications*, **8**(3), 559–568.

Chen, B., Chen, Z., Lv, S. 2011. A novel magnetic biochar efficiently sorbs organic pollutants and phosphate. *Bioresource Technology*, **102**(2), 716–723.

Chen, Y., Rekha, P., Arun, A., Shen, F., Lai, W.-A., Young, C. 2006. Phosphate solubilizing bacteria from subtropical soil and their tricalcium phosphate solubilizing abilities. *Applied Soil Ecology*, **34**(1), 33–41.

Choppala, G.K., Bolan, N.S., Megharaj, M., Chen, Z., Naidu, R. 2012. The Influence of Biochar and black carbon on reduction and bioavailability of chromate in soils. *Journal of Environmental Quality*, **41**(4), 1175–1184.

Clark, D.E., Sutton, W.H. 1996. Microwave processing of materials. *Annual Review of Materials Science*, **26**(1), 299–331.

Clough, T.J., Condron, L.M., Kammann, C., Müller, C. 2013. A review of biochar and soil nitrogen dynamics. *Agronomy (Basel)*, **3**, 275–293.

Cui, H.-J., Wang, M.K., Fu, M.-L., Ci, E. 2011. Enhancing phosphorus availability in phosphorus-fertilized zones by reducing phosphate adsorbed on ferrihydrite using rice straw-derived biochar. *Journal of Soils and Sediments*, **11**(7), 1135–1141.

Dong, X., Ma, L.Q., Zhu, Y., Li, Y., Gu, B. 2013. Mechanistic investigation of mercury sorption by Brazilian pepper biochars of different pyrolytic temperatures based on X-ray photoelectron spectroscopy and flow calorimetry. *Environmental Science & Technology*, **47**(21), 12156–12164.

Doydora, S.A., Cabrera, M.L., Das, K.C., Gaskin, J.W., Sonon, L.S., Miller, W.P. 2011. Release of nitrogen and phosphorus from poultry litter amended with acidified biochar. *International Journal of Environmental Research and Public Health*, **8**(5), 1491–1502.

Dungan, R., Frankenberger, W. 1999. Microbial transformations of selenium and the bioremediation of seleniferous environments. *Bioremediation Journal*, **3**(3), 171–188.

Evangelou, V., Zhang, Y. 1995. A review: Pyrite oxidation mechanisms and acid mine drainage prevention. *Critical Reviews in Environmental Science and Technology*, **25**(2), 141–199.

Fellet, G., Marchiol, L., Delle Vedove, G., Peressotti, A. 2011. Application of biochar on mine tailings: Effects and perspectives for land reclamation. *Chemosphere*, **83**(9), 1262–1267.

Fendorf, S., Wielinga, B.W., Hansel, C.M. 2000. Chromium transformations in natural environments: The role of biological and abiological processes in chromium (VI) reduction. *International Geology Review*, **42**(8), 691–701.

Glaser, B., Lehmann, J., Zech, W. 2002. Ameliorating physical and chemical properties of highly weathered soils in the tropics with charcoal—A review. *Biology and Fertility of Soils*, **35**(4), 219–230.

Hammerschmidt, C.R., Lamborg, C.H., Fitzgerald, W.F. 2007. Aqueous phase methylation as a potential source of methylmercury in wet deposition. *Atmospheric Environment*, **41**(8), 1663–1668.

Hindmarsh, J.T., McCurdy, R.F., Savory, J. 1986. Clinical and environmental aspects of arsenic toxicity. *Critical Reviews in Clinical Laboratory Sciences*, **23**(4), 315–347.

Hossain, M.K., Strezov, V., Yin Chan, K., Nelson, P.F. 2010. Agronomic properties of wastewater sludge biochar and bioavailability of metals in production of cherry tomato (*Lycopersicon esculentum*). *Chemosphere*, **78**(9), 1167–1171.

Jaisi, D.P., Blake, R.E., Kukkadapu, R.K. 2010. Fractionation of oxygen isotopes in phosphate during its interactions with iron oxides. *Geochimica et Cosmochimica Acta*, **74**(4), 1309–1319.

Jiang, J., Xu, R.-k., Jiang, T.-y., Li, Z. 2012. Immobilization of Cu (II), Pb (II) and Cd (II) by the addition of rice straw derived biochar to a simulated polluted Ultisol. *Journal of Hazardous Materials*, **229**, 145–150.

Johnson, D.B., Hallberg, K.B. 2005. Acid mine drainage remediation options: A review. *Science of the Total Environment*, **338**(1), 3–14.

Joseph, S., Taylor, P. 2014. The production and application of biochar in soils. In: *Advances in Biorefineries: Biomass and Waste Supply Chain Exploitation*, Waldron, K.W. (Ed), Elsevier, Cambridge, 525–555.

Kammann, C., Ratering, S., Eckhard, C., Müller, C. 2012. Biochar and hydrochar effects on greenhouse gas (carbon dioxide, nitrous oxide, and methane) fluxes from soils. *Journal of Environmental Quality*, **41**(4), 1052–1066.

Kappe, C.O. 2004. Controlled microwave heating in modern organic synthesis. *Angewandte Chemie International Edition*, **43**(46), 6250–6284.

Kappler, A., Wuestner, M.L., Ruecker, A., Harter, J., Halama, M., Behrens, S. 2014. Biochar as an electron shuttle between bacteria and Fe (III) minerals. *Environmental Science & Technology Letters*, **1**(8), 339–344.

Karami, N., Clemente, R., Moreno-Jiménez, E., Lepp, N.W., Beesley, L. 2011. Efficiency of green waste compost and biochar soil amendments for reducing lead and copper mobility and uptake to ryegrass. *Journal of Hazardous Materials*, **191**(1), 41–48.

Karunanithi, R., Szogi, A., Bolan, N., Naidu, R., Loganathan, P., Hunt, P.G., Vanotti, M.B., Saint, C.P., Ok, Y.S., Krishnamoorthy, S. 2014. Phosphorus recovery and reuse from waste streams. *Advances in Agronomy*, **131**, 1–78.

Keiluweit, M., Nico, P.S., Johnson, M.G., Kleber, M. 2010. Dynamic molecular structure of plant biomass-derived black carbon (biochar). *Environmental Science & Technology*, **44**(4), 1247–1253.

Kelly, C.N., Peltz, C.D., Stanton, M., Rutherford, D.W., Rostad, C.E. 2014. Biochar application to hardrock mine tailings: Soil quality, microbial activity, and toxic element sorption. *Applied Geochemistry*, **43**, 35–48.

Kesler, S.E. 1994. Mineral resources, economics and the environment, Macmillan Publishing Co, US.

Kim, J.G., Dixon, J.B., Chusuei, C.C., Deng, Y. 2002. Oxidation of chromium (III) to (VI) by manganese oxides. *Soil Science Society of America Journal*, **66**(1), 306–315.

Klass, D.L. 1998. *Biomass for renewable energy, fuels, and chemicals*. Academic Press, San Diego, CA.

Knowles, O.A., Robinson, B.H., Contangelo, A., Clucas, L. 2011. Biochar for the mitigation of nitrate leaching from soil amended with biosolids. *Science of the Total Environment*, **409**(17), 3206–3210.

Kwon, J.-S., Yun, S.-T., Lee, J.-H., Kim, S.-O., Jo, H.Y. 2010. Removal of divalent heavy metals (Cd, Cu, Pb, and Zn) and arsenic (III) from aqueous solutions using scoria: Kinetics and equilibria of sorption. *Journal of Hazardous Materials*, **174**(1), 307–313.

Laird, D.A., Fleming, P., Davis, D.D., Horton, R., Wang, B., Karlen, D.L. 2010. Impact of biochar amendments on the quality of a typical midwestern agricultural soil. *Geoderma*, **158**(3–4), 443–449.

Ledgard, S., Steele, K. 1992. Biological nitrogen fixation in mixed legume/grass pastures. *Plant and Soil*, **141**(1–2), 137–153.

Lehmann, J., John, G., Rondon, M. 2006. Bio-char sequestration in terrestrial ecosystems—A review. *Mitigation and Adaptation Strategies for Global Change*, **11**(2), 403–427.

Lehmann, J., Joseph, S. 2009. *Biochar for environmental management: Science and technology*. Earthscan, London.

Litaor, M., Seastedt, T., Walker, M., Carbone, M., Townsend, A. 2005. The biogeochemistry of phosphorus across an alpine topographic/snow gradient. *Geoderma*, **124**(1), 49–61.

Loganathan, P., Vigneswaran, S., Kandasamy, J., Bolan, N.S. 2014. Removal and recovery of phosphate from water using sorption. *Critical Reviews in Environmental Science and Technology*, **44**(8), 847–907.

Lv, P., Xiong, Z., Chang, J., Wu, C., Chen, Y., Zhu, J. 2004. An experimental study on biomass air–steam gasification in a fluidized bed. *Bioresource Technology*, **95**(1), 95–101.

Mahimairaja, S., Bolan, N.S., Adriano, D., Robinson, B. 2005. Arsenic contamination and its risk management in complex environmental settings. *Advances in Agronomy*, **86**, 1–82.

McLaughlin, M.J., Tiller, K., Naidu, R., Stevens, D. 1996. Review: The behaviour and environmental impact of contaminants in fertilizers. *Soil Research*, **34**(1), 1–54.

Melo, L.C., Coscione, A.R., Abreu, C.A., Puga, A.P., Camargo, O.A. 2013. Influence of pyrolysis temperature on cadmium and zinc sorption capacity of sugar cane straw–derived biochar. *BioResources*, **8**(4), 4992–5004.

Mukherjee, A., Zimmerman, A., Harris, W. 2011. Surface chemistry variations among a series of laboratory-produced biochars. *Geoderma*, **163**(3), 247–255.

Nagase, H., Ose, Y., Sato, T., Ishikawa, T. 1982. Methylation of mercury by humic substances in an aquatic environment. *Science of the Total Environment*, **25**(2), 133–142.

Namgay, T., Singh, B., Singh, B.P. 2010. Influence of biochar application to soil on the availability of As, Cd, Cu, Pb, and Zn to maize (*Zea maize* L.). *Soil Research*, **48**(7), 638–647.

Özçimen, D., Ersoy-Meriçboyu, A. 2010. Characterization of biochar and bio-oil samples obtained from carbonization of various biomass materials. *Renewable Energy*, **35**(6), 1319–1324.

Park, J., Song, W.Y., Ko, D., Eom, Y., Hansen, T.H., Schiller, M., Lee, T.G., Martinoia, E., Lee, Y. 2012. The phytochelatin transporters AtABCC1 and AtABCC2 mediate tolerance to cadmium and mercury. *The Plant Journal*, **69**(2), 278–288.

Ravichandran, M. 2004. Interactions between mercury and dissolved organic matter—A review. *Chemosphere*, **55**(3), 319–331.

Roberts, A., Longhurst, R., Brown, M. 1994. Cadmium status of soils, plants, and grazing animals in New Zealand. *New Zealand Journal of Agricultural Research*, **37**(1), 119–129.

Robinson, B., Greven, M., Green, S., Sivakumaran, S., Davidson, P., Clothier, B. 2006. Leaching of copper, chromium and arsenic from treated vineyard posts in Marlborough, New Zealand. *Science of the Total Environment*, **364**(1), 113–123.

Schulz, H., Glaser, B. 2012. Effects of biochar compared to organic and inorganic fertilizers on soil quality and plant growth in a greenhouse experiment. *Journal of Plant Nutrition and Soil Science*, **175**(3), 410–422.

Singh, B.P., Hatton, B.J., Singh, B., Cowie, A.L., Kathuria, A. 2010. Influence of biochars on nitrous oxide emission and nitrogen leaching from two contrasting soils. *Journal of Environmental Quality*, **39**(4), 1224–1235.

Sohi, S.P., Krull, E., Lopez-Capel, E., Bol, R. 2010. A review of biochar and its use and function in soil. *Advances in Agronomy*, **105**, 47–82.

Spokas, K.A. 2010. Review of the stability of biochar in soils: Predictability of O: C molar ratios. *Carbon Management*, **1**(2), 289–303.

Spokas, K.A., Koskinen, W.C., Baker, J.M., Reicosky, D.C. 2009. Impacts of woodchip biochar additions on greenhouse gas production and sorption/degradation of two herbicides in a Minnesota soil. *Chemosphere*, **77**(4), 574–581.

Spokas, K.A., Novak, J.M., Venterea, R.T. 2012. Biochar's role as an alternative N-fertilizer: Ammonia capture. *Plant and Soil*, **350**(1–2), 35–42.

Taghizadeh-Toosi, A., Clough, T.J., Condron, L.M., Sherlock, R.R., Anderson, C.R., Craigie, R.A. 2011. Biochar incorporation into pasture soil suppresses in situ nitrous oxide emissions from ruminant urine patches. *Journal of Environmental Quality*, **40**(2), 468–476.

Taghizadeh-Toosi, A., Clough, T.J., Sherlock, R.R., Condron, L.M. 2012. Biochar adsorbed ammonia is bioavailable. *Plant and Soil*, **350**(1–2), 57–69.

Thangarajan, R., Bolan, N.S., Tian, G., Naidu, R., Kunhikrishnan, A. 2013. Role of organic amendment application on greenhouse gas emission from soil. *Science of the Total Environment*, **465**, 72–96.

Uchimiya, M., Chang, S., Klasson, K.T. 2011. Screening biochars for heavy metal retention in soil: Role of oxygen functional groups. *Journal of Hazardous Materials*, **190**(1), 432–441.

Uchimiya, M., Lima, I.M., Thomas Klasson, K., Chang, S., Wartelle, L.H., Rodgers, J.E. 2010. Immobilization of heavy metal ions (CuII, CdII, NiII, and PbII) by broiler litter-derived biochars in water and soil. *Journal of Agricultural and Food Chemistry*, **58**(9), 5538–5544.

Van Zwieten, L., Kimber, S., Morris, S., Downie, A., Berger, E., Rust, J., Scheer, C. 2010. Influence of biochars on flux of N_2O and CO_2 from Ferrosol. *Australian Journal of Soil Research*, **48**(6–7), 555–568.

Van Zwieten, L., Singh, B., Joseph, S., Kimber, S., Cowie, A., Chan, K.Y. 2009. Biochar and emissions of non-CO_2 greenhouse gases from soil. In: *Biochar for environmental management: Science and technology*. Earthscan, London, 227–249.

Vaughan, N.E., Lenton, T.M. 2011. A review of climate geoengineering proposals. *Climatic Change*, **109**(3–4), 745–790.

Violante, A., Cozzolino, V., Perelomov, L., Caporale, A., Pigna, M. 2010. Mobility and bioavailability of heavy metals and metalloids in soil environments. *Journal of Soil Science and Plant Nutrition*, **10**(3), 268–292.

Vogeler, I., Vibart, R., Mackay, A., Dennis, S., Burggraaf, V., Beautrais, J. 2014. Modelling pastoral farm systems—Scaling from farm to region. *Science of The Total Environment*, **482**, 305–317.

Whitehead, D.C. 1990. Atmospheric ammonia in relation to grassland agriculture and livestock production. *Soil Use & Management*, **6**(2), 63–65.

Whitehead, D.C., Goulden, K.M., Hartley, R.D. 1986. Fractions of nitrogen, sulphur, phosphorus, calcium and magnesium in the herbage of perennial ryegrass as influenced by fertilizer nitrogen. *Animal Feed Science and Technology*, **14**(3–4), 231–242.

Wuana, R.A., Okieimen, F.E. 2011. Heavy metals in contaminated soils: A review of sources, chemistry, risks and best available strategies for remediation. *ISRN Ecology*, pp. 1–20.

Xu, R.-k., Zhao, A.-z. 2013. Effect of biochars on adsorption of Cu (II), Pb (II) and Cd (II) by three variable charge soils from southern China. *Environmental Science and Pollution Research*, **20**(12), 8491–8501.

Yanai, Y., Toyota, K., Okazaki, M. 2007. Effects of charcoal addition on N_2O emissions from soil resulting from rewetting air-dried soil in short-term laboratory experiments. *Soil Science and Plant Nutrition*, **53**(2), 181–188.

Yao, Y., Gao, B., Zhang, M., Inyang, M., Zimmerman, A.R. 2012. Effect of biochar amendment on sorption and leaching of nitrate, ammonium, and phosphate in a sandy soil. *Chemosphere*, **89**(11), 1467–1471.

Zhang, A., Cui, L., Pan, G., Li, L., Hussain, Q., Zhang, X., Zheng, J., Crowley, D. 2010. Effect of biochar amendment on yield and methane and nitrous oxide emissions from a rice paddy from Tai Lake plain, China. *Agriculture, Ecosystems & Environment*, **139**(4), 469–475.

6

Biochar for Organic Contaminant Management in Soil

Chapter 6
Biochar for Organic Contaminant Management in Soil

Xiaokai Zhang, Kim McGrouther, Lizhi He,
Huagang Huang, Kouping Lu, and Hailong Wang

Chapter Outline

6.1 Introduction

In recent years, the incidence of soil contamination is increasing globally (Bolan et al. 2004; Mench et al. 2010). Approximately 3.5 million sites in industrial and mine areas, landfills, energy production plants, and agricultural land are potentially contaminated in Europe (Petruzzelli 2012). Soil contamination with organic pollutants is usually caused by industrial activities or inadequate management of pesticides and chemicals in agricultural production and from household waste (Fabietti et al. 2010). Many organic pollutants are recalcitrant and may accumulate in soils where they can enter the human food chain or harm ecosystems.

Organic pollutants released from industrial and agricultural processes or household products include persistent organic pollutants (POPs), emerging organic pollutants, and some pesticides (Table 6.1). POPs of concern are from industrial processes and from the production of products that include chemical families such as polychlorinated biphenyls (PCBs), polychlorinated dibenzo-*p*-dioxins and dibenzofurans, and polycyclic aromatic hydrocarbons (PAHs) (WHO 2010). POPs can accumulate in soil and may be toxic to soil microorganisms (Masih and Taneja 2006). Emerging organic pollutants are a group of synthetic compounds that have been recently detected in soils. Typically, phthalate acid esters (PAEs) [e.g., dibutyl phthalate and di(2-ethylhexyl)phthalate,

TABLE 6.1
Classification of Organic Pollutants

Persistent Organic Pollutant	Emerging Organic Pollutant	Pesticide
Industrial Chemicals	**Hormones and Their Metabolites**	Aldrin
Polychlorinated	Estrogen	Chlordane
biphenyls	Progesterone	DDT
Polycyclic	Adrenocorticotropic hormone	Dieldrin
aromatic	Hormone norepinephrine	Endrin
hydrocarbons	Progesterone	Heptachlor
		Hexachlorobenzene
Industrial By-Products	**Industrial Chemicals**	Mirex
Dioxins	Phthalic acid esters	Toxaphene
Furan	Dyes	
Chlorophenols		
Toxaphene	**Pharmaceuticals and Personal Care Products**	
	Trimethoprim	
	Triclosan	
	Antibiotics	
	Nonylphenols	
	Trolamine	

pharmaceuticals, endocrine-disrupting chemicals, disinfection by-products, cyanotoxins, dioxane, and flame retardants] are considered within this group (Petrović et al. 2001; Zhang et al. 2014a). Many organic pollutants are currently, or were in the past, used as pesticides. Pesticides are a group of chemicals that are mainly used in agriculture. The main sources of pesticide pollution are wastewaters from agricultural industries and from pesticide-formulating or -manufacturing plants (Chiron et al. 2000). Pesticides do provide a variety of benefits to humans (Jeyaratnam 1990); however, excessive pesticide use that results in residues in water and soil may harm the environment.

Biochar has been used as an amendment to absorb and retain organic contaminants in soils (Zhang et al. 2013b). High porosity and large surface area enable biochars to adsorb organic pollutants, and adsorption onto biochars can reduce the bioavailability of organic pollutants (Sarmah et al. 2010; Beesley et al. 2011; Yuan and Xu 2011) and reduce the risk of these pollutants entering the human food chain or leaching into groundwater.

6.2 Effect of Biochar on Adsorption of Organic Pollutants

The organic matter fraction of soils can adsorb sparingly soluble organic compounds and thus plays an important role in the transport and bioavailability of organic pollutants in the soil. Soil organic carbon (SOC) can strongly influence organic pollutant sorption and mobility (Wauchope et al. 2002). Soils with high organic carbon have higher adsorption and weaker desorption capacity than the low organic carbon soils (Wang et al. 2010). However, soil organic matter can easily release most of the pollutants to the environment again. Addition of a soil amendment, such as biochar, may enhance the adsorption capacity of soil (Yu et al. 2006; Marín-Benito et al. 2013; Zhang et al. 2013a). Using biochar to absorb organic pollutants in soils has been widely studied (Table 6.2). Sorption and desorption equilibria are

TABLE 6.2
Effect of Biochar Application on Sorption of Organic Pollutants in Soils

Feedstock	Production Temperature (°C)	Contaminant	Effect	Reference
Wheat residue	Not clear	Diuron, bromoxynil, ametryne	Wheat char was found to be a highly effective sorbent for the pesticides, and its presence in soil contributed >70% to the pesticide sorption	Sheng et al. (2005)
Hardwood	450	PAHs	Pore water concentrations of PAHs were reduced by biochar, with a >50% decrease of the heavier, more toxicologically relevant PAHs	Beesley et al. (2010)
Bamboo	600	Pentachlorophenol	Biochar reduced PCP bioavailability in soil	Xu et al. (2011)
Eucalyptus wood chips	850	Diuron	Pesticide absorption increases with the biochar contact time with soil and application rate	Yu et al. (2011)
Wheat straw	500	CBs	Biochar amendment significantly reduced the bioavailability of CBs	Song et al. (2012b)
Rice straw	Not clear	Carbaryl	The biochar amendment of soils increased the irreversible sorption of pesticides	Qiu et al. (2013)

(Continued)

TABLE 6.2 (*Continued*)
Effect of Biochar Application on Sorption of Organic Pollutants in Soils

Feedstock	Production Temperature (°C)	Contaminant	Effect	Reference
Sewage sludge	550	PAHs	Biochar application was effective at significantly reducing the bioaccumulation of PAHs from contaminated soils into lettuce	Khan et al. (2013)
Wheat straw	300	4-Chloro-2-methylphenoxy	4-Chloro-2-methylphenoxy sorption by biochar and biochar-amended soil (1.0 wt% biochar) was 82 and 2.53 times higher than that by the nonamended soil, respectively	Tatarková et al. (2013)
Poplar wood	350	2,4-D and acetochlor	Biochar showed high sorption of the herbicides acetochlor and 2,4-D	Li et al. (2013)
Wheat straw, pine needles	550	PCBs	Application of 2% biochars to soils greatly reduced the bio-uptake of PCBs	Wang et al. (2013)

CB, chlorobenzene; PAH, polycyclic aromatic hydrocarbon; PCB, polychlorinated biphenyl; PCP, pentachlorophenol.

used to describe the adsorption and desorption processes of the organic pollutants with the biochar surface. Sheng et al. (2005) showed that wheat char was a highly effective sorbent for pesticides, and its presence in soil contributed >70% to the pesticide sorption.

The ability of biochar to adsorb organic pollutants is related to the feedstock biomass, the conditions under which the biochar is produced, and the types of organic pollutants present in the contaminated soils (Downie et al. 2009; Wu et al. 2012). These factors may consequently affect other processes, such as the bioavailability of contaminants.

6.2.1 Factors That Affect the Fate of Organic Pollutants in Soil

The adsorption capacity of biochar varies with the type of biochar, soil pH, and SOC content as well as with the type of organic pollutant.

6.2.1.1 Types of Biochar

Biochar adsorption of organic pollutants in the soil is greatly influenced by the biochar feedstock type and the pyrolysis temperature (Table 6.3). Cabrera et al. (2014) investigated the effect of various biochars on the sorption–desorption profiles of pesticides. They demonstrated that aminocyclopyrachlor and bentazone were almost completely sorbed by the soils amended with the biochars made from wood pellets. However, soils amended with biochar made from fast pyrolysis of macadamia nut shells had lower herbicide sorption compared to the unamended soil. This difference is mainly attributed to the competition between the dissolved organic carbon from the biochar and the herbicides for the sorption sites (Cabrera et al. 2014). Zhang et al. (2014b) investigated the effect of various biochars on the sorption–desorption profiles of diethyl phthalate (DEP). They showed that both wheat straw biochar and bamboo biochar can effectively adsorb DEP in different soil types. However, soils amended with rice straw biochar indicated lower sorption of the DEP compared to the bamboo biochar–amended soil (Figure 6.1). This difference was mainly attributed to the lignin content of the different biochar feedstocks.

TABLE 6.3
Effect of Pyrolysis (Production) Temperature and Feedstock Type on Adsorption of Organic Pollutants by Biochar

Feedstock	Production Temperature (°C)	Contaminant	Effect	Reference
Eucalyptus wood	450, 850	Diuron chlorpyrifos and carbofuran	Higher pyrolysis temperatures and higher rates of biochar applied to soils result in stronger adsorption and weaker desorption of pesticides	Yu et al. (2006)
Dairy manure	200, 350	Atrazine	At 200°C, partitioning of atrazine is positively related to biochar carbon content	Cao et al. (2009)
Pine wood	350, 700	Terbuthylazine	Soil sorption increased 2.7- and 63-fold in the BC350 and BC700 treatments, respectively	Wang et al. (2010)
Pine wood	350, 700	Phenanthrene	Biochar produced at 700°C showed a greater ability to enhance a soil's sorption ability than biochar produced at 350°C	Zhang et al. (2010)
Pine needles	100, 300, 400, 700	PAHs	Sorption capacity increased with pyrolysis temperature	Chen and Yuan (2011)
Swine manure	350, 700	Carbaryl	At low carbaryl concentrations, the sorption capacity of BC700 > BC350; similar sorption capacity at high carbaryl concentrations	Zhang et al. (2013a)

(Continued)

TABLE 6.3 (*Continued*)
Effect of Pyrolysis (Production) Temperature and Feedstock Type on Adsorption of Organic Pollutants by Biochar

Feedstock	Production Temperature (°C)	Contaminant	Effect	Reference
Sediment	200, 300, 400, 500	Phenanthrene, bisphenol A, sulfamethoxazol, norfloxacin, ofloxacin	Biochars derived from sediment show high sorption to organic contaminants	Wu et al. (2013)
Wheat straw	250, 300, 500	HCB	Biochar amendment of soil resulted in a rapid reduction in the bioavailability of HCB, even at 0.1% biochar application rate	Song et al. (2012a)
Shipping pallets, construction waste, softwood	450, 700	PCBs	Burt's biochar was more effective in reducing PCB uptake into plants and earthworms than blue leaf biochar	Denyes et al. (2013)
Wheat straw, bamboo	350, 650	DEP	Bamboo biochars were more effective than the straw biochars in improving soils' adsorption capacity of DEP	Zhang et al. (2014b)
Wood chip pellets, macadamia nutshells, hardwood	500, 540, 850	Bentazone, pyraclostrobin, aminocyclopyrachlor	Biochars with high surface areas and low dissolved organic carbon contents can increase the sorption of highly mobile pesticides in soil	Cabrera et al. (2014)

DEP, diethyl phthalate; HCB, hexachlorobenzene; PAH, polycyclic aromatic hydrocarbon; PCB, polychlorinated biphenyl.

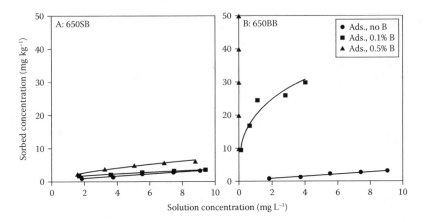

Figure 6.1 Effect of biochar (B) feedstock material on the adsorption (Ads.) of diethyl phthalate (DEP) in soil. (A) 650SB and (B) 650BB, where 650 is the pyrolysis temperature in °C, SB is rice straw biochar, and BB is bamboo biochar. Symbols are measured data, and the solid lines are Freundlich model fits.

Studies showed that the amount of aromatic carbon on the biochar surface plays an important role in the contaminant adsorption process and is mainly determined by the lignin content of the feedstock (Bornemann et al. 2007; Cesarino et al. 2012). Generally, the lignin content of bamboo at 37% (Deshpande et al. 2000) is considerably higher than that of rice straw, with values ranging from 5 to 24% (Binod et al. 2010). Another study demonstrated that Burt's biochar (made from waste wood) had higher adsorption capacity for PCBs than the blue leaf biochar (Denyes et al. 2013). Therefore, the feedstock material of biochar can strongly affect its ability to adsorb different organic pollutants.

Biochar's properties, such as specific surface area (SSA) and microporosity, can strongly affect its adsorption capacity. One of the important factors that affects SSA is the pyrolysis temperature. Chen and Yuan (2011) investigated PAH uptake isotherms of wood biochars, and they found that PAH sorption increases for materials with higher pyrolysis temperatures. James et al. (2005) also demonstrated that phenanthrene sorption increases for materials exposed to higher temperatures. They indicated that sorption increases also coincide with increases in the SSA of biochars produced at higher temperatures. Yu et al. (2006) also found that pesticide adsorption was stronger with biochars created at higher pyrolysis temperatures. Soil amended with this biochar showed a pronounced hysteresis in the desorption process of pesticides. Yu et al. (2006) noted that the degree of apparent hysteresis showed a good correlation with the micropore volume of the biochar-amended soils. They assessed the link between microporosity and the apparent sorption–desorption hysteresis and found a smaller increase in the hysteresis index value when the soil contained biochar with a lower pore volume. This result was associated with the presence of predominantly 450°C biochar and only small amounts of 850°C biochar. Biochar created at 450°C had a negligible proportion of micropores, whereas biochar created at 850°C contained a high proportion of micropores <1 nm in width. Similarly, bamboo biochar pyrolyzed at 650°C had a much higher adsorption capacity but a weaker desorption capacity of DEP in soils than the biochar produced at 350°C (Zhang et al. 2014b). Bornemann et al. (2007) showed higher adsorption capacity could be attributed to the greater microporosity created at higher pyrolysis temperatures. Pore-filling mechanisms were thought to be the dominating processes governing the sorption on these microporous materials (Bornemann et al. 2007). These studies indicate that adding biochar with high SSA and low organic carbon content (especially the soluble part) to different soils can enhance the adsorption capacity of the organic pollutants (Bornemann et al. 2007; Cabrera et al. 2014).

6.2.1.2 Soil pH

Soil pH is expected to influence the adsorption of organic pollutants onto biochars, in particular when the organic pollutants become dissociated or protonated at a certain pH range in soils (Sheng et al. 2005). Yang et al. (2004) indicated that a pH increase resulted in a decrease (20%/unit pH) in ametryne adsorption, due primarily to the development of a negative charge on the surface of the biochar. For other pollutants, such as atrazine and simazine, lower pHs can improve sorption affinity with biochar. Atrazine and simazine are weak-base pesticides, with pKa values of 1.7 and 1.6, respectively. Low pH solutions facilitate the formation of the two pesticides' triazine cations, and the occurrence of triazine cations in acidic media enhances the pesticide interaction with the negatively charged surface of the biochar (Liu et al. 2002; Zheng et al. 2010). Depending on pH, the carboxyl and phenolic groups on biochar may also be deprotonated and hence change biochar's properties.

Generally, at higher pH, the ester compounds (e.g., DEP) are likely to be easily ionized and become more polar and hydrophilic (Stales et al. 1997; Schaffer et al. 2012). Studies showed that biochar was more effective at adsorbing hydrophobic organic compounds than hydrophilic materials (Zhang et al. 2010). Hence, organic pollutants with higher water solubility and polarity are likely to be more difficult to be sorbed by biochar.

6.2.1.3 Effect of Other Factors on the Adsorption of Organic Pollutants

Other factors, such as the physicochemical properties of the organic pollutants, and the soil properties, also play important roles in determining the biochar adsorption–desorption processes (Table 6.4). Sorption of organic pollutants on biochar can occur by physical adsorption on a surface or by partitioning (dissolution) into a phase such as natural organic matter. Biochar is a porous material, and the dissolved organic matter in soil may block its pores, thereby reducing its adsorption capacity. Zhang et al. (2010) showed that the addition of biochar into soils enhanced their apparent sorption affinity for phenanthrene; however, the adsorption capacity of biochar decreased as the organic carbon content of these soils increased. Similarly, Wang et al. (2010) found that biochar-amended soil with high SOC had a lower adsorption capacity for pesticides than soil with low SOC (Wang et al. 2010). This difference may be attributed to the coating of biochar particles with dissolved organic matter, reducing the overall accessibility to the sorption sites (Zhang et al. 2010).

TABLE 6.4
Effect of Feedstock Type and Production Temperature on Adsorption of Organic Pollutants to Biochars

Feedstock	Production Temperature (°C)	Contaminant	Effect	Reference
Sediment	370	Phenanthrene, PAHs	Competition exists for BC sites between phenanthrene and native PAHs	Cornelissen and Gustafsson (2004)
Sediment	370	Phenanthrene, PAHs	Addition of PAHs significantly decreases biochar sorption of phenanthrene	Cornelissen and Gustafsson (2006)
Wood chips	500	Atrazine, acetochlor	Acetochlor adsorption increased 1.5 times, atrazine adsorption increased 1.38 times	Spokas et al. (2009)
Grass, wood	200, 300, 400, 500, 600, 700	PAEs	Grass chars displayed higher PAE sorption than wood chars	Sun et al. (2012)
Rice straw	500	1,2,4,5-Tetrachlorobenzene, 1,2,4-trichlorobenzene (1,2,4-TCB)	Effect of biochar content in soil on the bioavailability of CBs was more pronounced for 1,2,4-TCB relative to other CBs	Song et al. (2013)

1,2,4-TCB, 1,2,4-trichlorobenzene; CB, chlorobenzene; PAE, phthalate ester; PAH, polycyclic aromatic hydrocarbon.

Adsorption is the most important mode of interaction between biochar and organic pollutants; it controls the concentration of these pollutants in the soil liquid phase. The extent of adsorption mainly depends on the properties of organic pollutants, such as molecular structure, chemical functions, solubility, polarity, and polarizability. Wu et al. (2013) compared the adsorption capacities of biochar on different types organic pollutants, and they demonstrated that biochar had a different adsorptive ability for each of the following chemicals: phenanthrene (PHE), sulfamethoxazole (SMX), bisphenol A (BPA), norfloxacin (NOR), and ofloxacin (OFL). This difference was mainly due to SMX and PHE sorption being controlled via hydrophobic effects and hydrogen bonding playing an important role for the sorption of the chemicals with hydroxyl (for BPA) or carboxyl (for OFL and NOR) groups (Wu et al. 2013). However, Dechene et al. (2014) reported that the application of biochar did not affect the sorption or desorption of metazachlor oxalic acid, imazamox, and metazachlor sulfonic acid. These compounds are anionic and therefore unlikely to be adsorbed onto biochar because of its net negative surface charge (Dechene et al. 2014). Zhang et al. (2010) suggested that biochar application can be used to enhance soil sorption of hydrophobic organic compounds; hence, by inference, sorption of hydrophilic pesticides is less likely to be achieved by the addition of biochar to soil. Similarly, Sun et al. (2012) indicated that DEP (as the PAE with the greatest polarity and lowest hydrophobicity) was sorbed less on biochar than the less polar dibutyl phthalate and butyl benzyl phthalate. Therefore, it is important to know about the soil environment and the properties of organic pollutants when deciding whether to use biochar to adsorb specific soil pollutants.

6.3 Effect of Biochar on Organic Pollutant Leaching

Biochar addition to soils might increase the sorption of organic pollutants and therefore affect other sorption–desorption-related processes, such as leaching. Table 6.5 shows that

TABLE 6.5
Effect of Biochar on Leaching of Organic Pollutants

Feedstock	Pesticide	Effect	Reference
Wheat straw	MCPA	Approximately 35% of the applied MCPA was transported through biochar-amended soil	Tatarková et al. (2013)
Bamboo	PCP	Addition of 5% biochar decreased the cumulative leach-loss content of PCP by 42%	Xu et al. (2012)
Bamboo, Brazilian pepperwood, sugarcane bagasse, hickory wood	SMX	Only ~2–14% of the SMX was transported through biochar-amended soils, whereas 60% was found in the leachate of the unamended soils	Yao et al. (2012)
Hardwood	Simazine	Biochar suppressed simazine biodegradation and reduced simazine leaching	Jones et al. (2011)
Charcoal	Isoproturon	The amount of isoproturon in leachate decreased with the increase of the amount of charcoal addition to soil column	Si et al. (2011)
Wheat	Sulfosulfuron, isoproturon, propyzamid	For the biochar-amended clay soil, leaching was decreased by >93% for the moderately mobile pesticides isoproturon, imidacloprid, and propyzamid, whereas it was increased by 100% for the nonmobile pesticide pyraclostrobin	Larsbo et al. (2013)
Hardwood sawdust, macadamia nut shells, wood pellets	Fluometuron, 4-chloro-2-methylphen-oxyacetic acid	Lower leaching of pesticides was observed in the soils amended with the biochars with higher surface areas	Cabrera et al. (2011)

MCPA, 4-chloro-2-methylphenoxy; PCP, pentachlorophenol; SMX, sulfamethoxazole.

soil type, organic pollutants properties, and the types of biochar can strongly affect the leaching of organic pollutants in biochar-amended soil. Tatarková et al. (2013) reported that biochar-amended soil reduced 4-chloro-2-methylphenoxy leaching and dissipation. Approximately 35% of the applied 4-chloro-2-methylphenoxy was transported through biochar-amended soil, whereas almost 56% was recovered in the leachates transported through nonamended soil. The amount of pesticide that leaches through the soil may decrease as the biochar application rate is increased (Jones et al. 2011; Xu et al. 2012). Increasing the amount of biochar appears to increase retention and reduce pesticide leaching from the soil (Si et al. 2011; Yao et al. 2012). Similarly, Xu et al. (2012) showed that compared to the control, the addition of 5% bamboo biochar decreased the cumulative leach-loss content of pentachlorophenol by 42%; high biochar application rates led to lower pentachlorophenol content in leachates.

Soil types and organic pollutants types may significantly affect the leaching of organic pollutants. Larsbo et al. (2013) investigated the sorption and leaching of five pesticides with different sorption characteristics for two Swedish topsoils. They showed that biochar amendment to loam soil had no significant effects on leaching compared with unamended soil; however, for the clay soil, leaching was decreased by >93% for the moderately mobile pesticides (isoproturon, imidacloprid, and propyzamid) and leaching was increased by 100% for the nonmobile pesticide (pyraclostrobin) (Larsbo et al. 2013). This increase in leaching was thought to be facilitated by material released from the biochar. It is evident that biochar application to soil is not always a solution to reducing pesticide leaching from agricultural land. One study showed that amendment with some types of biochar resulted in enhanced leaching of the herbicides fluometuron and 4-chloro-2-methylphenoxyacetic acid (Cabrera et al. 2011). This result was thought to be due to some amendments releasing soluble organic compounds that compete or associate with the herbicide molecules, thereby enhancing their soil mobility (Cabrera et al. 2011). Therefore, factors such as the biochar raw material, biochar properties, and soil types as well as the type of chemical applied can play important roles in whether the chemical will leach.

6.4 Effect of Biochar on the Bioavailability of Organic Pollutants

The bioavailability of organic pollutants is determined by the accessibility of a chemical for biological uptake and possible toxicity. Previous studies demonstrated that biochar addition to soil can increase the absorption of a variety of organic contaminants, thereby reducing their uptake by plants. Even application of a small amount of biochar to soil can significantly reduce the bioavailability of nonpolar organic pollutants (Kookana 2010). Yang et al. (2006) demonstrated that increasing the biochar content in a soil reduced the bioavailability of herbicides; and that even at a low application rate [0.1% (w/w)], biochar in soil could significantly reduce the bioavailability of diuron. Nag et al. (2011) also showed that increasing biochar in the soil caused greater ryegrass biomass production and also reduced the bioavailability of chemicals that were applied to the soil. Similarly, Song et al. (2012a) investigated the effect of wheat straw biochar on the bioavailability of hexachlorobenzene, and they observed that addition of biochar reduced the volatilization and earthworm (*Eisenia foetida*) uptake of hexachlorobenzene from the soil.

Biochar can also effect pollutant partitioning and thereby affect its bioavailability. Reid et al. (2013) evaluated the influence of biochar on the loss, partitioning, and bioaccessibility of [14C]isoproturon (IPU); they showed that IPU extractability was reduced to <2% in biochar treatments. This result suggests that biochar application may reduce herbicide bioaccessibility. It is evident that the bioavailability of the organic pollutant may be determined by the properties of the amendment. A test showed that phenanthrene was quickly desorbed if the amendment did not have hydrophobic surfaces (Alexander 2000).

However, the aim of using agrochemicals in modern agricultural processes is to enhance crop yields (Reid et al. 2013). Biochar application may also reduce pesticide efficacy. Therefore, to keep good crop yields, the application rates of biochar should be carefully planned to avoid the unintended consequence of offsetting pesticide efficacy (Graber et al. 2012).

Biochar application can reduce the bioavailability of organic pollutants, mainly through sorption. Simultaneously, biological degradation of organic pollutants in soil can be substantially reduced (Kookana 2010; Sopeña et al. 2012). Ideally, for sustainable remediation of contaminated soils, organic contaminants in soil should be degraded. Some investigations have explored the feasibility of using biochar to accelerate the degradation of organic pollutants in soils. Some studies showed that biochar contains sorption sites and electron conductors (Kemper et al. 2008; Oh et al. 2012). These electron conductors could catalyze the reduction of some organic contaminants (e.g., nitroaromatic compounds), thereby enhancing their degradation. Yu et al. (2011) studied the reduction of nitrobenzenes to anilines by sulfides at room temperature. They reported that biochar can serve as a platform to accelerate the reduction of nitrobenzenes. Hydrolysis is a mechanism whereby compounds, such as pesticides adsorbed to biochar, can be degraded (Sarmah and Sabadie 2002; Zhang et al. 2013a).

Bioremediation is one of the commonly practiced technologies for cleaning up soils contaminated with organic pollutants (Smith et al. 2009; Chen et al. 2012). Chen et al. (2012) proposed the addition of concentrated microorganisms capable of decomposing certain types of organic pollutants (e.g., high-molecular-weight PAHs). They conducted a study on the dissipation of PAHs in a contaminated soil amended with immobilized bacteria, using biochar as a carrier (Su et al. 2006). It is important to select an appropriate biochar as an immobilization carrier to stimulate biodegradation of organic pollutants (Chen et al. 2012).

6.5 Effect of Aging of Biochar on Adsorption of Organic Pollutants

High SSA and microporosity make biochars very efficient sorbents to stabilize many organic pollutants. However, such behaviors and the properties of biochar applied to soil may change with time. This process is commonly referred to as aging (Kookana 2010). Studies have demonstrated that the interaction between biochars and soil natural organic molecules and clay minerals contributes to the aging of biochars (Uchimiya et al. 2010) and that the natural organic matter can block the micropores of biochars (Pignatello et al. 2006). He et al. (2014) demonstrated that the effect of biochar on imidacloprid adsorption was influenced by SOC content. They also showed that the adsorption capacity of aged biochar was significantly lower than that of biochar without aging. Zhang et al. (2010) found that the adsorption capacity of the pine biochar was reduced after the biochars were incubated with soil for 4 weeks (Figure 6.2). Martin et al. (2012) reported that the sorption–desorption properties of biochar were reduced at least by 47% for the herbicide diuron after aging under field conditions for 32 months. However, another study showed that hardwood-derived biochars aged in the field for 2 years

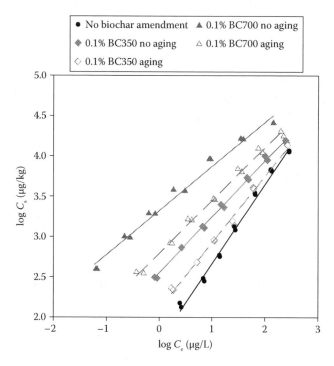

Figure 6.2 Sorption of phenanthrene in a sandy soil with biochar amendment. BC350 is biochar pyrolyzed at 350°C. (From Zhang, H. et al. 2010. *Environmental Pollution* 158, 2821–2825.)

had the same effect as fresh biochar on the sorption and mineralization of simazine (Jones et al. 2011). This suggests that the effects of biochar on herbicide behavior may be long lasting. Cheng et al. (2014) indicated that after rice straw biochar–amended sediment was aged for 98 days, pentachlorophenol was released more easily. Studies showed that not all aging processes caused a reduction of the adsorption capacity of biochar; the aging time, biochar type, and biochar application rate strongly affected the aging process.

6.6 Mechanisms for the Adsorption of Organic Pollutants

The studies referenced in the previous sections of this chapter showed that biochar can absorb a range of organic pollutants, thereby playing an important role in controlling the transport and the subsequent risk associated with release of pollutants (Bornemann et al. 2007; Chen et al. 2008). The mechanisms of adsorption have been investigated in many studies (Chun et al. 2004; Zhu et al. 2005; Nguyen et al. 2007). Nguyen et al. (2007) reported that a pore-filling mechanism mainly contributed to the sorption process for organic pollutants onto biochar, whereas Zhu et al. (2005) showed that π-electron interactions assisted with the adsorption. Yu et al. (2011) indicated that biochar could not only act as an adsorbent but also act as a platform to accelerate the reduction of nitrobenzenes. This study showed biochar can serve as a catalyst for the degradation of some organic pollutants.

These different adsorption mechanisms are related to the properties of the biochar. They are determined by the feedstock type and the pyrolysis temperature; thereby, different types of biochars could vary in their ability to stabilize certain types of organic pollutants. The mechanisms of adsorption of organic pollutants onto biochar include the following: (1) the combined mechanisms of direct electrostatic attraction and repulsion and intermolecular hydrogen bonding (Nguyen et al. 2007), (2) the π–π interactions between pollutant molecules and graphene layers of biochar (Zhu et al. 2005), and (3) electron donation for catalytic degradation of some organic pollutants (Yu et al. 2011).

6.7 Summary

Biochar can be an effective adsorbent for the adsorption of many types of organic pollutants in contaminated soils if the biochar, soil, and pesticide properties are aligned. The biochar sorption capacity is influenced by its feedstock biomass, pyrolysis conditions, types of pollutants, and soil properties. When biochar is applied to soil, its adsorption behavior may change with time, and this aging can result in a reduced sorption capacity of the biochar. In the process of using biochar to remediate soils contaminated with organic pollutants, an unintended adverse effect, such as reduced pesticide effectiveness, could also occur, so biochar application should be carefully planned.

References

Alexander, M. 2000. Aging, bioavailability, and overestimation of risk from environmental pollutants. *Environmental Science and Technology* 34, 4259–4265.

Beesley, L., Moreno-Jiménez, E., Gomez-Eyles, J.L. 2010. Effects of biochar and greenwaste compost amendments on mobility, bioavailability and toxicity of inorganic and organic contaminants in a multi-element polluted soil. *Environmental Pollution* 158, 2282–2287.

Beesley, L., Moreno-Jiménez, E., Gomez-Eyles, J.L., Harris, E., Robinson, B., Sizmur, T. 2011. A review of biochars' potential role in the remediation, revegetation and restoration of contaminated soils. *Environmental Pollution* 159, 3269–3282.

Binod, P., Sindhu, R., Singhania, R.R., Vikram, S., Devi, L., Nagalakshmi, S., Kurien, N., Sukumaran, R.K., Pandey, A. 2010. Bioethanol production from rice straw: An overview. *Bioresource Technology* 101, 4767–4774.

Bolan, N., Adriano, D., Mahimairaja, S. 2004. Distribution and bioavailability of trace elements in livestock and poultry manure by-products. *Critical Reviews in Environmental Science and Technology* 34, 291–338.

Bornemann, L.C., Kookana, R.S., Welp, G. 2007. Differential sorption behaviour of aromatic hydrocarbons on charcoals prepared at different temperatures from grass and wood. *Chemosphere* 67, 1033–1042.

Cabrera, A., Cox, L., Spokas, K., Hermosín, M.C., Cornejo, J., Koskinen, W.C. 2014. Influence of biochar amendments on the sorption–desorption of aminocyclopyrachlor, bentazone and pyraclostrobin pesticides to an agricultural soil. *Science of the Total Environment* 470–471, 438–443.

Cabrera, A., Cox, L., Spokas, K.A., Celis, R., Hermosín, M.C., Cornejo, J., Koskinen, W.C. 2011. Comparative sorption and leaching study of the herbicides fluometuron and 4-chloro-2-methylphenoxyacetic acid (MCPA) in a soil amended with biochars and other sorbents. *Journal of Agricultural and Food Chemistry* 59, 12550–12560.

Cao, X., Ma, L., Gao, B., Harris, W. 2009. Dairy-manure derived biochar effectively sorbs lead and atrazine. *Environmental Science & Technology* 43, 3285–3291.

Cesarino, I., Araújo, P., Domingues Júnior, A.P., Mazzafera, P. 2012. An overview of lignin metabolism and its effect on biomass recalcitrance. *Brazilian Journal of Botany* 35, 303–311.

Chen, B., Yuan, M. 2011. Enhanced sorption of polycyclic aromatic hydrocarbons by soil amended with biochar. *Journal of Soils and Sediments* 11, 62–71.

Chen, B., Yuan, M., Qian, L. 2012. Enhanced bioremediation of PAH-contaminated soil by immobilized bacteria with plant residue and biochar as carriers. *Journal of Soils and Sediments* 12, 1350–1359.

Chen, B., Zhou, D., Zhu, L., Shen, X. 2008. Sorption characteristics and mechanisms of organic contaminant to carbonaceous biosorbents in aqueous solution. *Science in China Series B: Chemistry* 51, 464–472.

Cheng, G., Zhu, L., Sun, M., Deng, J., Chen, H., Xu, X., Lou, L., Chen, Y. 2014. Desorption and distribution of pentachlorophenol (PCP) on aged black carbon containing sediment. *Journal of Soils and Sediments* 14, 344–352.

Chiron, S., Fernandez-Alba, A., Rodriguez, A., Garcia-Calvo, E. 2000. Pesticide chemical oxidation: State-of-the-art. *Water Research* 34, 366–377.

Chun, Y., Sheng, G., Chiou, C.T., Xing, B. 2004. Compositions and sorptive properties of crop residue-derived chars. *Environmental Science & Technology* 38, 4649–4655.

Cornelissen, G., Gustafsson, Ö. 2004. Sorption of phenanthrene to environmental black carbon in sediment with and without organic matter and native sorbates. *Environmental Science & Technology* 38, 148–155.

Cornelissen, G., Gustafsson, Ö. 2006. Effects of added PAHs and precipitated humic acid coatings on phenanthrene sorption to environmental black carbon. *Environmental Pollution* 141, 526–531.

Dechene, A., Rosendahl, I., Laabs, V., Amelung, W. 2014. Sorption of polar herbicides and herbicide metabolites by biochar-amended soil. *Chemosphere* 109, 180–186.

Denyes, M.J., Rutter, A., Zeeb, B.A. 2013. In situ application of activated carbon and biochar to PCB-contaminated soil and the effects of mixing regime. *Environmental Pollution* 182, 201–208.

Deshpande, A.P., Bhaskar Rao, M., Lakshmana Rao, C. 2000. Extraction of bamboo fibers and their use as reinforcement in polymeric composites. *Journal of Applied Polymer Science* 76, 83–92.

Downie, A., Crosky, A., Munroe, P. 2009. Physical properties of biochar. In: *Biochar for Environmental Management: Science and Technology*. London, Earthscan, (eds. J Lehmann, S Joseph) pp. 13–33.

Fabietti, G., Biasioli, M., Barberis, R., Ajmone-Marsan, F. 2010. Soil contamination by organic and inorganic pollutants at the regional scale: The case of Piedmont, Italy. *Journal of Soils and Sediments* 10, 290–300.

Graber, E., Tsechansky, L., Gerstl, Z., Lew, B. 2012. High surface area biochar negatively impacts herbicide efficacy. *Plant and Soil* 353, 95–106.

He, L., Zhang, X., Wang, H., Wu, H., Liu, H. 2015. Effect of bamboo and straw derived biochars on two soil adsorption and desorption of pesticide imidacloprid. *Acta Scientiae Circumstantiae* (in Chinese) 35(2), 535–540.

James, G., Sabatini, D., Chiou, C., Rutherford, D., Scott, A., Karapanagioti, H. 2005. Evaluating phenanthrene sorption on various wood chars. *Water Research* 39, 549–558.

Jeyaratnam, J. 1990. Acute pesticide poisoning: A major global health problem. *World Health Statistics Quarterly* 43, 139–144.

Jones, D.L., Edwards-Jones, G., Murphy, D.V. 2011. Biochar mediated alterations in herbicide breakdown and leaching in soil. *Soil Biology and Biochemistry* 43, 804–813.

Kemper, J.M., Ammar, E., Mitch, W.A. 2008. Abiotic degradation of hexahydro-1, 3, 5-trinitro-1, 3, 5-triazine in the presence of hydrogen sulfide and black carbon. *Environmental Science & Technology* 42, 2118–2123.

Khan, S., Wang, N., Reid, B.J., Freddo, A., Cai, C. 2013. Reduced bioaccumulation of PAHs by Lactuca satuva L. grown in contaminated soil amended with sewage sludge and sewage sludge derived biochar. *Environmental Pollution* 175, 64–68.

Kookana, R.S. 2010. The role of biochar in modifying the environmental fate, bioavailability, and efficacy of pesticides in soils: A review. *Soil Research* 48, 627–637.

Larsbo, M., Löfstrand, E., de Veer, D.v.A., Ulén, B. 2013. Pesticide leaching from two Swedish topsoils of contrasting texture amended with biochar. *Journal of Contaminant Hydrology* 147, 73–81.

Li, J., Li, Y., Wu, M., Zhang, Z., Lü, J. 2013. Effectiveness of low-temperature biochar in controlling the release and leaching of herbicides in soil. *Plant and Soil* 370, 333–344.

Liu, W., Zheng, W., Gan, J. 2002. Competitive sorption between imidacloprid and imidacloprid-urea on soil clay minerals and humic acids. *Journal of Agricultural and Food Chemistry* 50, 6823–6827.

Marín-Benito, J.M., Brown, C.D., Herrero-Hernández, E., Arienzo, M., Sánchez-Martín, M.J., Rodríguez-Cruz, M.S. 2013. Use of raw or incubated organic wastes as amendments in reducing pesticide leaching through soil columns. *Science of the Total Environment* 463–464, 589–599.

Martin, S.M., Kookana, R.S., Van Zwieten, L., Krull, E. 2012. Marked changes in herbicide sorption-desorption upon ageing of biochars in soil. *Journal of Hazardous Materials* 231–232, 70–78.

Masih, A., Taneja, A. 2006. Polycyclic aromatic hydrocarbons (PAHs) concentrations and related carcinogenic potencies in soil at a semi-arid region of India. *Chemosphere* 65, 449–456.

Mench, M., Lepp, N., Bert, V., Schwitzguébel, J.P., Gawronski, S.W., Schröder, P., Vangronsveld, J. 2010. Successes and limitations of phytotechnologies at field scale: Outcomes, assessment and outlook from COST Action 859. *Journal of Soils and Sediments* 10, 1039–1070.

Nag, S.K., Kookana, R., Smith, L., Krull, E., Macdonald, L.M., Gill, G. 2011. Poor efficacy of herbicides in biochar-amended soils as affected by their chemistry and mode of action. *Chemosphere* 84, 1572–1577.

Nguyen, T.H., Cho, H.-H., Poster, D.L., Ball, W.P. 2007. Evidence for a pore-filling mechanism in the adsorption of aromatic hydrocarbons to a natural wood char. *Environmental Science & Technology* 41, 1212–1217.

Oh, S.-Y., Son, J.-G., Lim, O.-T., Chiu, P.C. 2012. The role of black carbon as a catalyst for environmental redox transformation. *Environmental Geochemistry and Health* 34, 105–113.

Petrović, M., Eljarrat, E., López de Alda, M.J., Barceló, D. 2001. Analysis and environmental levels of endocrine-disrupting compounds in freshwater sediments. *TrAC Trends in Analytical Chemistry* 20, 637–648.

Petruzzelli, G. 2012. Soil Contamination and Remediation Strategies. Current research and future challenge. *EGU General Assembly Conference* 14, 7963.

Pignatello, J.J., Kwon, S., Lu, Y. 2006. Effect of natural organic substances on the surface and adsorptive properties of environmental black carbon (char): Attenuation of surface activity by humic and fulvic acids. *Environmental Science & Technology* 40, 7757–7763.

Qiu, Y., Wu, M., Jiang, J., Li, L., Daniel Sheng, G. 2013. Enhanced irreversible sorption of carbaryl to soils amended with crop-residue-derived biochar. *Chemosphere* 93, 69–74.

Reid, B., Pickering, F., Freddo, A., Whelan, M., Coulon, F. 2013. Influence of biochar on isoproturon partitioning and bioaccessibility in soil. *Environmental Pollution* 181, 44–50.

Sarmah, A.K., Sabadie, J. 2002. Hydrolysis of sulfonylurea herbicides in soils and aqueous solutions: A review. *Journal of Agricultural and Food Chemistry* 50, 6253–6265.

Sarmah, A.K., Srinivasan, P., Smernik, R.J., Manley-Harris, M., Antal, M.J., Downie, A., Van Zwieten, L. 2010. Retention capacity of biochar-amended New Zealand dairy farm soil for an estrogenic steroid hormone and its primary metabolite. *Soil Research* 48, 648–658.

Schaffer, M., Boxberger, N., Börnick, H., Licha, T., Worch, E. 2012. Sorption influenced transport of ionizable pharmaceuticals onto a natural sandy aquifer sediment at different pH. *Chemosphere* 87, 513–520.

Sheng, G., Yang, Y., Huang, M., Yang, K. 2005. Influence of pH on pesticide sorption by soil containing wheat residue-derived char. *Environmental Pollution* 134, 457–463.

Si, Y., Wang, M., Tian, C., Zhou, J., Zhou, D. 2011. Effect of charcoal amendment on adsorption, leaching and degradation of isoproturon in soils. *Journal of Contaminant Hydrology* 123, 75–81.

Smith, K.E., Thullner, M., Wick, L.Y., Harms, H. 2009. Sorption to humic acids enhances polycyclic aromatic hydrocarbon biodegradation. *Environmental Science & Technology* 43, 7205–7211.

Song, Y., Wang, F., Bian, Y., Kengara, F.O., Jia, M., Xie, Z., Jiang, X. 2012a. Bioavailability assessment of hexachlorobenzene in soil as affected by wheat straw biochar. *Journal of Hazardous Materials* 217, 391–397.

Song, Y., Wang, F., Kengara, F.O., Yang, X., Gu, C., Jiang, X. 2013. Immobilization of chlorobenzenes in soil using wheat straw biochar. *Journal of Agricultural and Food Chemistry* 61, 4210–4217.

Song, Y., Wang, F., Yang, X., Bian, Y., Gu, C., Xie, Z., Jiang, X. 2012b. Influence and assessment of biochar on the bioavailability of chlorobenzenes in soil. *Environmental Sciences* (in Chinese) 33, 169.

Sopeña, F., Semple, K., Sohi, S., Bending, G. 2012. Assessing the chemical and biological accessibility of the herbicide isoproturon in soil amended with biochar. *Chemosphere* 88 (1), 77–83.

Spokas, K., Koskinen, W., Baker, J., Reicosky, D. 2009. Impacts of woodchip biochar additions on greenhouse gas production and sorption/degradation of two herbicides in a Minnesota soil. *Chemosphere* 77, 574–581.

Stales, C.A., Peterson, D.R., Parkerton, T.F., Adams, W.J. 1997. The environmental fate of phthalate esters: A literature review. *Chemosphere* 35, 667–749.

Su, D., Li, P.-j., Frank, S., Xiong, X.-z. 2006. Biodegradation of benzo [a] pyrene in soil by Mucor sp. SF06 and Bacillus sp. SB02 co-immobilized on vermiculite. *Journal of Environmental Sciences* 18, 1204–1209.

Sun, K., Jin, J., Keiluweit, M., Kleber, M., Wang, Z., Pan, Z., Xing, B. 2012. Polar and aliphatic domains regulate sorption of phthalic acid esters (PAEs) to biochars. *Bioresource Technology* 118, 120–127.

Tatarková, V., Hiller, E., Vaculík, M. 2013. Impact of wheat straw biochar addition to soil on the sorption, leaching, dissipation of the herbicide (4-chloro-2-methylphenoxy)acetic acid and the growth of sunflower (Helianthus annuus L.). *Ecotoxicology and Environmental Safety* 92, 215–221.

Uchimiya, M., Lima, I.M., Klasson, K.T., Wartelle, L.H. 2010. Contaminant immobilization and nutrient release by biochar soil amendment: Roles of natural organic matter. *Chemosphere* 80, 935–940.

Wang, H., Lin, K., Hou, Z., Richardson, B., Gan, J. 2010. Sorption of the herbicide terbuthylazine in two New Zealand forest soils amended with biosolids and biochars. *Journal of Soils and Sediments* 10, 283–289.

Wang, Y., Wang, Y.-J., Wang, L., Fang, G.-D., Cang, L., Herath, H., Zhou, D.-M. 2013. Reducing the bioavailability of PCBs in soil to plant by biochars assessed with triolein-embedded cellulose acetate membrane technique. *Environmental Pollution* 174, 250–256.

Wauchope, R.D., Yeh, S., Linders, J.B.H.J., Kloskowski, R., Tanaka, K., Rubin, B., Katayama, A., Kördel, W., Gerstl, Z., Lane, M., Unsworth, J.B. 2002. Pesticide soil sorption parameters: Theory, measurement, uses, limitations and reliability. *Pest Management Science* 58, 419–445.

WHO. 2010. Persistent organic pollutants: Impact on child health. Geneva, Switzerland, ISBN 9789241501101.

Wu, M., Pan, B., Zhang, D., Xiao, D., Li, H., Wang, C., Ning, P. 2013. The sorption of organic contaminants on biochars derived from sediments with high organic carbon content. *Chemosphere* 90, 782–788.

Wu, W., Yang, M., Feng, Q., McGrouther, K., Wang, H., Lu, H., Chen, Y. 2012. Chemical characterization of rice straw-derived biochar for soil amendment. *Biomass and Bioenergy* 47, 268–276.

Xu, T., Lou, L., Luo, L., Cao, R., Duan, D., Chen, Y. 2011. Effect of bamboo biochar on pentachlorophenol leachability and bioavailability in agricultural soil. *Science of the Total Environment* 414, 727–731.

Xu, T., Lou, L., Luo, L., Cao, R., Duan, D., Chen, Y. 2012. Effect of bamboo biochar on pentachlorophenol leachability and bioavailability in agricultural soil. *Science of the Total Environment* 414, 727–731.

Yang, Y., Chun, Y., Sheng, G., Huang, M. 2004. pH-dependence of pesticide adsorption by wheat-residue-derived black carbon. *Langmuir* 20, 6736–6741.

Yang, Y., Sheng, G., Huang, M. 2006. Bioavailability of diuron in soil containing wheat-straw-derived char. *Science of the Total Environment* 354, 170–178.

Yao, Y., Gao, B., Chen, H., Jiang, L., Inyang, M., Zimmerman, A.R., Cao, X., Yang, L., Xue, Y., Li, H. 2012. Adsorption of sulfamethoxazole on biochar and its impact on reclaimed water irrigation. *Journal of Hazardous Materials* 209–210, 408–413.

Yu, X., Gong, W., Liu, X., Shi, L., Han, X., Bao, H. 2011. The use of carbon black to catalyze the reduction of nitrobenzenes by sulfides. *Journal of Hazardous Materials* 198, 340–346.

Yu, X.-Y., Ying, G.-G., Kookana, R.S. 2006. Sorption and desorption behaviors of diuron in soils amended with charcoal. *Journal of Agricultural and Food Chemistry* 54, 8545–8550.

Yuan, J., Xu, R. 2011. Progress of the research on the properties of biochars and their influence on soil environmental functions. *Ecology Environment Science* 20, 779–785.

Zhang, H., Lin, K., Wang, H., Gan, J. 2010. Effect of Pinus radiata derived biochars on soil sorption and desorption of phenanthrene. *Environmental Pollution* 158, 2821–2825.

Zhang, J., Sun, B., Xiong, X., Gao, N., Song, W., Du, E., Guan, X., Zhou, G. 2014a. Removal of emerging pollutants by Ru/TiO2-catalyzed permanganate oxidation. *Water Research* 63, 262–270.

Zhang, P., Sun, H., Yu, L., Sun, T. 2013a. Adsorption and catalytic hydrolysis of carbaryl and atrazine on pig manure–derived biochars: Impact of structural properties of biochars. *Journal of Hazardous Materials* 244–245, 217–224.

Zhang, X., He, L., Sarmah, A., Lin, K., Liu, Y., Li, J., Wang, H. 2014b. Retention and release of diethyl phthalate in biochar-amended vegetable garden soils. *Journal of Soils and Sediments* 14, 1790–1799.

Zhang, X., Wang, H., He, L., Lu, K., Sarmah, A., Li, J., Bolan, N.S., Pei, J., Huang, H. 2013b. Using biochar for remediation of soils contaminated with heavy metals and organic pollutants. *Environmental Science and Pollution Research* 20, 8472–8483.

Zheng, W., Guo, M., Chow, T., Bennett, D.N., Rajagopalan, N. 2010. Sorption properties of greenwaste biochar for two triazine pesticides. *Journal of Hazardous Materials* 181, 121–126.

Zhu, D., Kwon, S., Pignatello, J.J. 2005. Adsorption of single-ring organic compounds to wood charcoals prepared under different thermochemical conditions. *Environmental Science & Technology* 39, 3990–3998.

7

Biochar for Inorganic Contaminant Management in Waste and Wastewater

Chapter 7

Biochar for Inorganic Contaminant Management in Waste and Wastewater

Anitha Kunhikrishnan, Irshad Bibi, Nanthi Bolan, Balaji Seshadri, Girish Choppala, Nabeel Khan Niazi, Won-Il Kim, and Yong Sik Ok

Chapter Outline

7.1 Introduction

In many parts of the world, continuous withdrawal of fresh-water for various activities, including irrigation, have led to unsustainable rates of water consumption, which is not assisted by declining rainfall and increased rationing of water to the ecosystem services (Kunhikrishnan et al. 2012). Communities, particularly primary producers, are compelled to improve water-use efficiency and use alternative water supplies, including recycled wastewater sources, for irrigation. Using wastewater for irrigation raises concerns about public exposure to pathogens and contamination of soil, surface water, and groundwater. However, under controlled management, these water sources can be used safely and profitably for irrigation (Drechsel et al. 2010).

Wastewaters originate from several sources, including domestic sewage (municipal wastewater); agricultural, urban, and industrial effluents; and stormwater. Wastewater irrigation has many beneficial effects, such as groundwater recharging (Asano and Cotruvo 2004) and nutrient supply to plants (Anderson 2003). However, there are some detrimental effects, such as buildup of salts, pesticides, and heavy metal(loid)s in the soils irrigated with wastewater. At sites irrigated with wastewater, mobilization and transport of pesticides and heavy metal(loid)s into groundwater have been noted, which led to their enhanced bioavailability to soil biota and higher plants. For example, dissolved organic matter present in wastewater has been shown to facilitate the transport of pesticides and heavy metal(loid)s (Ashworth and Alloway 2004; Bolan et al. 2011; Kunhikrishnan et al. 2012; Müller et al. 2007). Wastewater irrigation has also been shown to act as a source of these contaminant input to soils (Eriksson and Donner 2009; Kunhikrishnan et al. 2012; Müller et al. 2007).

Heavy metal(loid)s reach the soil environment through both geogenic and anthropogenic processes or activities. Anthropogenic activities, primarily associated with the disposal of industrial and domestic waste materials, including wastewaters and biosolids, are the major sources of metal(loid) enrichment in soils and have potential to leach into groundwater (Adriano 2001; Bolan et al. 2014). Industries, such as mining and electroplating, discharge aqueous effluents containing high levels of heavy metal(loid)s, such as Cd, Cu, Hg, Pb, Zn, and U. For example, field studies showed that some soils in the northeastern China that received extensive wastewater irrigation contained as high as 24.6 mg·kg^{-1} As and 3.2 mg·kg^{-1} Cd; meanwhile, as much as 2.1 mg·L^{-1} Cd and 15.2 mg·L^{-1} As accumulates in groundwater, which poses an environmental risk for human and animal health (Guo and Zhou 2006; Wu et al. 2011).

Conventional methods and technologies that have been used for the removal of heavy metal(loid)s from wastewater include chemical precipitation, ion exchange, filtration, chemical oxidation and reduction, membrane technology (separation), reverse osmosis, electrochemical treatment, neutralization, electrodialysis, flotation, electrolytic recovery, evaporation removal, and adsorption on activated carbon (AC) (Barakat 2011; El-Ashtoukhy et al. 2008). Most of these methods, however, are often expensive, ineffective when heavy metal(loid) concentrations are low, specifically when they are <100 mg·L^{-1}, or both (Perez-Marin et al. 2008). In these cases, the most suitable and effective method for metal(loid) removal has been proven to be adsorption (Demirbas 2008). However, large-scale application of this technology is often limited by high capital and regeneration costs of the commercially available adsorbents, such as AC. Wastewater treatment requires vast quantities of AC. Improved, inexpensive, tailor-made, and readily regenerated adsorbents are required. This requirement has led to extensive research concerning the identification of suitable and relatively cheap materials for the production of low-cost adsorbents that are capable of removing significant quantities of metal(loid) ions from aqueous solutions (Liu and Zhang 2009).

A wide variety of materials, including agricultural by-products and wastes that have been examined to generate low-cost adsorbents, are converted to high value-added adsorbents that are also suitable for wastewater purification (Imamoglu and Tekir 2008). Some of the agricultural by-products that have been tested recently are wood and bark (Mohan et al. 2007), hazelnut husks (Imamoglu and Tekir 2008), rice husks (Cope et al. 2014), potato peel (Aman et al. 2008), orange waste (Pellera et al. 2012; Perez-Marin et al. 2008), pomegranate peel (El-Ashtoukhy et al. 2008), pinewood (Abdel-Fattah et al. 2014), sugar beet pulp (Aksu and Isoglu 2005), olive pomace (Pellera et al. 2012), and different types of plant wastes (Wan Ngah and Hanafiah 2008). These agricultural by-products have been used in their raw form or after some physical or chemical modification, or both. Hydrothermal treatment and conventional pyrolysis are two of the most commonly used methods to synthesize C-based products such as biochar from agricultural residues with high adsorption capacity (Liu and Zhang 2009; Mohan et al. 2007).

Biochar is a pyrogenic C-rich material, derived from thermal decomposition of biomass in a closed system with little or no O supply (Lehmann et al. 2006). Biochar has been extensively studied in the past few years for its potential for C sequestration and for its ability to enhance the nutrient level of soils (Schulz and Glaser 2012) as well as for its positive effect on plant growth (Sohi et al. 2010). Biochar is now widely studied for its metal(loid) sorption efficiency in soils and water. The use of biochar as a low-cost sorbent to remove organic and inorganic contaminants from aqueous solutions is an emerging and promising wastewater treatment technology, and it has already been demonstrated in previous studies (Ahmad et al. 2014; Cao et al. 2009; Mohan et al. 2007, 2014a, 2014b; Uchimiya et al. 2011).

This chapter aims to describe different sources of wastewater and heavy metal(loid) input to water. It discusses the methods to remove metal(loid)s from wastewater. The role of biochar as a low-cost adsorbent of metal(loid)s has been elaborated; details of its synthesis and characterization and studies involving the remediation of metal(loid)s from wastewater by biochars have been reviewed. Future research needs of biochars and its potential implementation are also highlighted.

7.2 Sources of Wastewater Streams and Their Heavy Metal(loid)s Distribution

7.2.1 Municipal Wastewater and Stormwater

Municipal wastewater consists of discharges from households, institutions, and commercial buildings. Secondary treated wastewater typically contains low levels of contaminants as these tend to settle under gravitation with solid fractions in the treatment lagoons. Settling of suspended solids also lowers both the chemical and biochemical oxygen demand. Municipal wastewater contains high concentrations of nutrients, especially N and P; trace elements, such as Fe and Mn; and dissolved salts, particularly Na, Cl, and, in some cases, bicarbonates. These parameters, including heavy metal(loid) contents, are critical when wastewater is reused in agriculture. For example, in a recent study in Zambia, Kapungwe (2013) observed higher than acceptable limits of Cr, Co, Cu, Pb, and Ni in domestic sewage and industrial wastewater that was used to irrigate crops.

Water that falls on roofs or collects on paved areas such as driveways, roads, or footpaths is called stormwater. The stormwater system runs from outdoor drains down the gutters and untreated into our natural waterways (creeks, rivers, groundwaters, wetlands, and oceans). In Australia, for example, ~10,300 million liter of stormwater is generated annually (Laurenson et al. 2010). In many cases, urban stormwater runoff contains a broad range of pollutants that are transported to natural water systems (Aryal et al. 2010). Pollutants include pesticides; herbicides; oil; grease; and heavy metal(loid)s, such as Cd, Cr, Cu, Ni, Pb, and Zn (Wong et al. 2000). These metal(loid)s are either dissolved in the stormwater or bound to particulates; the degree of binding is a function of pH, average pavement residence time, and the nature and quantity of solids present (Sansalone and Buchberger 1997). This partition between the solid and aqueous phase has a major effect on the occurrence, transport, fate, and biological effects of heavy metal(loid)s in aquatic systems (Ran et al. 2000).

Nutrients such as N and P are also important pollutants in stormwater. If accumulated in great quantities, they can eventually cause toxic algal blooms and other pollution problems in the waterways. Hence, urban stormwater harvesting has emerged in recent years as a viable option to reduce pressures on existing water sources and to alleviate adverse environmental impacts associated with stormwater runoff (Roy et al. 2008). The extent to which runoff from storms must be retained depends upon the nature and magnitude of the water pollution that might result from the discharge. Other variables include rainfall distribution and land management practices.

7.2.2 Farm Wastewater

Farm effluents such as those emanating from dairy sheds and piggeries are being increasingly used as sources of irrigation water and nutrients (Bolan et al. 2009). For example, in New Zealand, dairy and piggery effluents generate annually ~9,000 Mg of N, 1,250 Mg of P, and 14,000 Mg of K (Bolan et al. 2004a). Effluents from farms differ in their composition depending on the animal production system from which they are derived (poultry, pigs, beef, dairy). Generally, farm wastewater is rich in organic and inorganic components (Wang et al. 2004). For example, Cu and Zn are commonly used as feed additives, growth promoters, for disease prevention or treatment, and their concentration in the final wastewater can be significant (Bolan et al. 2004b). Recently, Abe et al. (2012) measured the daily output of Zn and Cu in wastewater from livestock farms to aquatic environments because waste from animal husbandry operations contains high levels of these metal(loid)s. They surveyed 21 pig farms and 6 dairy farms. The unit (i.e., per head) output load from piggery wastewater treatment facilities ranged from 0.13 to 17.8 mg·head^{-1} day^{-1} for Zn and from 0.15 to 9.4 mg·head^{-1} day^{-1} for Cu. For dairy farms, the unit output load from wastewater treatment facilities was estimated to be 1.8–3.6 mg·head^{-1} day^{-1} for Zn and 0.6 mg·head^{-1} day^{-1} for Cu.

Irrigation of farm effluent onto pasture is increasingly being recognized as a means for biological treatment in many countries, including Australia and New Zealand, and they acknowledge the fact that farm effluent is a resource to be used for its nutrient content rather than a waste for disposal. Traditionally, the farm effluents are treated biologically using two-pond systems and then discharged to land or streams. Bolan et al. (2009) have suggested that land application of farm effluent facilitates the recycling of valuable nutrients, C and water, and if managed well, helps to mitigate surface water pollution.

7.2.3 Industrial Wastewater

Use of metal(loid)s in industry is widespread, and the type and quantity of metal(loid) discharges from industries depend on many factors, including industrial type, process variables, and pollution abatement practices (Barakat 2011). Chromium, for example, is widely used in electroplating, metal(loid) finishing, magnetic tapes, pigments, leather tanning, wood protection, chemical manufacturing, brass, electrical and electronic equipment, and catalysis. The volume and characteristics of wastewater streams derived from various industries such as tanneries depend on the processes adopted for water consumption. For example, in Kanpur, India, a cluster of >60 tanneries are situated on the bank of the Ganga River. Very high Cr concentrations, in the order of $16.3 \text{ mg} \cdot \text{L}^{-1}$, were found in these waters compared to the permissible concentrations of $0.05 \text{ mg} \cdot \text{L}^{-1}$ recommended for drinking water (Singh et al. 2009). Another significant source of heavy metal(loid) wastes result from printed circuit board manufacturing. Tin, Pb, and Ni solder plates are the most widely used resistant over-plates. Other sources for the metal(loid) wastes include the wood processing industry, which uses chromated copper-arsenate for treating wood; inorganic pigment manufacturing units that contain Cr compounds and cadmium sulfide; and petroleum refining, which generates conversion catalysts contaminated with Ni, V, and Cr. All of these sources produce a large quantity of wastewaters, residues, and sludges that can be categorized as hazardous wastes requiring extensive waste treatment (Sorme and Lagerkvist 2002).

The pulp and paper industry is the third largest industrial polluter of water, air, and land (Rout 2008). Pulp and paper mill effluent either from thermomechanical pulp mill or chemi-thermomechanical pulp mill is often irrigated to land after primary treatment (Smith et al. 2003; Wang et al. 1999). Pulp mill effluent has high chemical and biochemical oxygen demand, some wood-derived organic compounds, metal(loid)s, fatty acids, and resins, with relatively high C:N ratios (Gaete et al. 2000; Kannan and Oblisami 1990; Mishra et al. 2013); hence, consequent land application of pulp mill effluents is becoming a common method for recycling nutrients (Rubilar et al. 2008). Pathak et al. (2013) concluded that an undiluted paper mill effluent irrigation for 60 days increased the nutrient status of the soil, including an increase in metal(loid)s such as Ni and Cd that, in turn, was responsible for the metal(loid) increase in leaves and roots of spinach (*Spinacia oleracea*). Ramola and Singh (2013) studied the concentrations of some heavy metal(loid)s (Cd, Cr, Pb, Ni, Cu, and Zn) in the effluents of pharmaceutical industries operating in Dehradun, India. They noticed that Cr, Cd, Pb, and Ni concentrations were above the permissible limit recommended by World Health Organization standards. In a recent study, Samuel et al. (2014) analyzed the presence of metal(loid)s in sugar mill effluent and detected slightly elevated levels of As, Cd, Cu, Cr, Hg, Pb, and Zn. Similarly, results of a study by Abdalla et al. (2013) indicated Zn and Cu in the wastewater of all analyzed sugarcane plants.

7.3 Techniques of Heavy Metal(loid)s Removal from Wastewater Streams

7.3.1 Chemical Precipitation

Precipitation is the formation of solid(s), either in solution or on a surface, during the sorption process. Precipitation is used to remove inorganic contaminants from wastewater using coagulants such as lime (calcium carbonate), alum, Fe salts [iron(II) chloride, iron(III) chloride], and organic polymers. The conventional chemical precipitation process includes hydroxide precipitation and sulfide precipitation. Hydroxide precipitation is widely used due to its relative simplicity, cost-effectiveness, and its ease of pH control during the process (Huisman et al. 2006).

Mirbagheri and Hosseini (2005) used sodium hydroxide (NaOH) and calcium hydroxide to remove Cu and chromate [Cr(VI)] ions from wastewater. This technique is limited due to its excessive sludge production and the usage of large amount of chemicals to reduce metal(loid)s to an acceptable level for discharge. In sulfide precipitation process, its precipitates are not amphoteric in nature, and the solubility of metal(loid)-sulfide precipitates is substantially less than that of hydroxide precipitates. However, the major limitations are generation of toxic hydrogen sulfide fumes and the discharge of treated wastewater containing residual levels of sulfide (Fu and Wang 2011).

7.3.2 Ion Exchange

The ion exchange process has been extensively used for the remediation of wastewater given to its several merits, including the large metal(loid) removal efficiency and rapid kinetics (Barakat 2011). Ion exchange resins have the potential to provide its positively charged (cationic) surface sites for the retention of various metal(loid)s in wastewater. The weakly acid resins with carboxylic groups (–COOH) and strongly acidic resins with sulfonic groups (–SO_3H) are considered the most common cation exchange resins. The hydrogen ions in the –SO_3H or –COOH functional groups of the resin material could offer the surface to remove the metal(loid) cations by exchange process and the metal(loid) anions by ligand exchange mechanism (Barakat 2011; Lakherwal 2014). Abo-Farha et al. (2009) studied the effect of ionic charge on the sorptive removal of Fe^{3+} and Pb^{2+} from contaminated water by using the cation-exchange resin purolite. Naturally occurring materials, such as zeolites and silicate minerals, have also been used for the removal of metal(loid)s from wastewater due to their low cost and abundance in the environment (Doula 2009).

7.3.3 Membrane Filtration

Membrane filtration techniques are highly capable of metal(loid) removal from wastewaters. The membrane filtration processes used are ultrafiltration, reverse osmosis, nanofiltration, and electrodialysis.

Ultrafiltration uses small transmembrane pressures to remove finely dissolved and colloidal particles. The dissolved metal(loid) ions (as hydrated ions or low-molecular-weight complexes) can easily flow through the ultrafiltration membranes. Micellar-enhanced ultrafiltration uses surfactants such as sodium dodecyl sulfate, and polymer-enhanced ultrafiltration uses water-soluble polymers to effectively separate metal(loid)s from wastewater (Lakherwal 2014; Landaburu-Aguirre et al. 2009). Aroua et al. (2007) noticed up to 100% rejection rate for Cr(III) species at pH >7.0 in contaminated water, for water-soluble polymers, chitosan, polyethyleneimine, and pectin. Reverse osmosis uses a semipermeable membrane that allows the fluid to pass through it, while holding back a majority of the contaminants. Mohsen-Nia et al. (2007) efficiently removed Cu^{2+} and Ni^{2+} ions from wastewater by using reverse osmosis, and the removal efficiency increased >99% in the presence of Na-EDTA.

Nanofiltration is intermediate between ultrafiltration and reverse osmosis. It is a promising technique for the removal of heavy metal(loid)s such as Ni (Murthy and Chaudhari 2008), Cr (Muthukrishnan and Guha 2008), Cu (Ahmad and Ooi 2010), and As (Figoli et al. 2010) from wastewater. Electrodialysis process separates ions across charged membranes from one solution to another by using an electric field as the driving force. Cifuentes et al. (2009) and Lambert et al. (2006) found that electrodialysis proved very effective in removing Cu and Fe, and Cr(III) ions, respectively. The membrane filtration techniques are highly efficient, easy to operate, and space saving. However, they are limited by the expensive nature of the nanofiltration systems on a large scale and high power consumption and restoration of membranes in the case of reverse osmosis.

7.3.4 Coagulation and Flocculation

Coagulation refers to the charge neutralization of particles. Widely used coagulant materials, such as aluminium, ferrous sulfate, and ferric chloride salts, successfully remove the particulate matter in wastewater by charge neutralization of particles, thereby forming aggregates with amorphous metal(loid)

(Fe/Al) hydroxide precipitates (Fu and Wang 2011). Chang and Wang (2007) used sodium xanthogenate group in conjunction with polyethyleneimine to remove both soluble metal(loid)s and insoluble substances efficiently by coagulation. Flocculation is the action of polymers to form bridges between the flocs and bind the particles into large agglomerates or clumps. Flocculants of mercaptoacetyl chitosan (Chang et al. 2009), Konjac-graft-poly (acrylamide)-co-sodium xanthate (Duan et al. 2010), and polyampholyte chitosan derivatives-N-carboxyethylated chitosans (Bratskaya et al. 2009) were used to remove metal(loid)s. Although the coagulation–flocculation technique is simple and cost-effective and it enhances filtration, time consumption and the formation of sludge that needs subsequent treatment are potential drawbacks.

7.3.5 Flotation and Electrochemical Treatment

Flotation separates the metal(loid)s from contaminated wastewater by using the bubble attachment mechanism. The process of ion flotation is based on imparting hydrophobic particles to the ionic metal(loid) species in wastewaters by using surfactant materials, followed by the removal of these hydrophobic species by air bubbling. For example, 96% of Cr(III) (~pH 8) was removed from the wastewater by precipitate flotation, where an anionic collector was used for agglomeration and ethanol as a frother (Medina et al. 2005). In the electrochemical technique, plating-out of metal(loid) ions is accomplished on a cathode surface, and metal(loid)s are then recovered in their elemental state. The metal(loid) ions are hydrolyzed and coprecipitated as hydroxides. However, this technique requires huge capital investment and extensive supply of electricity.

7.3.6 Adsorption

Adsorption is the transfer of ions from the solution phase to the solid phase whereby ions are bound by physical interactions, chemical interactions, or both (Babel and Kurniawan 2003). Recently, adsorption has become one of the alternative treatment techniques for wastewater contaminated with heavy metal(loid)s. Various low-cost adsorbents, derived from agricultural wastes, industrial by-products, natural materials, or modified biopolymers, have been recently developed and applied for the removal of heavy metal(loid)s from wastewater.

AC removes metal(loid)s from wastewater due to its large micro- and mesopore volumes that provide the AC with a high surface area, thus offering more surface sites for metal(loid) binding. For example, Kongsuwan et al. (2009) reported that the maximum adsorption capacity of eucalyptus bark–derived AC for Cu and Pb was 0.44 and 0.52 mmol·g^{-1}, respectively. Guo et al. (2010) reported that poultry litter–based AC demonstrated significantly greater adsorption capacity for metal(loid)s than commercial AC derived from bituminous coal and coconut shells. However, several studies have demonstrated outstanding removal capacities for metal(loid)s using low-cost adsorbents such as chitosan, zeolites, natural or modified clays, waste slurry, and lignin compared to AC (Babel and Kurniawan 2003; Baker et al. 2009; Srinivasan 2011). Jiang et al. (2010) noticed that the removal rate of Pb, Cd, Ni, and Cu ions from wastewater by kaolinite clay was fast, with maximum sorption attained in first 30 min.

Biosorption of metal(loid)s from wastewater is becoming apparent recently, and the merits of using biosorbents include their high efficiency, cost-effectiveness, and eco-friendly nature. The biosorbent materials could be derived from (i) nonliving biomass, for example, bark, peels of various fruits, krill, squid, and crab shells; (ii) algal dead biomass; and (iii) microorganism biomass, for example, fungi and bacteria (Barakat 2011; Fu and Wang 2011). For example, Pavan et al. (2006) used mandarin peel and obtained sorption capacities of 1.92, 1.37, and 1.31 mmol·g^{-1} for Ni, Co, and Cu, respectively. Ajjabi and Chouba (2009) observed that at the optimum biosorbent dosage (20 g·L^{-1}) and initial solution pH 5, marine green macroalga (*Chaetomorpha linum*) removed maximum Cu and Zn ions, with sorption capacities of 1.46 and 1.97 mmol·g^{-1}, respectively. Bhainsa and D'Souza (2008) noticed that the maximum removal capacity for Cu by viable and pretreated NaOH-treated fungal (*Rhizopus oryzae*) biomass was 19.4 and 43.7 mg·g^{-1}, respectively.

Biochar can be prepared using an extensive range of feedstock types, including agricultural and forest residues; industrial by-products and wastes; municipal solid waste materials; and nonconventional materials, such as waste tires, papers, and even bones. The sorption of metal(loid)s by biochars can be affected by several factors, including feedstock types, pyrolysis conditions, and modification and activation methods. The strong sorption ability of biochars is attributed to their surface properties originating from the feedstock materials. The domestic animal waste materials are another major source of potential feedstock for biochar production. Biochars produced from poultry litter generally have high contents of P and ash, making them effective adsorbents for removal of metal(loid)s from wastewater.

7.4 Biochar Production and Characterization

7.4.1 Production Methods of Biomass Pyrolysis

Biochar is produced as a charred material by pyrolysis, or the process of thermochemical decomposition of organic material at elevated temperatures in the absence of oxygen. There are three product streams from pyrolysis: (i) noncondensable gases; (ii) a combustible bio-oil representing the condensable liquids (tars); and (iii) biochar, a solid residual coproduct. Pyrolysis technology can be differentiated by the residence time, pyrolytic temperature (e.g., slow and fast pyrolysis processes), pressure, size of adsorbent, and the heating rate and method. Slow pyrolysis technique maximizes biochar yield, but the other variants of hydrothermal carbonization (HTC) and microwave-assisted pyrolysis are also appealing due to their ability to handle wet biomass sources, which reduces biomass drying costs. Current biochar production is focused on advanced pyrolysis systems that allow precise control of operating conditions, and when coupled with feedstock selection, can regulate the physical and chemical properties of biochar (Lima et al. 2010; Zhang et al. 2007). Different processes describing the production of biochar are detailed below.

7.4.1.1 Slow and Fast Pyrolysis

The most efficient process for char production under dry conditions is slow pyrolysis. Biomass is heated slowly to about 500°C in the absence of air over a long period, and ultimately leads to biochar with 30–45% C (Bruun et al. 2012). Although slow pyrolysis yields a high-C, energy-dense solid char product, the coproducts are a watery, low-molecular-weight acidic liquid called pyroligneous acid or wood tar, and a low-energy, combustible gas. Fast pyrolysis requires dry biomass (<10 wt% moisture), rapid heat transfer ($\sim$100–1000°C$\cdot$s^{-1}), fast temperature increase by heating small biomass particles (1–2 mm) to 400–500°C and vapor residence times of 1 s (maximum 5 s) (Lima et al. 2010).

The characteristics of biochar are affected by the extent of pyrolysis (pyrolytic temperature and residence pressure), biomass size, and kiln or furnace residence time (Nartey and Zhao 2014). Particle size and shape, physical properties, composition (e.g., lignin, cellulose, hemicellulose), and ash content are the most important feedstock properties that influence the properties of the final product (Joseph et al. 2009). The high sorption potential of pyrochars compared to hydrothermal biochars is due to their physicochemical properties, that is, a large surface area and microporosity, and a high aromaticity linked to the presence of O-containing functional groups (Liu and Zhang 2011). Many studies have shown that biochar produced at higher temperatures has higher surface area and pore volume (Ahmad et al. 2012; Ghani et al. 2013).

7.4.1.2 Torrefaction

Torrefaction is a thermochemical treatment process that is carried out at an operating temperature range of 200–300°C in a nonoxidative environment to improve the physical and chemical characteristics of biomass and also to aid in further conversion to biofuels (Sadaka and Negi 2009). This is achieved by facilitating decomposition of easily degraded volatile matter (hemicellulose) and repolymerization of cellulose and lignin (Hill et al. 2013). The torrefied biomass resembles the original material in shape and size but is darker, friable, and hydrophobic with a higher calorific content (Lipinsky et al. 2002). Torrefaction is an energy densification process that uses mild pyrolysis. Depending on the torrefaction temperature and biomass residence time in the reactor, hemicellulose, cellulose, and lignin content of biomass are partly decomposed.

The solid products obtained from torrefaction have some advantageous properties: (i) higher energy density and heating value, (ii) reduced transport cost due to reduced moisture of the end product, (iii) higher resistivity of torrefied biomass to fungal attack due to the hydrophobic nature, (iv) reduced grinding energy requirements, and (v) creation of a more uniform fuel for gasification or cofiring for electricity. Despite the benefits, torrefaction does not create adsorbent chars because only partial biomass decomposition occurs to prevent decay and induce some water loss (Mohan et al. 2014a). However, Nhuchhen et al. (2014) recently suggested that torrefaction can provide a good alternative to the traditional means of biochar production mainly because of its low temperature (200–300°C), leading to a larger fraction of biomass char retained in solid form.

7.4.1.3 Gasification

Gasification takes place at much higher temperatures than pyrolysis and torrefaction, and it produces a clean gas that can be used in internal combustion engines or to produce electricity. In this process, the energy in biomass or any other organic matter is converted to combustible gases (mixture of carbon monoxide [CO], methane [CH_4], and hydrogen gas [H_2]) at temperatures ranging from 600°C to 1000°C, with char, water, and condensable tar as minor products. In contrast to pyrolysis, biomass gasification is performed under a partially oxidizing atmosphere. Initially, in the first step called pyrolysis, the organic matter is decomposed by heat into gaseous and liquid volatile materials and char. In the second step, the hot char reacts with the gases (mainly carbon dioxide [CO_2] and water), leading to product gases, namely, CO, H_2, and CH_4. This producer gas or syngas mixture leaves the reactor with some heavy hydrocarbons called tar. Separation and elimination of tar from the syngas account for a significant portion of fuel production costs (Baldwin et al. 2012). Brewer et al. (2009) characterized the chars from slow and fast pyrolysis and gasification of switchgrass and corn stover. Higher char aromaticity was obtained in slow pyrolysis than in fast pyrolysis or gasification. However, although the size of fused aromatic ring compounds was similar (~7–8 rings per compound) in fast and slow pyrolysis chars, the gasification char was more highly condensed (~17 rings per compound).

7.4.1.4 Hydrochar

Hydrochar refers to the solid product from HTC of C-rich biomass in the presence of water, which is also called hydrous pyrolysis or wet pyrolysis (Hu et al. 2010; Sevilla and Fuertes 2009). The HTC process usually takes place at relatively low temperatures (150–350°C) and under high pressure (~2 MPa) and can be applied directly to wet feedstocks, such as wet animal manures, sewage sludge, and algae (Xue et al. 2012). As a result, this process does not require an energy-intensive predrying step, in contrast to pyrolysis (Mumme et al. 2011). In addition, HTC has been reported to have a higher yield (30–60 wt%) than fast or slow pyrolysis (Funke and Ziegler 2010; Sevilla and Fuertes 2009).

The hydrochar process is eco-friendly because it does not generate any hazardous chemical waste or by-products as does dry pyrolysis (Hu et al. 2010). The effect of hydrochar on wastewater remediation will likely differ as result of their varying physicochemical properties arising from specific production conditions (Libra et al. 2011). For example, the surface area of pyrochars varies considerably and reaches values up to 1000 $m^2 \cdot g^{-1}$ (Qiu et al. 2009; Uchimiya et al. 2011). But for hydrochars, average surface area values of 8 $m^2 \cdot g^{-1}$ have been reported (Schimmelpfennig and Glaser 2012). Studies report that the adsorption capacity of hydrochar to metal(loid)s seems to be much lower than that of dry thermally produced biochars or other common adsorbents (Mohan et al. 2007; Uchimiya et al. 2010), probably because hydrochar contains fewer O-containing surface functional groups than the dry thermal biochars (Uchimiya et al. 2011).

7.4.2 Characterization of Biochar

7.4.2.1 Physical Characterization

Process temperature is the main factor-governing surface area, increasing from 120 $m^2 \cdot g^{-1}$ at 400°C to 460 $m^2 \cdot g^{-1}$ at 900°C (Day et al. 2005). Kim et al. (2013) investigated the feasibility of biochar produced from the grass *Miscanthus sacchariflorus* via slow pyrolysis at 300, 400, 500, and 600°C for removing Cd from aqueous solution. The surface area increased greatly in biochar produced at a pyrolytic temperature of ≥500°C, which increased Cd sorption capacity (Figure 7.1). The pore space increased with increasing pyrolytic temperature due to the escape of volatile substances such as cellulose and hemicelluloses and the formation of channel structures during pyrolysis (Ahmad et al. 2012).

Cope et al. (2014) produced biochar from rice husks at 550°C to create a high specific surface area (SSA) rice husk biochar (RHB). The RHB was then amended with iron oxide by using dissolved ferric nitrate to provide a surface

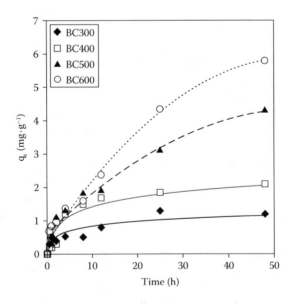

Figure 7.1 Kinetics of cadmium sorption onto biochars (1 $g \cdot L^{-1}$) produced by slow pyrolysis at 300, 400, 500, and 600°C. (From Kim, W. et al. 2013. *Bioresour. Technol.* 138: 266–270.)

chemistry conducive to As adsorption. The amended RHB's SSA was nearly 2.5 orders of magnitude higher and the arsenate [As(V)] adsorptive level was nearly 2 orders of magnitude higher than values reported for iron oxide–amended sand. Rice husks were then pyrolyzed at temperatures ranging from 450°C to 1050°C to create an even higher SSA material. The 950°C RHB was chosen as an amendment with iron oxide due to its high SSA and feasibility of being produced in the field. The sorption maximum (Q_{max}) values demonstrated that the 950°C iron oxide–amended RHB (1.46 mg·g^{-1}) significantly improved the As(V) adsorption capacity compared to 550°C iron oxide–amended RHB (1.15 mg·g^{-1}). Further study demonstrated that postamendment mesoporous volume and mesoporous surface area appear to be better indicators of As adsorptive capacity than SSA.

7.4.2.2 Chemical Characterization

Elemental ratios of O:C, O:H, and C:H have been found to provide a reliable measure of both the extent of pyrolysis and the level of oxidative alteration of biochar. Various spectroscopic techniques, such as diffuse reflectance infrared Fourier transform (FTIR) infra red spectroscopy, x-ray photoelectron spectroscopy (XPS), energy dispersive x-ray spectroscopy (EDX), and near-edge x-ray absorption fine structure spectroscopy, have been used to examine the surface chemistry of biochar in more detail (Brewer et al. 2009; Singh et al. 2014; Sohi et al. 2010). These analyses provide qualitative information that may reveal the mechanisms behind aging and functionalization of biochar. Kim et al. (2013) noticed an increase in pH and surface area at pyrolytic temperatures ≤500°C, which increased Cd sorption capacity. Ahmad et al. (2012) also reported that biochar produced at higher temperatures exhibited a higher pH. Kim et al. (2013) observed that the C content in the biochars increased from 68.48% to 90.71% as the temperature increased from 300°C to 600°C, whereas the volatile matter content decreased from 41.87% to 7.70%, mainly due to decomposition of hemicelluloses and cellulose (Lee et al. 2012). Hydrogen and oxygen content decreased as pyrolysis temperature increased, resulting in a decrease in the H:C and O:C molar ratios.

However, nitrogen content was not dependent on pyrolytic temperature. These findings suggest that the higher temperature produced more aromatic and less hydrophilic biochars, and their FTIR spectra results agree with the changes in elemental composition (Figure 7.2).

Brewer et al. (2009) characterized the biochar produced from fast pyrolysis and gasification of switchgrass and corn stover and noticed that the most dramatic change was the O–H stretch peak around 3400^{-1}, which was almost absent in the gasification char spectrum, implying the presence of high amounts of O-containing functional groups in fast pyrolysis biochars (Figure 7.3a and b). The other important peaks in the biochar spectra were the aliphatic C–H stretch (3000–2860 cm^{-1}), the aromatic C–H stretch (3060 cm^{-1}), the carboxyl (C=O) stretch (1700 cm^{-1}), and the various aromatic ring modes at 1590 and 1515 cm^{-1}. Tytłak et al. (2014) investigated the potential of biochars produced by thermal decomposition for removing Cr(VI) ions from wastewater, and XPS studies confirmed that Cr(III) ions were the most abundant Cr species on the biochars' surfaces.

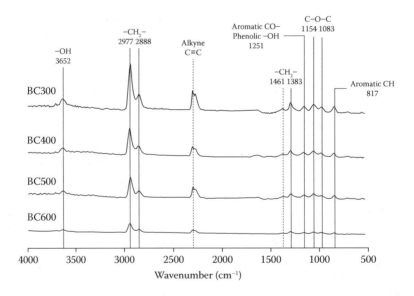

Figure 7.2 Fourier transform infrared spectra of biochars produced from *Miscanthus sacchariflorus* by slow pyrolysis at 300, 400, 500, and 600°C. (From Kim, W. et al. 2013. *Bioresour. Technol.* 138: 266–270.)

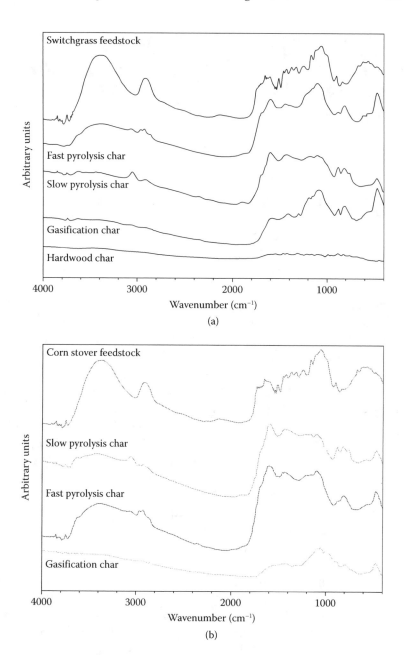

Figure 7.3 Fourier transform infrared photoacoustic spectra of (a) switchgrass feedstock and chars and a commercial hardwood char and (b) corn stover feedstock and biochars. (From Brewer, C.E. et al. 2009. *Environ. Prog. Sustain. Energy* 28: 386–396.)

7.5 Remediation of Heavy Metal(loid)s–Contaminated Wastewater by Using Biochar

7.5.1 Remediation by Using Biochars

7.5.1.1 Slow and Fast Pyrolysis Biochars

Xu et al. (2013) determined the effectiveness of dairy manure (DM) biochar through slow pyrolysis at 200 and 350°C as a sorbent in removing Cd, Cu, and Zn from aqueous solutions. DM350 biochar was more effective in sorbing all three metal(loid)s, with both biochars having the highest affinity for Cu. The maximum sorption capacities of Cu, Zn, and Cd were 48.4, 31.6, and 31.9 $mg \cdot g^{-1}$, respectively, for DM200 and 54.4, 32.8, and 51.4 $mg \cdot g^{-1}$, respectively, for DM350. Sorption of metal(loid)s by biochar was mainly attributed to their precipitation with phosphate ions or carbonate ions originating in biochar, with less to the surface complexation through hydroxyl (–OH) groups or delocalized π electrons. Tong and Xu (2013) examined the removal efficiency of Cu from an acidic electroplating effluent by biochars at 300, 400, and 500°C generated from canola, rice, soybean, and peanut straws. The biochars simultaneously removed Cu from the effluent, mainly through adsorption and precipitation, and neutralized acidity. The removal efficiency of Cu by the biochars followed the order: peanut straw char > soybean straw char > canola straw char > rice straw char, and the optimum temperature and reaction time were 400°C and 8 h, respectively.

Abdel-Fattah et al. (2014) noted that pinewood biochar via slow pyrolysis provided an effective means of Cr(VI) removal from leather tanning wastewater (76–84%) and seawater (70–83%). The biochar was rich in O-functional groups with O:C ratio 0.19 and had a relatively low (4.587 $m^2 \cdot g^{-1}$) Brunauer–Emmett–Teller surface area with the abundance of micropores. Roberts et al. (2015) used an iron-based sorbent produced from farmed seaweed (*Gracilaria*; Rhodophyta treated with ferric solution), and then converted to biochar through slow pyrolysis to remove Se from wastewater. The resulting sorbent was capable of binding both selenite [Se(IV)] and selenate [Se(VI)] from wastewater. The rate of Se(VI) sorption was minimally affected by temperature and the capacity of biosorbent for Se (i.e., Q_{max} was unaffected by pH). The Q_{max} values for the optimized iron-based biochar ranged from 2.60 to 2.72 mg Se(VI)$\cdot$g^{-1} biochar between pH 2.5 and 8.0. Biochar produced from malt spent rootlets was used for the removal of Hg(II) from pure aqueous solutions by Boutsika et al. (2014). After a 24-h contact time at biochar concentrations of 0.3 and 1 g$\cdot$L^{-1}, the Hg(II) removal was 71 and 100%, respectively. The biochar sorption capacity for Hg reached its maximum after 2 h; 33% of Hg(II) was removed within the first 5 min.

7.5.1.2 Hydrochars

Hydrochar is porous and has reactive, functionalized aromatic surfaces (Kumar et al. 2011), making hydrochar a potential alternative low-cost adsorbent to remove contaminants from water. The removal of heavy metal(loid)s by hydrochar has been suggested to be mainly controlled by interactions between metal(loid) ions in solution and O-containing functional groups on hydrochar's surface (Liu and Zhang 2009). Several studies suggested that hydrochar may contain few metal(loid)-reactive surface functional groups, such as $-OH$ and $-COOH$ groups (Liu and Zhang 2009; Sevilla et al. 2011).

Liu and Zhang (2009) measured the sorption capacity of Pb^{2+} on hydrochars and compared with pyrochars prepared from pinewood (P300) and rice husk (R300). They found that the maximum Pb adsorption capacity was ~4.25 (P300) and 2.40 (R300) mg$\cdot$g^{-1}, values that are lower than that of pyrochar. Both batch and column experiments were used in a study by Kumar et al. (2011) to assess the removal of uranium [U(VI)] from aqueous solution by switch grass hydrochar at 300°C. Although it was found that the switch grass hydrochar could be used as a reactive barrier medium to remove U(VI) from groundwater, its maximum sorption capacity was only ~2–4 mg·g^{-1} under acid or neutral pH conditions. In another study, Pellera et al. (2012) studied the adsorption of Cu from aqueous solutions by agricultural by-product chars, such as rice husks, olive pomace, and orange waste, as well as compost, by using pyrolysis (300 and 600°C) and hydrothermal treatment (300°C). Although the highest Cu adsorption capacity for rice husks and olive pomace was achieved by only pyrochars at 300°C, for orange waste and compost, the highest adsorption capacity was obtained with both hydrochars and pyrochars at 300°C.

To increase the effectiveness of hydrochar as an alternative adsorbent for water purification or remediation, it may be necessary to modify or activate the hydrochar surface to enhance its ability to remove the metal(loid)s. Xue et al. (2012) examined the effect of hydrogen peroxide (H_2O_2) treatment on hydrochar from peanut hull to remove aqueous heavy metal(loid)s (Pb^{2+}, Cu^{2+}, Ni^{2+}, and Cd^{2+}). H_2O_2 modification increased the O-containing functional groups, particularly –COOH groups, on the hydrochar surfaces. As a result, the modified hydrochar showed enhanced Pb sorption ability, with a sorption capacity of 22.82 mg·g^{-1}, a value that was comparable to that of commercial AC and was >20 times of that of untreated hydrochar (0.88 mg·g^{-1}). Model results indicated that the heavy metal(loid) removal ability of the modified hydrochar followed the order of Pb^{2+} > Cu^{2+} > Cd^{2+} > Ni^{2+} (Table 7.1). Regmi et al. (2012) used switchgrass hydrochar to remove Cu and Cd from aqueous solution. The cold activation process using potassium hydroxide at room temperature was developed to enhance the porous structure and sorption properties of the hydrochar, and its sorption efficiency was compared with AC. The activated hydrochar exhibited a higher adsorption potential for Cu and Cd than normal hydrochar and AC. At an initial metal(loid) concentration of 40 mg·L^{-1} at pH 5.0 and contact time of 24 h, 100% Cu and Cd removal by activated hydrochar at 2 g·L^{-1} was observed, a value that was much greater than normal hydrochar (16 and 5.6%) and AC (4 and 7.7%). The adsorption capacities of activated hydrochar for Cd and Cu removal were 34 and 31 mg·g^{-1}, respectively.

TABLE 7.1
Selected Literature on the Sorption Capacity of Different Biochars for Heavy Metal(loid)s Removal in Wastewater

Biomass	Biochar	Heavy Meta(loid)	Pyrolysis Temperature (°C)	Maximum Sorption Capacity		Reference
				Unmodified (mg · g⁻¹)	Modified (mg · g⁻¹)	
Sewage sludge	Hydrochar (KOH)	Cu	—	15.67	30.0	Spataru (2014)
Peanut hull	Hydrochar (H_2O_2)	Pb	300	0.88	22.82	Xue et al. (2012)
Grass straw	Magnetic: Fe_3O_4 fabricated with grass straw biochar (unmodified) Fe_3O_4 loaded on grass straw and pyrolyzed (modified)	As(V), As(III)	500	As(V): 1.42 As(III): 1.75	As(V): 3.1 As(III): 2.0	Baig et al. (2014)
Oak bark	Magnetic	Cd, Pb	450	Cd: 5.4, Pb: 13.1	Cd: 7.4, Pb: 30.2	Mohan et al. (2014b)
Pinewood	Magnetic (hematite + pinewood)	As(V)	600	265.2	428.7	Wang et al. (2015)
Pinewood	Modified [Fe(NO$_3$)$_3$]	Cr(VI)	200	31.96	53.45	Li et al. (2010)

(Continued)

TABLE 7.1 (Continued)
Selected Literature on the Sorption Capacity of Different Biochars for Heavy Metal(loid)s Removal in Wastewater

Biomass	Biochar	Heavy Metal(loid)	Pyrolysis Temperature (°C)	Maximum Sorption Capacity		Reference
				Unmodified ($mg \cdot g^{-1}$)	Modified ($mg \cdot g^{-1}$)	
Pine cone	Modified [$Zn(NO_3)_2$]	As(III)	500	0.0057	0.007	Van Vinh et al. (2014)
Municipal solid waste	Modified (KOH)	As(V)	500	24.49	30.98	Jin et al. (2014)
Brewers draff	Modified (KOH)	Cu	650	8.77	10.3	Trakal et al. (2014)
Cassava stem	Modified (KOH)	Cd	300	10.46	24.88	Prapagdee et al. (2014)
Cocoa husk	Modified (chitosan)	Cr(VI)	500	136.98	333	Okoya et al. (2014)
Rice husk	Modified (polyacrylamide hydrogel biochar composite + N,N'-methylene bisacrylamide)	As	—	—	28.32	Sanyang et al. (2014)

Fe_3O_4, magnetite; $Fe(NO_3)_3$, iron(III) nitrate; H_2O_2, hydrogen peroxide; KOH, potassium hydroxide; $Zn(NO_3)_2$, zinc nitrate.

7.5.1.3 *Magnetic Biochars*

Increasingly recognized as a multifunctional material, biochar has being explored for agricultural and environmental applications. However, powdered biochar, like powdered AC, is difficult to separate from the aqueous solution. Introducing magnetic medium (e.g., magnetite or maghemite [γ-Fe_2O_3]) to sorbents (e.g., AC and carbon nanotubes [CNTs]) by chemical coprecipitation is an efficient method to separate sorbents by magnetic separation (Šafařik et al. 1997; Zhang et al. 2007) (Table 7.1). Furthermore, the combined magnetic medium offers a potential to add the functions of bulk magnetic sorbent (Wiatrowski et al. 2009), such as the strong sorption affinity of Se (Loyo et al. 2008) and organic As (Lim et al. 2009) with magnetic iron oxide.

Zhou et al. (2014) tested a novel environmental sorbent that combines the advantages of biochar, chitosan, and zerovalent iron (ZVI). Chitosan was used as a dispersing and soldering reagent to attach fine ZVI particles onto bamboo biochar surfaces. The ZVI–chitosan–biochar composites (BBCF) showed enhanced ability to sorb Pb^{2+}, Cr(VI), and As(V) from aqueous solutions (Figure 7.4). The removal of Pb^{2+} and Cr(VI) by

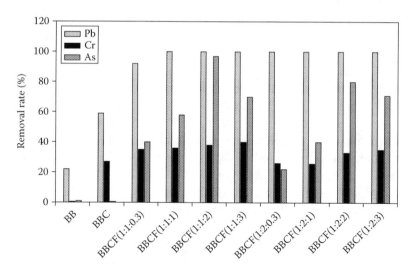

Figure 7.4 Removal of lead (Pb^{2+}), chromium [Cr(VI)], and arsenic [As(V)] from aqueous solution by bamboo biochar (BB), biochar coated with chitosan (BBC), and biochar coated with chitosan and zerovalent iron (BBCF). (From Zhou, Y. et al. 2014. *Bioresour. Technol.* 152: 538–542.)

the chitosan–biochar-supported ZVI was mainly controlled by both the reduction and surface adsorption mechanisms, and the removal of As(V) was likely controlled by electrostatic attraction with the iron particles on the BBCF surfaces. The contaminant-laden BBCF was removed from aqueous solution easily by magnetic attraction.

Mubarak et al. (2013) compared the adsorption capacity of functionalized CNTs and magnetic biochar prepared using empty fruit bunch for Zn^{2+} removal, and the maximum Zn^{2+} adsorption capacities were 1.05 and 1.18 $mg \cdot g^{-1}$ for CNTs and magnetic biochar, respectively. Zhang et al. (2013) fabricated a magnetic biochar with colloidal or nanosized γ-Fe_2O_3 particles embedded in the porous matrix via iron(III) chloride–treated cottonwood pyrolysis at 600°C. A large quantity of γ-Fe_2O_3 particles with sizes from hundreds of nanometers to several micrometers grew within the porous biochar. Its sorption capacity for As(V) removal was 3147 $mg \cdot kg^{-1}$.

7.5.1.4 Chemically Modified Biochars

A biochar/AlOOH nano-flakes nanocomposite was fabricated from aluminium chloride–pretreated biomass through slow pyrolysis, which was a highly effective adsorbent to remove As and phosphate (Zhang and Gao 2013). The Langmuir adsorption capacity of phosphate and As on the biochar/AlOOH was ~1,35,000 and ~17,410 $mg \cdot kg^{-1}$, respectively. Sanyang et al. (2014) studied hydrogel-biochar composite (HBC-RH) prepared using acrylamide as a monomer, with N,N'-methylenebisacrylamide as a cross-linker, ammonium persulfate as an initiator, and RHB for the removal of Zn from wastewater. They noticed that the maximum monolayer sorption capacity of HBC-RH for Zn was 35.75 $mg \cdot g^{-1}$. Nguyen and Lee (2014) treated biochar with nitrogen-containing surface groups (ammonia [NH_3]) to remove Cu from aqueous solutions. The alkaline modification substantially increased surface areas and the amount of nitrogen functional groups introduced onto the structure of adsorbent. Surface area (275.1 $m^2 \cdot g^{-1}$) of the biochar treated with NH_3 were much higher (14.5 ± 1.3 times) than those of the untreated biochar. They observed a maximum Cu sorption capacity of 37.5 $mg \cdot g^{-1}$ and attributed the possible adsorption mechanism to be the interactions between metal(loid) ions and nitrogen functional groups on the biochar surface.

A chemically modified biochar with abundant amino groups for the removal of Cr(VI) from aqueous solution was prepared using polyethylenimine (PEI) by Ma et al. (2014). The maximum adsorption capacity of modified biochar was 435.7 mg·g⁻¹, which was much higher than that of pristine biochar (23.09 mg·g⁻¹) (Figure 7.5). Results also indicated that the removal of Cr(VI) by the PEI-modified biochar depended on solution pH, and a low pH value was favorable for Cr(VI) removal. Biochar was modified as a high efficient and selective absorbent for Cu^{2+} ions by nitration and reduction by Yang and Jiang (2014). The results demonstrated that the amino-modified biochar exhibited excellent adsorption performance through strong complexation for Cu^{2+} as identified by XPS and FTIR spectroscopy. The adsorption capacity and bed volume of the modified biochar were five- and eightfold of the pristine biochar, respectively. Biochar/MnOx composite was successfully synthesized via potassium permanganate modification of corn straw biochar under high temperature (600°C)

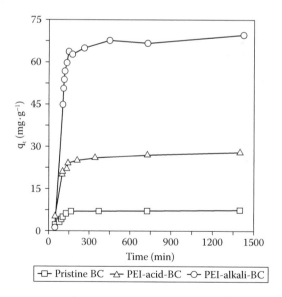

Figure 7.5 Adsorption kinetics of Cr(VI) by pristine and modified (polyethylenimine) biochars. The initial Cr(VI) concentration and content of biochar was 100 mg·L⁻¹ and 1 g·L⁻¹, respectively. (From Ma, Y. et al. 2014. *Bioresour. Technol.* 169: 403–408.)

to remove Cu^{2+} (Song et al. 2014). The MnOx-loaded biochars exhibited much higher adsorption capacity to Cu^{2+} relative to original biochar, with the maximum adsorption capacity as high as 160 mg·g^{-1}. This was mainly due to the formation of inner-sphere complexes with MnOx and O-containing groups.

7.5.2 Mechanisms of Heavy Metal(loid)s Removal

The main mechanisms of biochar interactions with metal(loid)s include electrostatic interaction, ion exchange, chemical precipitation, and complexation with functional groups on the biochar surface (Table 7.2 and Figure 7.6). Sorption may include electrostatic attraction and inner-sphere complexation with –COOH, alcoholic –OH, or phenolic –OH functional groups on the biochar surface as well as coprecipitation. For example, mechanisms for Pb sorption by a sludge-derived biochar in aqueous solution were explained by Lu et al. (2012). Four different possible mechanisms were proposed: (i) electrostatic outer-sphere complexation due to metal(loid) exchange with K^+ and Na^+ available in the biochar, (ii) coprecipitation and inner-sphere complexation of metal(loid)s with organic matter and mineral oxides of the biochar, (iii) surface complexation with active –COOH and –OH functional groups of the biochar, and (iv) precipitation as Pb–phosphate–silicate ($5PbO \cdot P_2O_5 \cdot SiO_2$). In the case of Cr, sorption on biochars has been attributed to binding with negatively charged biochar active sites after its reduction to Cr(III) due to O-containing functional groups (Dong et al. 2011). The main mechanisms for metal(loid) removal from wastewater by using biochars are detailed below.

7.5.2.1 Adsorption

Surface functional groups containing oxygen play a major role in the adsorption process by initiating the chemical bonding between adsorbent and adsorbate species. The negative charge and chemical functional groups on the biochar surface determine its capacity for metal(loid) sorption (Pan et al. 2013). The sorption efficiency of biochar depends on the polarity, surface functional groups, pore size distribution, and surface area (Kolodynska et al. 2012). The advantage of biochar over

TABLE 7.2
Selected Literature on the Effect of Biochars on Remediation of Heavy Metal(loid)s in Wastewater

Biochar	Pyrolysis Temperature (°C)	Heavy Metal(loid)	Mode of Action	Reference
Hardwood and corn straw	450, 600	Cu, Zn	Endothermic adsorption	Chen et al. (2011)
Crop straw	400	Cr(III)	Specific adsorption	Pan et al. (2013)
Soybean stalk	300–700	Hg	Precipitation, complexation, and reduction	Kong et al. (2011)
Pecan shell	800	Cu	Sorption on humic acid at pH 6; precipitation of azurite or tenorite at pH 7, 8, and 9	Ippolito et al. (2012)
Orchard pruning biomass	500	Pb, Cr	Surface electrostatic interaction and surface complexation	Caporale et al. (2014)
Aromatic spent (eucalptus [*Eucalyptus citriodora*] and rose [*Rosa damascena*])	450	Cd, Cr, Cu, Pb	Surface adsorption followed by intraparticle diffusion	Khare et al. (2013)
Sugar beet tailing	300	Cr	Electrostatic attraction; reduction of Cr(VI) to Cr(III); complexation	Dong et al. (2011)
Sugarcane bagasse	600	Pb	Surface adsorption and precipitation	Inyang et al. (2011)
Sugarcane bagasse	250–600	Pb	Complexation, precipitation, and intraparticle diffusion	Ding et al. (2014)
Dairy manure	200, 350	Cd, Cu, Zn	Precipitation and surface complexation	Xu et al. (2013)
Anaerobically digested garden waste	400	Cu, Zn	Chemisorption	Zhang and Luo (2014)
Sewage sludge	550	Pb	Adsorption due to cation release, functional groups complexation, and surface precipitation	Lu et al. (2012)

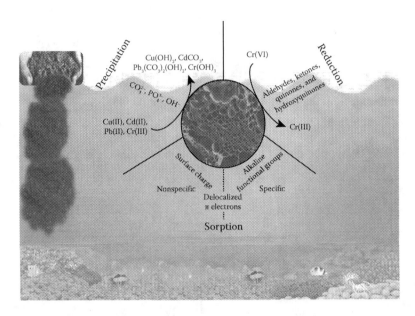

Figure 7.6 Schematic representation of mechanisms involved in metal(loid)s removal from wastewater by biochar.

other biosorbents is the stable form of C structure that makes them hard to degrade. Zhang and Luo (2014) noticed that the kinetic adsorption characteristics that were well described by the pseudo-second-order model indicated that chemisorption mechanism controls the metal(loid) adsorption onto anaerobically digested garden waste biochar. Similarly, Hu et al. (2015) suggested that As sorption on the Fe-impregnated hickory chip biochar was mainly controlled by chemisorption.

Samsuri et al. (2013) suggested that the mechanism responsible for the adsorption of As(III) and As(V) could be the formation of surface complexes between the functional groups of biochars produced from empty fruit bunch and rice husk, and As(III) and As(V). Copper adsorption by biochars prepared from three crop straws at 400°C was investigated by Tong et al. (2011) under acidic conditions. The adsorption capacity followed the order: peanut straw char > soybean straw char > canola straw char, whereas desorption of preadsorbed Cu^{2+} followed a reverse trend. The more negative surface charge on biochars from canola straw led to more electrostatic adsorption of Cu^{2+} compared to the other two biochars (Figure 7.7a).

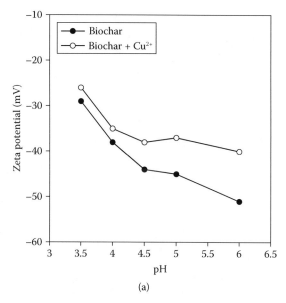

(a)

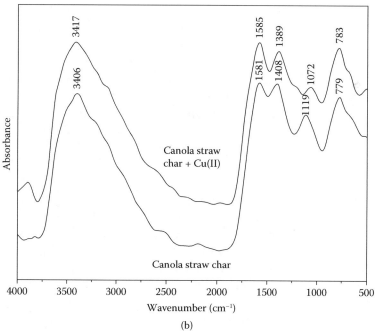

(b)

Figure 7.7 Zeta potential (a) and Fourier transform infrared photoacoustic spectra (b) of canola straw biochar before and after copper (Cu^{2+}) adsorption. (From Tong, S.J. et al. 2011. *Chem. Eng. J.* 172: 828–834.)

The FTIR spectroscopy data revealed that adsorption of Cu^{2+} caused an apparent shift of the vibrational bands assigned to the –COOH and phenolic –OH groups along with less negative zeta potential of the biochars that adsorbed Cu^{2+}, which suggested that the Cu^{2+} was adsorbed specifically through the formation of surface complexes (Figure 7.7b). Dong et al. (2013) observed that the irreversible sorption of Hg was via complexation with phenolic –OH and –COOH functional groups in low-temperature Brazilian pepper biochars (BP300 and BP450) and graphite-like structure in high-temperature biochar (BP600).

7.5.2.2 Precipitation

Inyang et al. (2012) found that the removal of Pb from aqueous solution by anaerobically digested dairy waste and whole sugar beet biochars was mainly through surface precipitation as $PbCO_3$ (cerrusite), $Pb_3(CO_3)_2(OH)_2$ (hydrocerrusite), and $Pb_5(PO_4)_3Cl$ (pyromorphite), which was confirmed by scanning electron microscopy-energy-dispersive x-ray spectroscopy, x-ray powder diffraction, and FTIR spectroscopy measurements. A similar phenomenon was observed for Pb removal by biochars made from anaerobically digested bagasse in which precipitation as hydrocerrusite was shown to be the dominant sorption mechanism (Inyang et al. 2011). Previous studies have demonstrated that slow release of negatively charged ions, such as carbonate and phosphate, from biochars can precipitate metal(loid) ions, particularly Pb (Cao et al. 2009; Inyang et al. 2011).

Agrafioti et al. (2014a) investigated the removal of As(V) and Cr(VI) from water by using biochars derived from rice husk, sewage sludge, and solid wastes. The results indicated that biochar derived from sewage sludge was the most efficient and removed 89% of Cr(VI) and 53% of As(V). They associated this to the biochar's high Fe_2O_3 content in ash, causing enhanced metal(loid) adsorption via precipitation. Agrafioti et al. (2014b) conducted another similar study for the removal of As and Cr by biochar, but modified the biochar (rice husk) with Ca and Fe (CaO, Fe^0, and Fe^{3+}). The modified biochars exhibited high As(V) removal capacity (>95%), except for rice husk impregnated with Fe^0, whose removal capacity reached only 58%.

Although all modified biochars exhibited much better As(V) removal capacity compared to the nonimpregnated biochars, the Cr(VI) removal rates were not as high as the As(V) rates. The results suggested that the main mechanisms of As(V) and Cr(VI) removal were possibly metal(loid) precipitation and electrostatic interactions between the modified biochars and the adsorbate. The precipitation of Cr(III) hydroxide and the formation of Cr^{3+} surface complexes with the functional groups on biochar were the main mechanisms for Cr(III) removal as observed in a study by Pan et al. (2014).

7.5.2.3 Reduction

It has been considered that the most dominant mechanism of Cr(VI) removal is the surface reduction of Cr(VI) to Cr(III), followed by adsorption of Cr(III) (Saha and Orvig 2010). Based on Saha and Orvig (2010), there are four potential mechanisms explaining Cr(VI) sorption by biosorbents: (i) anionic adsorption, (ii) adsorption-coupled reduction, (iii) anionic and cationic adsorption, and (iv) reduction and cationic adsorption. When the solution pH is between 7 and 9.5, Cr(VI) reduction is unlikely because Cr(VI) gets reduced at low pH values.

Park et al. (2006) proposed that adsorption coupled with reduction is the main mechanism of Cr(VI) removal by biomaterials. Due to its high redox potential value (+1.3 V), Cr(VI) is easily reduced to Cr(III) under acid conditions in the presence of organic matter. Dong et al. (2011) studied the removal of Cr(VI) from aqueous solutions by using sugar beet tailing biochar. They hypothesized that sugar beet biochar effectively removed Cr(VI) via electrostatic attraction of Cr(VI) coupled with Cr(VI) reduction to Cr(III) and Cr(III) complexation. Initially, under strongly acidic condition, the negatively charged Cr(VI) species migrated to the positively charged surfaces of biochar (protonated −COOH, alcohol and −OH groups) with the help of electrostatic driving forces; with the participation of hydrogen ions and the electron donors from biochar, Cr(VI) was then reduced to Cr(III); and finally, part of the converted Cr(III) was released to the medium, with the rest being complexed with functional groups on the sugar beet biochar (Figure 7.8). The potential of two biochars produced by

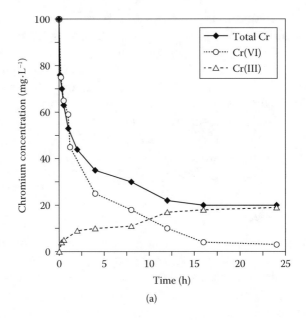

(a)

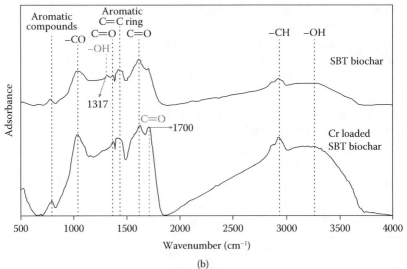

(b)

Figure 7.8 (a) Effect of reaction time on Cr removal by sugar beet biochar (SBT) and (b) Fourier transform infrared spectra of SBT biochar before and after reaction with 100 mg·L^{-1} Cr(VI) for 24 h at pH 2.0. (From Dong, X. et al. 2011. *J. Hazard. Mater.* 190: 909–915.)

the thermal decomposition of wheat straw (BCS) and wicker (BCW) for the removal of Cr(VI) ions in wastewater was investigated by Tytłak et al. (2014). The optimal adsorption capacities were obtained at pH 2 and were 24.6 and 23.6 $mg \cdot g^{-1}$ for BCS and BCW, respectively. The results indicated that the sorption mechanism of Cr(VI) on biochar involves anionic and cationic adsorption combined with Cr(VI) species reduction.

7.6 Conclusions and Future Research Needs

Municipal wastewater is increasingly being used as a valuable resource for irrigation in urban and peri-urban agriculture because of its availability and thus has partially solving the problem of effluent disposal. Although the wastewater-borne nutrients reduce the need for inputs of mineral fertilizers or manures and thus reduce production costs, pathogens and contaminants, including heavy metal(loid)s, in wastewater pose a health risk for the producers and consumers of the crops. Most of the wastewater purification methods currently in use, including precipitation, membrane filtration, and flotation, require high operating costs and often generate chemical sludge that itself is a disposal problem. Therefore, adsorption has become one of the best alternative treatment techniques for wastewater laden with heavy metal(loid)s. Although AC has undoubtedly been the most popular and widely used adsorbent in wastewater treatment, it remains an expensive material because the higher the quality of AC, the greater its cost. Therefore, the feasibility of using various low-cost, locally available adsorbents derived from agricultural waste or industrial by-products has been investigated for the removal of metal(loid) ions from wastewater.

Using biochar as a low-cost sorbent is an emerging technology with promising potential to remove metal(loid)s from aqueous solutions (Ahmad et al. 2014; Mohan et al. 2014a; Uchimiya et al. 2011). Biochars converted from agricultural residues, animal wastes, and woody materials have been tested for their ability to sorb various metal(loid)s, including As, Pb, Cu, Cr, and Cd (Agrafioti et al. 2014a, 2014b; Cao et al. 2009;

Dong et al. 2011; Mohan et al. 2014b; Uchimiya et al. 2011). The adsorption capacity of biochar differs with physicochemical properties such as pH, surface area, and functional groups that depend on biomass and production methods (Kolodynska et al. 2012). Biochar converted from anaerobically digested biomass could be used as a means of biological activation to make high-quality biochar-based sorbents. Also, chemically modified biochars are gaining attention because of their high sorption capacity compared to unmodified biochars. Recently explored hydrochars are eco-friendly because they do not generate any hazardous chemical wastes or by-products as do the dry pyrolysis biochars. Magnetic biochar is a potential metal(loid) sorbent and is advantageous because magnetic materials and amorphous biochar can be easily recollected by a magnet in aqueous solution. Few recent studies showed that Fe-treated biochar and raw biochar produced from macroalgae are effective biosorbents of metalloids and metals, respectively (Kidgell et al. 2014a; Roberts et al. 2015). However, the treatment of complex effluents that contain both metalloid and metal contaminants presents a challenging scenario. Kidgell et al. (2014b) tested a multiple-biosorbent (*Oedogonium* sp.) approach to bioremediation by using Fe-biochar and biochar to remediate both metalloids and metals in effluent from a coal-fired power station. The most effective treatment was the sequential use of Fe-biochar to remove metalloids from the wastewater, followed by biochar to remove metals.

All forms of biochars, that is, in raw form prepared through a dry (pyrochars) or a hydrothermal (hydrochars) method, modified or activated by chemicals and magnetic materials, are proving to be highly efficient in the removal of heavy metal(loid)s from wastewater. However, more studies are warranted, and future biochar research should address the following:

- Improved energy efficiency and emission control
- Higher biochar yields
- Competitive sorption of metal(loid)s should be taken into account before choosing the biochar for treating mixed wastewaters
- More activation techniques for biochars and algal-based biochars should be evaluated, with a range of

different feedstocks for their potential to sorb various heavy metal(loid)s

- The produced biochar should only contain a limited and acceptable amount of heavy metal(loid)s, allowing its approval and reuse
- Studies related to potential risk of biochars to aquatic organisms should be conducted and addressed

Acknowledgments

Postdoctoral fellowship program PJ009828 at the National Academy of Agricultural Science, Rural Development Administration, Republic of Korea, supported Dr. Kunhikrishnan. Dr. Nabeel Niazi thanks the Grand Challenges Canada–Stars in Global Health, Round 5 (0433-01) for the financial support at the University of Agriculture Faisalabad (Pakistan) and the Australian Research Council for supporting the postdoctoral research fellowship at the Southern Cross University through ARC Discovery Project DP140100012. Dr. Irshad Bibi was funded by the Endeavour Postdoctoral Research Fellowship through the Australian Government at the Southern Cross University and by the University of Agriculture, Faisalabad. Dr. Balaji Seshadri was financially supported by Cooperative Research Centre for Contamination Assessment and Remediation of the Environment (CRC CARE), Australia, in collaboration with University of South Australia. Dr. Girish Choppala would like to thank Southern Cross GeoScience (http://scu.edu.au/geoscience/) for providing resources to write this chapter.

References

Abdalla, M.O.M., Hassabo, A.A., and Elsheikh, N.A.H. 2013. Assessment of some heavy metals in waste water and milk of animals grazed around sugarcane plants in Sudan. *Livest. Res. Rural Dev.* 25. http://www.lrrd.org/lrrd25/12/abda25212.htm.

Abdel-Fattah, T.M., Mahmoud, M.E., Ahmed, S.B., Huff, M.D., Lee, J.W., and Kumar, S. 2014. Biochar from woody biomass for removing metal contaminants and carbon sequestration. *J. Ind. Eng. Chem.* 25: 103–109. doi: http://dx.doi.org/10.1016/j.jiec.2014.06.030.

Abe, K., Waki, M., Suzuki, K., Kasuya, M., Suzuki, R., Itahashi, S., and Banzai, K. 2012. Estimation of Zn and Cu unit output loads from animal husbandry facilities. *Wat. Sci. Technol.* 66: 653–658.

Abo-Farha, S.A., Abdel-Aal, A.Y., Ashourb, I.A., and Garamon, S.E. 2009. Removal of some heavy metal cations by synthetic resin purolite C100. *J. Hazard. Mater.* 169: 190–194.

Adriano, D.C. 2001. *Trace Elements in Terrestrial Environments. Biogeochemistry, Bioavailability and Risks of Metals*, 2nd ed. New York: Springer.

Agrafioti, E., Kalderis, D., and Diamadopoulos, E. 2014a. Arsenic and chromium removal from water using biochars derived from rice husk, organic solid wastes and sewage sludge. *J. Environ. Manag.* 133: 309–314.

Agrafioti, E., Kalderis, D., and Diamadopoulos, E. 2014b. Ca and Fe modified biochars as adsorbents of arsenic and chromium in aqueous solutions. *J. Environ. Manag.* 146: 444–450.

Ahmad, M., Lee, S.S., Dou, X., Mohan, D., Sung, J.K., Yang, J.E., and Ok, Y.S. 2012. Effects of pyrolysis temperature on soybean stover- and peanut shell-derived biochar properties and TCE adsorption in water. *Bioresour. Technol.* 118: 536–544.

Ahmad, A.L. and Ooi, B.S. 2010. A study on acid reclamation and copper recovery using low pressure nanofiltration membrane. *Chem. Eng. J.* 56: 257–263.

Ahmad, M., Rajapaksha, A.U., Lim, J.E., Zhang, M., Bolan, N.S., Mohan, D., Vithanage, M., Lee, S.S., and Ok, Y.S. 2014. Biochar as a sorbent for contaminant management in soil and water: A review. *Chemosphere* 99: 19–33.

Ajjabi, L.C., and Chouba, L. 2009. Biosorption of Cu^{2+} and Zn^{2+} from aqueous solutions by dried marine green macroalga *Chaetomorpha linum*. *J. Environ. Manag.* 90: 3485–3489.

Aksu, Z., and Isoglu, I.A. 2005. Removal of copper(II) ions from aqueous solution by biosorption onto agricultural waste sugar beet pulp. *Process Biochem.* 40: 3031–3044.

Aman, T., Kazi, A.A., Sabri, M.U., and Bano, Q. 2008. Potato peels as solid waste for the removal of heavy metal copper(II) from waste water/industrial effluent. *Colloids Surf. B Biointerfaces* 63: 116–121.

Anderson, J. 2003. The environmental benefits of water recycling and reuse. *Wat. Sci. Technol.* 3: 1–10.

Aroua, M.K., Zuki, F.M., and Sulaiman, N.M. 2007. Removal of chromium ions from aqueous solutions by polymer-enhanced ultrafiltration. *J. Hazard. Mater.* 147: 752–758.

Aryal, R., Vigneswaran, S., Kandasamy, J., and Naidu, R. 2010. Urban stormwater quality and treatment. *Kor. J. Chem. Eng.* 27: 1343–1359.

Asano, T., and Cotruvo, J.A. 2004. Groundwater recharge with reclaimed municipal wastewater: Health and regulatory considerations. *Water Res.* 38: 1941–1951.

Ashworth, D.J., and Alloway, B.J. 2004. Soil mobility of sewage sludge-derived dissolved organic matter, copper, nickel and zinc. *Environ. Pollut.* 127: 137–144.

Babel, S., and Kurniawan, T.A. 2003. Low-cost adsorbents for heavy metals uptake from contaminated water: A review. *J. Hazard. Mater.* 97: 219–243.

Baig, S.A., Zhu, J., Muhammad, N., Sheng, T., and Xu, X. 2014. Effect of synthesis methods on magnetic Kans grass biochar for enhanced As(III, V) adsorption from aqueous solutions. *Biomass Bioenerg.* 71: 299–310.

Baker, H., Massadeh, A., and Younes, H. 2009. Natural Jordanian zeolite: Removal of heavy metal ions from water samples using column and batch methods. *Environ. Monit. Assess.* 157: 319–330.

Baldwin, R.M., Magrini-Bair, K.A., Nimlos, M.R., Pepiot, P., Donohoe, B.S., Hensley, J.E., and Phillips, S.D. 2012. Current research on thermochemical conversion of biomass at the National Renewable Energy Laboratory. *Appl. Catal. B* 115–116: 320–329.

Barakat, M.A. 2011. New trends in removing heavy metals from industrial wastewater. *Arabian J. Chem.* 4: 361–377.

Bhainsa, K.C., and D'Souza, S.F. 2008. Removal of copper ions by the filamentous fungus, *Rhizopus oryzae* from aqueous solution. *Bioresour. Technol.* 99: 3829–3835.

Bolan, N.S., Adriano, D.C., Kunhikrishnan, A., James, T., McDowell, R., and Senesi, N. 2011. Dissolved organic matter: Biogeochemistry, dynamics and environmental significance in soils. *Adv. Agron.* 110: 1–75.

Bolan, N.S., Adriano, D.C., and Mahimairaja, S. 2004b. Distribution and bioavailability of trace elements in livestock and poultry manure by-products. *Crit. Rev. Environ. Sci. Technol.* 34: 291–338.

Bolan, N.S., Kunhikrishnan, A., Thangarajan, R., Kumpiene, J., Park, J.H., Makino, T., Kirkham, M.B., and Schekel, K. 2014. Remediation of heavy metal(loid)s contaminated soils—To mobilize or to immobilize? *J. Hazard. Mater.* 266: 141–166.

Bolan, N.S., Laurenson, S., Luo, J., and Sukias, J. 2009. Integrated treatment of farm effluents in New Zealand's dairy operations. *Bioresour. Technol.* 100: 5490–5497.

Bolan, N.S., Wong, L., and Adriano, D.C. 2004a. Nutrient removal from farm effluents. *Bioresour. Technol.* 94: 251–260.

Boutsika, L.G., Karapanagioti, H.K., and Manariotis, I.D. 2014. Aqueous mercury sorption by biochar from malt spent rootlets. *Water Air Soil Pollut.* 225: 1805–1814.

Bratskaya, S.Y., Pestov, A.V., Yatluk, Y.G., and Avramenko, V.A. 2009. Heavy metals removal by flocculation/precipitation using N-(2-carboxyethyl) chitosans. *Colloids Surf. A* 339: 140–144.

Brewer, C.E., Schmidt-Rohr, K., Satrio, J.A., and Brown, R.C. 2009. Characterization of biochar from fast pyrolysis and gasification systems. *Environ. Prog. Sustain. Energy* 28: 386–396.

Bruun, E.W., Ambus, P., Egsgaard, H., and Hauggaard-Nielsen, H. 2012. Effects of slow and fast pyrolysis biochar on soil C and N turnover dynamics. *Soil Biol. Biochem.* 46: 73–79.

Cao, X.D., Ma, L.N., Gao, B., and Harris, W. 2009. Dairy-manure derived biochar effectively sorbs lead and atrazine. *Environ. Sci. Technol.* 43: 3285–3291.

Caporale, A.G., Pigna, M., Sommella, A., and Conte, P. 2014. Effect of pruning-derived biochar on heavy metals removal and water dynamics. *Biol. Fert. Soils* 50: 1211–1222.

Chang, Q., and Wang, G. 2007. Study on the macromolecular coagulant PEX which traps heavy metals. *Chem. Eng. Sci.* 62: 4636–4643.

Chang, Q., Zhang, M., and Wang, J.X. 2009. Removal of Cu^{2+} and turbidity from wastewater by mercaptoacetyl chitosan. *J. Hazard. Mater.* 169: 621–625.

Chen, X., Chen, G., Chen, L., Chen, Y., Lehmann, J., McBride, M.B., and Hay, A.G. 2011. Adsorption of copper and zinc by biochars produced from pyrolysis of hardwood and corn straw in aqueous solution. *Bioresour. Technol.* 102: 8877–8884.

Cifuentes, L., García, I., Arriagada, P., and Casas, J.M. 2009. The use of electrodialysis for metal separation and water recovery from $CuSO_4$-H_2SO_4-Fe solutions. *Sep. Purif. Technol.* 68: 105–108.

Cope, C.O., Webster, D.S., and Sabatini, D.A. 2014. Arsenate adsorption onto iron oxide amended rice husk char. *Sci. Total Environ.* 488–489: 554–561.

Day, D., Evans, R.J., Lee, J.W., and Reicosky, D. 2005. Economical CO_2, SOx, and NOx capture from fossil–fuel utilization with combined renewable hydrogen production and large-scale carbon sequestration. *Energy* 30: 2558–2579.

Demirbas, A. 2008. Heavy metal adsorption onto agro-based waste materials: A review. *J. Hazard. Mater.* 157: 220–229.

Ding, W., Dong, X., Ime, I.M., Gao, B., and Ma, L.Q. 2014. Pyrolytic temperatures impact lead sorption mechanisms by bagasse biochars. *Chemosphere* 105: 68–74.

Dong, X., Ma, L.Q., and Li, Y. 2011. Characteristics and mechanisms of hexavalent chromium removal by biochar from sugar beet tailing. *J. Hazard. Mater.* 190: 909–915.

Dong, X., Ma, L.Q., Zhu, Y., Li, Y., and Gu, B. 2013. Mechanistic investigation of mercury sorption by Brazilian pepper biochars of different pyrolytic temperatures based on X-ray photoelectron spectroscopy and flow calorimetry. *Environ. Sci. Technol.* 47: 12156–12164.

Doula, M.K. 2009. Simultaneous removal of Cu, Mn and Zn from drinking water with the use of clinoptilolite and its Fe-modified form. *Water Res.* 43: 3659–3672.

Drechsel, P., Scott, C.A., Raschid-Sally, L., Redwood, M., and Bahri, A. 2010. *Wastewater Irrigation and Health: Assessing and Mitigation Risks in Low-Income Countries.* UK: Earthscan-IDRC-IWMI. www.idrc.ca/en/ev-149129-201-1-DO_TOPIC.html.

Duan, J.C., Lu, Q., Chen, R.W., Duan, Y.Q., Wang, L.F., Gao, L., and Pan, S.Y. 2010. Synthesis of a novel flocculant on the basis of crosslinked Konjac glucomannan-graftpolyacrylamide-co-sodium xanthate and its application in removal of Cu^{2+} ion. *Carbohydr. Polym.* 80: 436–441.

El-Ashtoukhy, E.S.Z., Amin, N.K., and Abdelwahab, O. 2008. Removal of lead (II) and copper (II) from aqueous solution using pomegranate peel as a new adsorbent. *Desalination* 223: 162–173.

Eriksson, E., and Donner, E. 2009. Metals in greywater: Sources, presence and removal efficiencies. *Desalination* 248: 271–278.

Figoli, A., Cassano, A., Criscuoli, A., Mozumder, M.S.I., Uddin, M.T., Islam, D.A., and Drioli, E. 2010. Influence of operating parameters on the arsenic removal by nanofiltration. *Water Res.* 44: 97–104.

Fu, F., and Wang, Q. 2011. Removal of heavy metal ions from wastewaters: A review. *J Environ. Manag.* 92: 407–418.

Funke, A., and Ziegler, F. 2010. Hydrothermal carbonization of biomass: A summary and discussion of chemical mechanisms for process engineering. *Biofuels Bioprod. Biorefin.* 4: 160–177.

Gaete, H., Larrain, A., Bay-Schmith, E., Baeza, J., and Rodriguez, J. 2000. Ecotoxicological assessment of two pulp mill effluents, Biobio river basin Chile. *Bull. Environ. Contam. Toxicol.* 65: 183–189.

Ghani, W.A.W.A.K., Mohd, A., da Silva, G., Bachmann, R.T., Taufiq-Yap, Y.H., Rashid, U., and Al-Muhtaseb, A.H. 2013. Biochar production from waste rubber-wood-sawdust and its potential use in C sequestration: Chemical and physical characterization. *Ind. Crops Prod.* 44: 18–24.

Guo, G.L., and Zhou, Q.X. 2006. Evaluation of heavy metal contamination in Phaeozem of northeast China. *Environ. Geochem. Health.* 28: 331–340.

Guo, M.X., Qiu, G.N., and Song, W.P. 2010. Poultry litter-based activated carbon for removing heavy metal ions in water. *Waste Manag.* 30: 308–315.

Hill, H.J., Grigsby, W.J., and Hall, P.W. 2013. Chemical and cellulose crystallite changes in *Pinus radiata* during torrefaction. *Biomass Bioenerg.* 56: 92–98.

Hu, X., Ding, Z., Zimmerman, A.R., Wang, S., and Gao, B. 2015. Batch and column sorption of arsenic onto iron-impregnated biochar synthesized through hydrolysis. *Water Res.* 68: 206–216.

Hu, B.B., Wang, K., Wu, L., Yu, S.H., Antonietti, M., and Titirici, M.M. 2010. Engineering carbon materials from the hydrothermal carbonization process of biomass. *Adv. Mater.* 22: 813–828.

Huisman, J.L., Schouten, G., and Schultz, C. 2006. Biologically produced sulphide for purification of process streams, effluent treatment and recovery of metals in the metal and mining industry. *Hydrometallurgy* 83: 106–113.

Imamoglu, M., and Tekir, O. 2008. Removal of copper (II) and lead (II) ions from aqueous solutions by adsorption on activated carbon from a new precursor hazelnut husks. *Desalination* 228: 108–113.

Inyang, M., Gao, B., Ding, W., Pullammanappallil, P., Zimmerman, A.R., and Cao, X. 2011. Enhanced lead sorption by biochar derived from anaerobically digested sugarcane bagasse. *Sep. Sci. Technol.* 46: 1950–1956.

Inyang, M., Gao, B., Yao, Y., Xue, Y.W., Zimmerman, A.R., Pullammanappallil, P., and Cao, X.D. 2012. Removal of heavy metals from aqueous solution by biochars derived from anaerobically digested biomass. *Bioresour. Technol.* 110: 50–56.

Ippolito, J.A., Strawn, D.G., Scheckel, K.G., Novak, J.M., Ahmedna, M., and Niandou, M.A.S. 2012. Macroscopic and molecular investigations of copper sorption by a steam-activated biochar. *J. Environ. Qual.* 41: 150–156.

Jiang, M.Q., Jin, X.Y., Lu, X.Q., and Chen, Z.L. 2010. Adsorption of Pb(II), Cd(II), Ni(II) and Cu(II) onto natural kaolinite clay. *Desalination* 252: 33–39.

Jin, H., Capareda, S., Chang, Z., Gao, J., Xu, Y., and Zhang, J. 2014. Biochar pyrolytically produced from municipal solid wastes for aqueous As(V) removal: Adsorption property and its improvement with KOH activation. *Bioresour. Technol.* 169: 622–629.

Joseph, S., Peacocke, C., Lehmann, J., and Munroe, P. 2009. Developing a biochar classification and test methods. In: *Biochar for Environmental Management Science and Technology*, J. Lehmann, and S. Joseph (eds.). London: Earthscan, pp. 107–112.

Kannan, K., and Oblisami, G. 1990. Effect of pulp and paper mill effluent irrigation on carbon dioxide evolution in soils. *J. Agron. Crop Sci.* 164: 116–119.

Kapungwe, E.M. 2013. Heavy metal contaminated water, soils and crops in peri urban wastewater irrigation farming in Mufulira and Kafue towns in Zambia. *J. Geogr. Geol.* 5: 55–72.

Khare, P., Dilshad, U., Rout, P.K., Yadav, V., and Jain, S. 2013. Plant refuses driven biochar: Application as metal adsorbent from acidic solutions. *Arabian J. Chem.* doi: http://dx.doi.org/10.1016/j.arabjc.2013.11.047.

Kidgell, J.T., de Nys, R., Hu, Y., Paul, N.A., and Roberts, D.A. 2014a. Bioremediation of a complex industrial effluent by biosorbents derived from freshwater macroalgae. *PLoS One* 9: e94706. doi: 10.1371/journal.pone.0094706.

Kidgell, J.T., de Nys, R., Paul, N.A., and Roberts, D.A. 2014b. The sequential application of macroalgal biosorbents for the bioremediation of a complex industrial effluent. *PLoS One* 9: e101309. doi: 10.1371/journal.pone.0101309.

Kim, W., Shim, T., Kim, Y.S., Hyun, S., Ryu, C., Park, Y.K., and Jung, J. 2013. Characterization of cadmium removal from aqueous solution by biochar produced from a giant *Miscanthus* at different pyrolytic temperatures. *Bioresour. Technol.* 138: 266–270.

Kolodynska, D., Wnetrzak, R., Leahy, J.J., Hayes, M.H.B., Kwapinski, W., and Hubicki, Z. 2012. Kinetic and adsorptive characterization of biochar in metal ions removal. *Chem. Eng. J.* 197: 295–305.

Kong, H., He, J., Gao, Y., Wu, H., and Zhu, X. 2011. Cosorption of phenanthrene and mercury(II) from aqueous solution by soybean stalk-based biochar. *J. Agr. Food Chem.* 59: 12116–12123.

Kongsuwan, A., Patnukao, P., and Pavasant, P. 2009. Binary component sorption of Cu(II) and Pb(II) with activated carbon from *Eucalyptus camaldulensis* Dehn bark. *J. Ind. Eng. Chem.* 15: 465–470.

Kumar, S., Loganathan, V.A., Gupta, R.B., and Barnett, M.O. 2011. An assessment of U(VI) removal from groundwater using biochar produced from hydrothermal carbonization. *J. Environ. Manag.* 92: 2504–2512.

Kunhikrishnan, A., Bolan, N.S., Müller, K., Laurenson, S., Naidu, R., and Kim, W.I. 2012. The influence of wastewater irrigation on the transformation and bioavailability of heavy metal(loid)s in soil. *Adv. Agron.* 115: 215–297.

Lakherwal, D. 2014. Adsorption of heavy metals: A review. *Int. J. Environ. Res. Dev.* 4: 41–48.

Lambert, J., Avila-Rodriguez, M., Durand, G., and Rakib, M. 2006. Separation of sodium ions from trivalent chromium by electrodialysis using monovalent cation selective membranes. *J. Memb. Sci.* 280: 219–225.

Landaburu-Aguirre, J., García, V., Pongrácz, E., and Keiski, R.L. 2009. The removal of zinc from synthetic wastewaters by micellar-enhanced ultrafiltration: Statistical design of experiments. *Desalination* 240: 262–269.

Laurenson, S., Kunhikrishnan, A., Bolan, N.S., Naidu, R., McKay, J., and Keremane, G. 2010. Management of recycled water for sustainable production and environmental protection: A case study with Northern Adelaide Plains recycling scheme. *Int. J. Environ. Res. Dev.* 1: 176–180.

Lee, Y., Eum, P.R.B., Ryu, C., Park, Y.K., Jung, J., and Hyun, S. 2012. Characteristics of biochar produced from slow pyrolysis of Geodae-Uksae 1. *Bioresour. Technol.* 130: 345–350.

Lehmann, J., Gaunt, J., and Rondon, M. 2006. Bio-char sequestration in terrestrial ecosystems—A review. *Mitig. Adapt. Strategies Glob. Change* 11: 403–427.

Li, Z., Katsumi, T., Inui, T., and Imaizumi, S. 2010. Woods charred at low temperatures and their modification for the adsorption of Cr(VI) ions from aqueous solution. *Adsorpt. Sci. Technol.* 28: 419–435.

Libra, J.A., Ro, K.S., Kammann, C., Funke, A., Berge, N.D., Neubauer, Y., Titirici, M.M., Führner, C., Bens, O., Kern, J., and Emmerich, K.H. 2011. Hydrothermal carbonization of biomass residuals: A comparative review of the chemistry, processes and applications of wet and dry pyrolysis. *Biofuels* 2: 71–106.

Lim, S.F., Zheng, Y.M., and Chen, J.P. 2009. Organic arsenic adsorption onto a magnetic sorbent. *Langmuir* 25: 4973–4978.

Lima, I.M., Boateng, A.A., and Klasson, K.T. 2010. Physicochemical and adsorptive properties of fast-pyrolysis bio-chars and their steam activated counterparts. *J. Chem. Technol. Biotechnol.* 85: 1515–1521.

Lipinsky, E.S., Arcate, J.R., and Reed, T.B. 2002. Enhanced wood fuels via torrefaction. *Fuel Chem. Div. Preprints* 47: 408–410.

Liu, Z., and Zhang, F. 2009. Removal of lead from water using bio-chars prepared from hydrothermal liquefaction of biomass. *J. Hazard. Mater.* 167: 933–939.

Liu, Z., and Zhang, F.S. 2011. Removal of copper (II) and phenol from aqueous solution using porous carbons derived from hydrothermal chars. *Desalination* 267: 101–106.

Loyo, R.L., Nikitenko, S.I., Scheinost, A.C., and Simonoff, M. 2008. Immobilization of selenite on Fe_3O_4 and Fe/Fe_3C ultra small particles. *Environ. Sci. Technol.* 42: 2451–2456.

Lu, H., Zhang, W., Yang, Y., Huang, X., Wang, S., and Qiu, R. 2012. Relative distribution of Pb^{2+} sorption mechanisms by sludge-derived biochar. *Water Res.* 46: 854–862.

Ma, Y., Liu, W.J., Zhang, N., Li, Y.S., Jiang, H., and Sheng, G.P. 2014. Polyethylenimine modified biochar adsorbent for hexavalent chromium removal from the aqueous solution. *Bioresour. Technol.* 169: 403–408.

Medina, B.Y., Torem, M.L., and de Mesquita, L.M.S. 2005. On the kinetics of precipitate flotation of Cr III using sodium dodecylsulfate and ethanol. *Miner. Eng.* 18: 225–231.

Mirbagheri, S.A., and Hosseini, S.N. 2005. Pilot plant investigation on petrochemical wastewater treatment for the removal of copper and chromium with the objective of reuse. *Desalination* 171: 85–93.

Mishra, S., Mohanty, M., Pradhan, C., Patra, H.K., Das, R., and Sahoo, S. 2013. Physico-chemical assessment of paper mill effluent and its heavy metal remediation using aquatic macrophytes—A case study at JK Paper mill, Rayagada, India. *Environ. Monit. Assess.* 185: 4347–4359.

Mohan, D., Kumar, H., Sarswat, A., Alexandre-Franco, M., and Pittman, C.U., Jr. 2014b. Cadmium and lead remediation using magnetic oak wood and oak bark fast pyrolysis bio-chars. *Chem. Eng. J.* 236: 513–528.

Mohan, D., Pittman, C.U., Bricka, M., Smith, F., Yancey, B., Mohammad, J., Steele, P.H., Alexandre-Franco, M.F., Gomez-Serrano, V., and Gong, H. 2007. Sorption of arsenic, cadmium, and lead by chars produced from fast pyrolysis of wood and bark during bio-oil production. *J. Colloid Interface Sci.* 310: 57–73.

Mohan, D., Sarswat, A., Ok, Y.S., and Pittman, C.U., Jr. 2014a. Organic and inorganic contaminants removal from water with biochar, a renewable, low cost and sustainable adsorbent— A critical review. *Bioresour. Technol.* 160: 191–202.

Mohsen-Nia, M., Montazeri, P., and Modarress, H. 2007. Removal of Cu^{2+} and Ni^{2+} from wastewater with a chelating agent and reverse osmosis processes. *Desalination* 217: 276–281.

Mubarak, N.M., Alicia, R.F., Abdullah, E.C., Sahu, J.N., Haslija, A.B.A., and Tan, J. 2013. Statistical optimization and kinetic studies on removal of Zn^{2+} using functionalized carbon nanotubes and magnetic biochar. *J. Environ. Chem. Eng.* 1: 486–495.

Müller, K., Magesan, G.N., and Bolan, N.S. 2007. A critical review on the influence of effluent irrigation on the fate of pesticides in soil. *Agric. Ecosyst. Environ.* 120: 93–116.

Mumme, J., Eckervogt, L., Pielert, J., Diakité, M., Rupp, F., and Kern, J. 2011. Hydrothermal carbonization of anaerobically digested maize silage. *Bioresour. Technol.* 102: 9255–9260.

Murthy, Z.V.P., and Chaudhari, L.B. 2008. Application of nanofiltration for the rejection of nickel ions from aqueous solutions and estimation of membrane transport parameters. *J. Hazard. Mater.* 160: 70–77.

Muthukrishnan, M., and Guha, B.K. 2008. Effect of pH on rejection of hexavalent chromium by nanofiltration. *Desalination* 219: 171–178.

Nartey, O.D., and Zhao. B. 2014. Biochar preparation, characterization, and adsorptive capacity and its effect on bioavailability of contaminants: An overview. *Adv. Mater. Sci. Eng.* 2014: Article ID 715398, pp. 1–12, doi: 10.1155/2014/715398.

Nguyen, M.V., and Lee, B.K. 2014. Cu^{2+} ion adsorption from aqueous solutions by amine activated poultry manure biochar. *J Selçuk Uni. Natl. Appl. Sci. Digital Proceeding of The ICOEST'2014*, SIDE Side, Turkey, May 14–17.

Nhuchhen, D.R., Basu, P., and Acharya, B. 2014. A comprehensive review on biomass torrefaction. *Int. J. Ren. Energ. Biofuels*, 2014: Article ID 506376, pp. 1–56, doi: 10.5171/2014.506376.

Okoya, A.A., Akinyele, A.B., Ofoezie, I.E., Amuda, O.S., Alayande, O.S., and Makinde, O.W. 2014. Adsorption of heavy metal ions onto chitosan grafted cocoa husk char. *Afr. J. Pure Appl. Chem.* 8: 147–161.

Pan, J., Jiang, J., and Xu, R. 2013. Adsorption of Cr(III) from acidic solutions by crop straw derived biochars. *J. Environ. Sci.* 25: 1957–1965.

Pan, J.J., Jiang, J., and Xu, R.K. 2014. Removal of Cr(VI) from aqueous solutions by $Na_2SO_3/FeSO_4$ combined with peanut straw biochar. *Chemosphere* 101: 71–76.

Park, D., Yun, Y.S., and Park, J.M. 2006. Comment on the removal mechanism of hexavalent chromium by biomaterials or biomaterial-based activated carbons. *J. Ind. Chem. Res.* 45: 2405–2407.

Pathak, C., Chopra, A.K., and Srivastava, S. 2013. Accumulation of heavy metals in *Spinacia oleracea* irrigated with paper mill effluent and sewage. *Environ. Monit. Assess.* 185: 7343–7352.

Pavan, F.A., Lima, I.S., Lima, E.C., Airoldi, C., and Gushikem, Y. 2006. Use of Ponkan mandarin peels as biosorbent for toxic metals uptake from aqueous solutions. *J. Hazard. Mater.* 137: 527–533.

Pellera, F.M., Giannis, A., Kalderis, D., Anastasiadou, K., Stegmann, R., Wang, J.Y., and Gidarakos, E. 2012. Adsorption of Cu(II) ions from aqueous solutions on biochars prepared from agricultural by-products. *J. Environ. Manag.* 96: 35–42.

Perez-Marin, A.B., Ballester, A., Gonzalez, F., Blazquez, M.L., Munoz, J.A., Saez, J., and Zapata, V.M. 2008. Study of cadmium, zinc and lead biosorption by orange wastes using the subsequent addition method. *Bioresour. Technol.* 99: 8101–8106.

Prapagdee, S., Piyatiratitivorakul, S., and Petsom, A. 2014. Activation of cassava stem biochar by physico-chemical method for stimulating cadmium removal efficiency from aqueous solution. *Environ. Asia* 7: 60–69.

Qiu, Y., Zheng, Z., Zhou, Z., and Sheng, G.D. 2009. Effectiveness and mechanisms of dye adsorption on a straw-based biochar. *Bioresour. Technol.* 100: 5348–5351.

Ramola, B., and Singh, A. 2013. Heavy metal concentrations in pharmaceutical effluents of industrial area of Dehradun (Uttarakhand), India. *J. Environ. Anal. Toxicol.* 3: 173–176.

Ran, Y., Fu, J.M., Sheng, G.Y., Beckett, R., and Hart, B.T. 2000. Fractionation and composition of colloidal and suspended particulate materials in rivers. *Chemosphere* 41: 33–43.

Regmi, P., Moscoso, J.L.G., Kumar, S., Cao, X.Y., Mao, J.D., and Schafran, G. 2012. Removal of copper and cadmium from aqueous solution using switchgrass biochar produced via hydrothermal carbonization process. *J. Environ. Manag.* 109: 61–69.

Roberts, D.A., Paul, N.A., Dworjanyn, S.A., Hu, Y., Bird, M.I., and de Nys, R. 2015. Gracilaria waste biomass (*sampah rumput laut*) as a bioresource for selenium biosorption. *J. Appl. Phycol.* 27: 611–620.

Rout, D.C. 2008. Managing the water resource at J.K. paper mill. *Ecovision* 2: 28–30.

Roy, A.H., Wenger, S.J., Fletcher, T.D., Walsh, C.J., Ladson, A.R., Shuster, W.D., Thurston, H.W., and Brown, R.R. 2008. Impediments and solutions to sustainable, watershed-scale urban stormwater management: Lessons from Australia and the United States. *Environ. Manag.* 42: 344–359.

Rubilar, O., Diez, M.C., and Gianfreda, L. 2008. Transformation of chlorinated phenolic compounds by white rot fungi. *Crit. Rev. Env. Sci. Technol.* 38: 227–268.

Sadaka, S., and Negi, S. 2009. Improvements of biomass physical and thermochemical characteristics via torrefaction process. *Environ. Prog. Sustain. Energy.* 28: 427–434.

Šafařik, I., Nymburska, K., and Šafaříková, M. 1997. Adsorption of water-soluble organic dyes on magnetic charcoal. *J. Chem. Technol. Biotechnol.* 69: 1–4.

Saha, B., and Orvig, C. 2010. Biosorbents for hexavalent chromium elimination from industrial and municipal effluents. *Coord. Chem. Rev.* 254: 2959–2972.

Samsuri, A.W., Sadegh-Zadeh, F., and Seh-Bardan, B.J. 2013. Adsorption of As(III) and As(V) by Fe coated biochars and biochars produced from empty fruit bunch and rice husk. *J. Environ. Chem. Eng.* 1: 981–988.

Samuel, F.A., Mohan, V., and Rebecca, L.J. 2014. Physicochemical and heavy metal analysis of sugar mill effluent. *J. Chem. Pharm. Res.* 6: 585–587.

Sansalone, J.J. and Buchberger, S.G. 1997. Characterization of solid and metal element distributions in urban highway stormwater. *Water Sci. Technol.* 36: 155–160.

Sanyang, M.L., Azlina, W., Ghani, W.A.K., Idris, A., and. Ahmad, M.B. 2014. Hydrogel biochar composite for arsenic removal from wastewater. *Desalin. Water Treat.* doi: 10.1080/19443994.2014. 989412.

Schimmelpfennig, S., and Glaser, B. 2012. One step forward toward characterization: Some important material properties to distinguish biochars. *J. Environ. Qual.* 41: 1001–1013.

Schulz, H., and Glaser, B. 2012. Effects of biochar compared to organic and inorganic fertilizers on soil quality and plant growth in a greenhouse experiment. *J. Plant. Nutr. Soil. Sci.* 175: 410–422.

Sevilla, M., and Fuertes, A.B. 2009. The production of carbon materials by hydrothermal carbonization of cellulose. *Carbon.* 47: 2281–2289.

Sevilla, M., Macia-Agullo, J.A., and Fuertes, A.B. 2011. Hydrothermal carbonization of biomass as a route for the sequestration of CO_2: Chemical and structural properties of the carbonized products. *Biomass. Bioenerg.* 35: 3152–3159.

Singh, B., Fang, Y., Cowie, B.C.C., and Thomsen, L. 2014. NEXAFS and XPS characterization of carbon functional groups of fresh and aged biochars. *Org. Geochem.* 77: 1–10.

Singh, R.K., Sengupta, B., Bali, R., Shukla, B.P., Gurunadharao, V.V.S., and Srivatstava, R. 2009. Identification and mapping of chromium (VI) plume in groundwater for remediation: A case study at Kanpur, Uttar Pradesh. *J. Geol. Soc. Ind.* 74: 49–57.

Smith, C.T., Carnus, J.M., Wang, H., Gielen, G.J.H.P., Stuthridge, T.R. and Tomer, M.D. 2003. Land application of chemi-thermo mechanical pulp and paper mill effluent in New Zealand: From research to practice. In: *Environmental Impacts of Pulp and Paper Waste Streams*, T. Stuthridge, M. van den Heuval, N. Marvin, A. Slade, and J. Gifford (eds.), Pensacola, FL: Society of Environmental Toxicology and Chemistry, pp. 177–187.

Sohi, S.P., Krull, E., Lopez-Capel, E., and Bol, R. 2010. A review of biochar and its use and function in soil. *Adv. Agron.* 105: 47–82.

Song, Z., Lian, F., Yu, Z., Zhu, L., Xing, B., and Qiu, W. 2014. Synthesis and characterization of a novel MnOx-loaded biochar and its adsorption properties for Cu^{2+} in aqueous solution. *Chem. Eng. J.* 242: 36–42.

Sorme, L., and Lagerkvist, R. 2002. Sources of heavy metals in urban wastewater in Stockholm. *Sci. Total. Environ.* 298: 131–145.

Spataru, A. 2014. The use of hydrochar as a low cost adsorbent for heavy metal and phosphate removal from wastewater. *Master thesis.* Ghent University, Belgium.

Srinivasan. R. 2011. Advances in application of natural clay and its composites in removal of biological, organic, and inorganic contaminants from drinking water. *Adv. Mater. Sci. Eng.* 2011: Article ID 872531, pp. 1–17, doi: 10.1155/2011/872531.

Tong, S.J., Li, J.Y., Yuan, J.H., and Xu, R.K. 2011. Adsorption of Cu(II) by biochars generated from three crop straws. *Chem. Eng. J.* 172: 828–834.

Tong, X., and Xu, R. 2013. Removal of Cu(II) from acidic electroplating effluent by biochars generated from crop straws. *J. Environ. Sci.* 25: 652–658.

Trakal, L., Šigut, R., Šillerová, H., Faturíková, D. and Komárek, M. 2014. Copper removal from aqueous solution using biochar: Effect of chemical activation. *Arabian J. Chem.* 7: 43–52.

Tytłak, A., Oleszczuk, P., and Dobrowolski, R. 2014. Sorption and desorption of Cr(VI) ions from water by biochars in different environmental conditions. *Environ. Sci. Pollut. Res.* 22: 5985–5994. DOI: 10.1007/s11356-014-3752-4.

Uchimiya, M., Chang, S., and Klasson, K.T. 2011. Screening biochars for heavy metal retention in soil: Role of oxygen functional groups. *J. Hazard. Mater.* 190: 432–441.

Uchimiya, M., Lima, I.M., Klasson, K.T., Chang, S.C., Wartelle, L.H., and Rodgers, J.E. 2010. Immobilization of heavy metal ions (Cu-II, Cd-II, Ni-II, and Pb-II) by broiler litter-derived biochars in water and soil. *J. Agric. Food Chem.* 58: 5538–5544.

Van Vinh, N., Zafar, M., Behera, S.K., and Park, H.S. 2014. Arsenic(III) removal from aqueous solution by raw and zinc loaded pine cone biochar: Equilibrium, kinetics, and thermodynamics studies. *Int. J. Environ. Sci. Technol.* 12: 1283–1294. doi: 10.1007/s13762-014-0507-1.

Wan Ngah, W.S., and Hanafiah, M.A.K.M. 2008. Removal of heavy metal ions from wastewater by chemically modified plant wastes as adsorbents: A review. *Bioresour. Technol.* 99: 3935–3948.

Wang, H., Gielen, G.J.H.P., Judd, M.L., Stuthridge T.R., Blackwell, B., Tomer, M.D., and Perace, S. 1999. Treatment efficiency of land application for thermo-mechanical pulp mill (TMP) effluent constituents. *Appita J.* 52: 383–386.

Wang, H., Magesan, G.N., and Bolan, N.S. 2004. An overview of the environmental effects of land application of farm effluents. *N. Z. J. Agr. Res.* 47: 389–403.

Wang, S., Gao, B., Zimmerman, A.R., Li, Y., Ma, L., Harris, W.G., and Migliaccio, K.W. 2015. Removal of arsenic by magnetic biochar prepared from pinewood and natural hematite. *Bioresour. Technol.* 175: 391–395.

Wiatrowski, H.A., Das, S., Kunkkadapu, R., Ilton, E. S., Barkay, T., and Yee, N. 2009. Reduction of Hg(II) to Hg(0) by magnetite. *Environ. Sci. Technol.* 43: 5307–5313.

Wong, T.H.F., Breen, P.F., and Lloyd, S.D. 2000. *Water Sensitive Road Design—Design Options for Improving Stormwater Quality of Road Runoff.* Melbourne: Co-operative Research Centre for Catchment Hydrology.

Wu, X.L., Yang, Y.L., Tang, Q.F., Xu, Q., Liu, X.D., Huang, Y.Y., and Yin, X.C. 2011. Ecological risk assessment and source analysis of heavy metals in river water, groundwater along river banks and river sediments in Shenyang. *Chin. J. Ecol.* 30: 438–447.

Xu, X.Y., Cao, X.D., Zhao, L., Wang, H.L., Yu, H.R., and Gao, B. 2013. Removal of Cu, Zn, and Cd from aqueous solutions by the dairy manure-derived biochar. *Environ. Sci. Pollut. Res.* 20: 358–368.

Xue, Y., Gao, B., Yao, Y., Inyang, M., Zhang, M., Zimmerman, A.R., and Ro, K.S. 2012. Hydrogen peroxide modification enhances the ability of biochar (hydrochar) produced from hydrothermal carbonization of peanut hull to remove aqueous heavy metals: Batch and column tests. *Chem. Eng. J.* 200: 673–680.

Yang G.X., and Jiang, H. 2014. Amino modification of biochar for enhanced adsorption of copper ions from synthetic wastewater. *Water Res.* 48: 396–405.

Zhang, G., Qu, J., Liu, H., Cooper, A.T., and Wu, R. 2007. CuFe$_2$O$_4$/activated carbon composite: A novel magnetic adsorbent for the removal of acid orange II and catalytic regeneration. *Chemosphere* 68: 1058–1066.

Zhang, M., and Gao, B. 2013. Removal of arsenic, methylene blue, and phosphate by biochar/AlOOH nanocomposite. *Chem. Eng. J.* 226: 286–292.

Zhang, M., Gao, B., Varnoosfaderani, S., Hebard, A., Yao, Y., and Inyang, M. 2013. Preparation and characterization of a novel magnetic biochar for arsenic removal. *Bioresour. Technol.* 130: 457–462.

Zhang, Y., and Luo, W. 2014. Adsorptive removal of heavy metal from acidic wastewater with biochar produced from anaerobically digested residues: Kinetics and surface complexation modeling. *Bioresources* 9: 2484–2499.

Zhou, Y., Gao, B., Zimmerman, A.R., Chen, H., Zhang, M., and Cao, X. 2014. Biochar-supported zerovalent iron for removal of various contaminants from aqueous solutions. *Bioresour. Technol.* 152: 538–542.

8

Biochar for Organic Contaminant Management in Water and Wastewater

Chapter 8

Biochar for Organic Contaminant Management in Water and Wastewater

Ming Zhang and Li Lu

8.1 Organic Contaminants in Water and Wastewater

The fast development of our global society and its rapidly growing economics in the past century has led to the discharge of a large quantity of anthropogenic chemicals into aquatic environments from industries (e.g., dyes), households (e.g., detergents, pharmaceuticals, and personal care products), and agriculture (e.g., pesticides, herbicides). More than 700 pollutants, both organic and inorganic, have been detected in water (Ali and Gupta 2007). Most of anthropogenic chemicals are hydrophobic and persistent and are well-known as hydrophobic organic contaminants (HOCs) or persistent organic pollutants (POPs). They are resistant to chemical degradation, biological degradation, or both, and the activated sludge process in conventional wastewater treatment plants (WWTPs) is not effective in removing these xenobiotic organic compounds (Petrasek et al. 1983). Consequently, WWTPs are considered as one of the important sources of these chemicals into aquatic environments (Katsoyiannis and Samara 2005; Pham and Proulx 1997). These xenobiotic organic compounds in aquatic environments may accumulate in plant and animal tissues and translocate to humans via water consumption and through the food chain, thereby posing long-term threats to human health and the safety of the ecosystem, even when introduced at extremely low concentrations (Katsoyiannis and Samara 2005).

Therefore, effective removal of xenobiotic organic contaminants from water and wastewater is crucial to ensure water safety. Sorption is recognized as a fast, effective, low-cost, and nonselective universal approach for HOC removal. Activated carbon (AC) has long been preferred as the sorbent for its excellent performance in removing chemicals from water and wastewater due to its large surface area and developed pore volume (Chern and Chien 2003; Jung et al. 2001). However, large-scale wastewater treatment application of AC is not applicable due to its high cost of production and regeneration (Ali and Gupta 2007). Developing a low-cost sorbent instead of AC is key in the adsorption approach in water and wastewater treatment. Studies have examined zeolites (Bosso and Enzweiler 2002), clays (Shen 2002), agricultural wastes (Robinson et al. 2002), and artificial polymeric materials (Azanova and Hradil 1999) as alternative sorbents. Carbonaceous sorbents, such as biochar (BC), have gained increasing attention due to their close physical and chemical properties to AC and their lower cost.

8.2 Sorption of Organic Contaminants by BC

Because BC serves as a popular soil amendment in agriculture, the features of BC itself are well understood. In addition

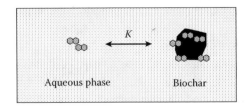

Figure 8.1 Sorption of organic contaminants by biochar particles. ∞ stands for organic contaminants. K is the distribution coefficient of chemicals between the biochar and the aqueous phase.

to its benefit for soil carbon balance and soil microbial activity, BC is recently recognized as a widespread, highly efficient, and low-cost environmentally engineered sorbent. Sorption of organic contaminants by BC is in essence the distribution of these chemicals between BC and water in an aqueous environment (Figure 8.1).

Some studies have been carried out to investigate the application prospects of BC in water and wastewater treatment. For example, BCs prepared from straw, anaerobic digestion residue, palm bark, and eucalyptus were used to sorb cationic dyes from different dye-contaminated wastewater as substitutes for AC; >80% of dyes can be removed using such BCs (Qiu et al. 2009; Sun et al. 2013). Cao et al. (2009) developed a dairy manure–derived BC that performed as an effective sorbent for atrazine. Yao et al. (2012) tested the removal of sulfamethoxazole from reclaimed water by BC and suggested that the BC may be used as a low-cost alternative sorbent for treating effluent from wastewater plants, and thereby prevent soil and groundwater contamination by reclaimed water irrigation.

8.2.1 Sorption-Driven Force in BC-Water System

BC produced at a high temperature is feasible for sorbing organic contaminants due to its high surface area and developed pore structure. Chen et al. (2008) summarized the sorption of organic contaminants to BC as a two-part process: adsorption on carbonized fractions and partition in noncarbonized fractions of BC. Ionic electrostatic attraction, hydrogen bonding, π–π electron–donor–acceptor (π–π EDA) interaction, and hydrophobic interaction are considered as the main driving forces of adsorption, and dissolution of organic contaminants in noncarbonized organic fractions of BC is commonly termed *partition* (Ahmad et al. 2014).

8.2.1.1 Ionic Electrostatic Attraction

As described in former chapters the surface of BC is negatively charged, and there is thus considerable exchange of cationic ions, such as K^+, Na^+, and Ca^{2+}, in the BC matrix. This property facilitates the sorption of cationic organic contaminants by ion exchange and subsequently electrostatic attraction. Cationic dyes are a group of typical organic contaminants that can be sorbed by BC through ionic electrostatic attraction. For example, Rhodamine B (RB), a popular cationic dye, in water was effectively removed by both BC and AC. However, BC was slightly more effective due to the RB–BC electrostatic attraction (Qiu et al. 2009). Figure 8.2 illustrates the interactions between cationic organic contaminants and the negatively charged BC surface. In contrast, BC is not as effective at sorption of anionic organic contaminants because of electrostatic repulsion.

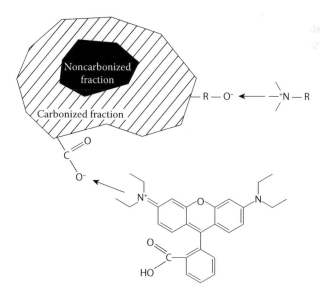

Figure 8.2 Scheme of interactions between cationic organic contaminants and the negatively charged biochar surface (ionic electrostatic attraction).

For example, consider sulfamethoxazole (SMX), an antibiotic. The sorption of SMX by BC was greatly decreased with pH increase from 3.0 to 8.0 because anionic SMX⁻ species increased and gradually became dominant at pH > 6.0 (Zheng et al. 2013). However, the negatively charged BC still sorbed SMX⁻ via negative charge–assisted hydrogen bonding.

8.2.1.2 Hydrogen Bonding

Qiu et al. (2009) studied the interaction of brilliant blue (KNR) and straw-based BC by using Fourier transform infrared (FTIR) spectroscopy, and they observed intermolecular hydrogen bonding (O=H–O bonds), an important mechanisms for the sorption of KNR on BC. Teixidó et al. (2011) reported strong H bonding between anionic sulfamethazine and carboxylate or phenolate group on hardwood BC produced at 600°C in an alkaline environment, indicating an H-bonding-induced effective sorption, although there existed an electrostatic repulsion between anionic sulfamethazine and the negatively charged BC surface. Sun et al. (2011) found that BC produced at lower temperature was more effective for adsorption of the polar compounds norflurazon and fluridone; this influence was mainly attributed to the H bonding between the O-containing groups of the BC and the polar chemicals (Figure 8.3).

8.2.1.3 π–π EDA

With an increase in the pyrolytic temperature, more aromatic structures form in BC, accompanied by loss of H- and O-containing groups, indicated by the higher C/H ratio. Sun et al. (2012) found that both electron-rich and -poor functional groups are present in high-temperature-derived BCs that are capable of interacting with both electron donors and electron acceptors. For example, the graphene-like surface of BC acts as an π-electron donor and the organic contaminants acts as an π-electron acceptor; their interactions greatly promoted the adsorption of aromatic organic contaminants by the BC. Jia et al. (2013) found a slight shift to higher wavenumbers from 1598 and 1512 cm⁻¹ to 1606 and 1516 cm⁻¹ of maize-straw BC after

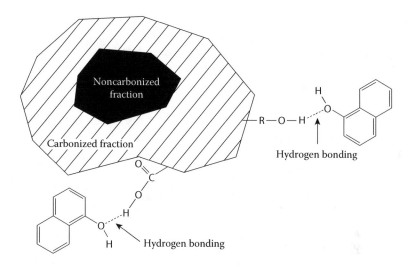

Figure 8.3 Scheme of the hydrogen bonding interaction between adsorbent (biochar) and aromatic organic contaminants.

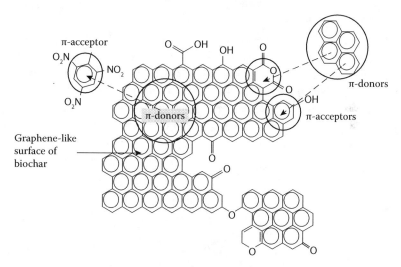

Figure 8.4 Scheme of the π–π electron–donor–acceptor interaction between adsorbent (biochar) and aromatic organic contaminants.

oxytetracycline (OTC) sorption by using FTIR spectroscopy, which may be attributed to the π–π EDA interaction between graphene sheets in the BC and OTC structures (Ji et al. 2011). Figure 8.4 shows a brief scheme of the π–π EDA interaction between adsorbent (BC) and aromatic organic contaminants.

8.2.1.4 Partition

The term *partition* refers to a process similar to solution in which an organic molecule penetrates into the entire network of the solid phase. In fact, this process was first used by Chiou and coworkers (Chiou 2002; Chiou et al. 1979, 1983) to interpret the sorption of HOCs by soils, where the soil organic matter behaves as a partition medium. The noncarbonized organic matter of BCs also serves as a similar partition phase for nonpolar organic chemicals. The process might be partly interpreted by a liquid-dissolved process in organic matter similar to the partition process of the solvent extraction from water (Figure 8.5).

8.2.2 Sorption Isotherms and Models

The sorption process is generally studied by plotting the equilibrium concentration of a compound in the sorbent as a function of its equilibrium concentration in solution at a given temperature. The result of sorption experiment at constant temperature is termed *isotherm*. According to the shape of the isotherm curve and the corresponding sorption theory, the isotherm has been classified into several general modes, and each isotherm has been interpreted by a specific mathematical model.

The main driving forces of the sorption of organic contaminants by BC have been discussed above, and different sorption

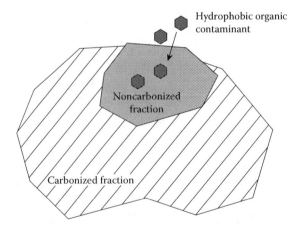

Figure 8.5 Scheme of the partition of organic contaminants between the noncarbonized fraction of biochar and water.

mechanisms have also been proposed depending on the nature of the sorbent. For BC, it is generally not fully carbonized, and the BC bulk consists of the carbonized fraction and the non-carbonized organic matter. The carbonized fraction serves as an adsorbent, whereas the noncarbonized organic matter serves as a partition phase (Chen et al. 2008; Chun et al. 2004; Nguyen et al. 2007; Zhu et al. 2005). Therefore, here we mainly discuss the sorption isotherm and model in three phenomena: adsorption, partition, and the dual-mode sorption.

8.2.2.1 Adsorption and Nonlinear Sorption Model

Adsorption is mainly a surface-based phenomenon that can be defined as the accumulation of sorbates (i.e., contaminants) on the surface or inner surface of sorbents from an aqueous or gaseous phase (i.e., BC) (Ali and Gupta 2007; Faust and Aly 1983). According to this definition, the above-mentioned electrostatic attraction, hydrogen bonding, and π–π EDA interactions between carbonized BC and organic contaminants can be the main mechanisms of adsorption.

For adsorption theory, there are basically two well established nonlinear isotherms: the Freundlich model and the Langmuir model.

8.2.2.1.1 Freundlich Model

The shape of the graph of the amount adsorbed per unit weight of adsorbent versus the concentration in the aqueous form was often well fitted the empirical Freundlich equation (Freundlich 1906) that is mathematically expressed as Equation 8.1:

$$Q_e = K_f C_e^{1/n} \tag{8.1}$$

where Q_e is the adsorption amount (mg/kg) at equilibrium concentration (C_e, mg/L), K_f is adsorption coefficient, and n is an index of nonlinearity of the isotherms.

The logarithmic variant of this equation can be described as Equation 8.2:

$$\log Q_e = \log K_f + \frac{1}{n}\log C_e \tag{8.2}$$

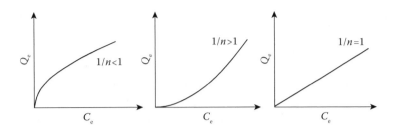

Figure 8.6 Typical Freundlich isotherms describing sorption of organic compounds in water by sorbents.

The coefficients K_f and n can be obtained from intercept and slope from linear regression of $\log Q_e$ and $\log C_e$.

A $1/n>1$ indicates a curved upward, S-type isotherm, and a $1/n<1$ indicates a curved downward, L-type isotherm. A value of $1/n=1$ represents a linear adsorption (Figure 8.6). As for its physical significance, the value of $1/n$ is related to the magnitude and diversity of energies associated with a particular sorption process (Weber et al. 1992). Generally, L-type is more usual for the case with less competition for adsorption sites, and S-type is more characteristic for the cooperative sorption (Site 2001).

8.2.2.1.2 Langmuir Model

The Langmuir equation of isotherm for adsorption at the solid–liquid interface is initially derived from the theoretical description of single molecules on solid surfaces (Langmuir 1916). It is based on the assumption that the number of adsorption sites on the adsorbent surface is definite and that each adsorption site can adsorb only one molecule of chemical. In contrast to the empirical Freundlich equation, the constant parameters of Langmuir equation have a strictly defined physical meaning. The Langmuir equation of isotherm for adsorption at the solid–liquid interface can be expressed as Equation 8.3:

$$Q_e = \frac{Q_m K_a C_e}{1 + K_a C_e} \tag{8.3}$$

where K_a is Langmuir equilibrium constant, Q_e is the sorption amount of sorbates on sorbents at equilibrium concentration of C_e, and Q_m is the maximum adsorbed amount (total monolayer coverage theoretically).

This equation can be transformed to reciprocal form as Equation 8.4:

$$\frac{1}{Q_e} = \frac{1}{Q_m K_a C_e} + \frac{1}{Q_m} \qquad (8.4)$$

The regression of $1/Q_e$ versus $1/C_e$ indicates a straight line with a slope of $1/Q_m K_a$ and an intercept of $1/Q_m$ (Figure 8.7).

Sun et al. (2013) used BCs prepared from anaerobic digestion residue, palm bark, and eucalyptus as substitutes for AC to remove cationic dyes from different dye-contaminated wastewater. They proposed that compared with the Freundlich isotherm model, the Langmuir isotherm model was more suitable for describing the adsorption of the cationic methylene blue on the BCs.

Because of its basic assumption of monolayer adsorption, the Langmuir model is commonly thought of as an ideal localized model; however, it has several limitations, one of which is that it neglects the interactions among already adsorbed molecules. The possibility of large molecules occupying more than one site also is not considered. Many attempts were made to modify the Langmuir model by taking into account of the interactions among adsorbed molecules and the surface heterogeneity of the solid.

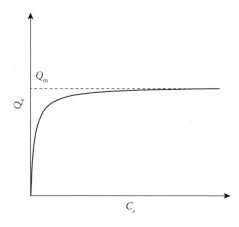

Figure 8.7 Typical Langmuir isotherms describing sorption of organic compounds in water by sorbents.

8.2.2.2 Partition and Linear Sorption Model

The accumulation of sorbates (i.e., contaminants) in non-carbonized organic matter through a dissolve-like process is termed *partition*. The partition theory was first proposed by Chiou et al. (1979) and is now widely accepted as a useful linear model describing the transfer of nonionic organic compounds from water to organic matter.

8.2.2.2.1 Linear Partition Model

The linear partition model can be expressed as Equation 8.5:

$$Q_e = K_d C_e \tag{8.5}$$

where K_d was the partition coefficient of organics between sorbent and water, and Q_e is the sorption amount of sorbent at equilibrium concentration C_e.

Partition is one of the important sorption mechanisms of organic contaminants on the noncarbonized biomass of BC. For example, the sorption of methyl *tert*-butyl ether to BC pyrolyzed at low temperature (200 and 300°C) was proved to be dominated by partitioning into the noncarbonized part of BC (Xiao 2014).

8.2.2.3 Dual-Mode Sorption and Model

Some studies on the sorption of organic contaminants by BC have shown that because of the heterogeneous nature of BC, its sorption includes both adsorption and partition. The total sorption is the sum of sorption in the partition–dissolution domain and adsorption in the hole-filling domain. The model is thus named dual-mode model, as suggested by Xing et al. (1996) and Xing and Pignatello (1997), and is expressed as Equation 8.6:

$$Q_e = K_p C_e + \frac{Q_0 b C_e}{1 + b C_e} \tag{8.6}$$

where Q_e is the sorption amount of sorbent at equilibrium concentration C_e, K_p is the partition coefficient, Q_0 is the maximum sorption capacity, and b is the affinity constant.

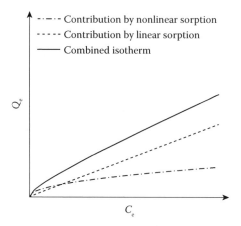

Figure 8.8 Linear and nonlinear sorption isotherms in dual-mode sorption model.

The combined isotherm shows apparent linearity at relatively high concentrations of an organic compound and nonlinearity at relatively low concentrations. Figure 8.8 interpreted the linear and nonlinear parts of dual-mode sorption isotherm.

Chen et al. (2012) compared the sorption behaviors of organic contaminants on BCs with different pyrolytic temperature. The sorption of naphthalene and 1-naphthol by the BCs with higher pyrolytic temperature can be described by the Langmuir model and are well fitted, whereas the isotherms for BCs with lower pyrolytic temperature were more adequately described by the dual-mode model.

8.2.3 Factors That Influence BC Sorption

8.2.3.1 Solution Chemistry

Conditions in solution, such as pH, ion strength and temperature may also lead to the property change of both sorbents and sorbates, thus affect the sorption behavior.

8.2.3.1.1 pH Values

The pH of an aqueous solution can significantly affect the sorption of organic contaminants on BC, especially of ionogenic organic contaminants. Jia et al. (2013) found that sorption of oxytetracycline (OTC) on maize straw–derived BC was highly pH dependent. Sorption capability initially increased and then decreased with a rise in pH, and pH 5.5 was the best for OTC sorption. The electronic repulsion between anionic OTC^- and OTC^{2-} at high pH values >5.5 and the highly negatively charged BC surface may decrease the sorption of OTC on BC. The same pattern was also observed at high pH values for the sorption of sulfonamides to demineralized and original pine-wood BC and was described as mainly due to the repulsion between deprotonated sulfonamide species and deprotonated O-functionalities of BC at higher pH (Xie et al. 2014).

However, for cationic contaminants, an increase in pH may enhance their sorption capacity on BC. For example, sorption of methyl violet on BC increased sharply from pH 7.7 to 8.7. It is inferred that the dissociation of phenolic hydroxyl groups on the BC surface increased the net negative charge on surface, thereby increasing the electrostatic attraction to methyl violet (Xu et al. 2011).

The properties of some chemicals at different pHs may also lead to different sorption mechanisms. The solubility of neutral sulfamethoxazole (SMX) species (at pH 3.22) and anionic SMX species (at pH 7.5) is 281 17,900 mg/L, respectively (Lian et al. 2014). Therefore, much higher hydrophobic partitioning of SMX into amorphous BC structure occurred at the lower pH range.

8.2.3.1.2 Ion Strength

In a real water and a wastewater environment, ion strength would vary significantly due to the source of wastewater. Ion strength has been observed to affect the sorption of organic contaminants to BC. For example, sorption of OTC decreased greatly with an increase of ion strength, as indicated by potassium chloride concentration (Jia et al. 2013). It is suggested that K^+ might compete with OTC for sorption sites on the BC surface, leading to a decrease of OTC sorption. In contrast, an increase of brilliant blue dye adsorption on BC was observed when ion strength (Na^+) increased. The neutralization of the negative charge of the BC surface by Na^+ and consequently the compression of electrical double layer near the BC surface was believed to be the main mechanism (Qiu et al. 2009). However, it was also observed that the presence of sodium chloride (0.02–0.1 mol/L) did not affect the sorption of SMX on BC (Lian et al. 2014).

8.2.3.1.3 Temperature

Sorption energy is an important parameter to use to evaluate the sorption characteristics of a sorbent to certain organic contaminants. There were few comprehensive studies until recently that focused on the effect of temperature on the sorption of organic contaminants to BC. For polar and ionic organic contaminant SMX, the increase of temperature enhanced the sorption of SMX on BC, and the positive enthalpy change (ΔH) values suggested an endothermic process. The negative Gibbs free energy (ΔG) implied a thermodynamic favorable and spontaneous sorption of SMX on BC. Lower temperature pyrolyzed BC had the lowest ΔG value, indicating more favorable sorption of SMX (Lian et al. 2014), perhaps due to its more polar structure to polar SMX, or its higher noncarbonized fraction in BC as partition phase to SMX. Similar trends were observed when the sorption of basic dye methylene blue on acid-treated kenaf fiber char (Mahmoud et al. 2012).

8.2.3.2 Coexisting Chemicals

8.2.3.2.1 Bivalent Metals

With its significantly negatively charged surface, BC has a strong potential to interact with bivalent metals and other positively charged contaminants. Therefore, the coexisting bivalent metals in solution may have a pronounced effect on the sorption of organics onto BC. It has been observed that Cu^{2+} inhibited >50% sorption of SMX and >20% sorption of sulfapyridine (Xie et al. 2014). The strong hydration of Cu^{2+} and subsequently the formation of Cu^{2+} complexes on the BC surface was believed to form hydration shells, thereby blocking the available sorption areas on the BC surface (Chen et al. 2007). Competition of O-containing functional groups on the BC surface by Cu^{2+} may hinder the formation of charge-assisted H bonds with sulfonamides (Teixidó et al. 2011). Pb^{2+} was observed to suppress the sorption of OTC under acidic conditions (Jia et al. 2013); however, in the same study, Cu^{2+} significantly enhanced the sorption of OTC at all tested pH values. FTIR characterization showed that Cu^{2+} reacted with carboxylic and phenolic groups of BC and acted as a bridge between BC and OTC, finally forming BC–O≡Cu–OTC complexes.

Cd^{2+} had no effect on OTC sorption because there was almost no strong interaction between Cd^{2+} and the BC surface functional groups (Jia et al. 2013).

8.2.3.2.2 Dissolved Humic Acid

Dissolved humic acid (dissolved organic carbon [DOC]) commonly exists in water, especially in a natural water environment. Significant inhibition of sulfonamides by BC and hydrogenated BC has been reported previously (Xie et al. 2014), and a similar phenomenon was observed by Pignatello et al. (2006) and Ji et al. (2009) in the sorption of organics by black carbon and carbon nanotubes. The competition of DOC on the BC surface with organic contaminants through π–π electron coupling or π–π EDA interactions was the main mechanism for the sorption inhibition.

8.2.3.2.3 Organic Contaminants

Zhang et al. (2012) studied the competitive sorption of metsulfuron-methyl (ME) and tetracycline (TC) on corn straw BC and found that ME and TC affect their sorption on BC mutually. TC exhibited weak competition with ME in low-temperature-derived BC (200–300°C) because partitioning into the noncarbonized part of BC dominated the sorption of ME, whereas sorption of TC for all BC and sorption of ME at high-temperature-derived BC (400–600°C) were dominated by adsorption, leading to the mutual competition of ME and TC adsorption for BC surface-specific sites. Zheng et al. (2010) also found a significant mutual sorption inhibition between two triazine pesticides, atrazine and simazine, when they coexist in the aqueous environment was due to competition of sorption sites. These results implied that when BC is applied in an aqueous environment for removal of organic contaminants, its sorption capacity to a certain organic contaminant is expected to be lower when some other organics coexist.

8.2.3.3 Aging

In most of the current studies on BC sorption to both organic and inorganic contaminants, newly pyrolyzed BCs have been used and characterized. However, the morphologic and property changes of BC have occurred via long-term aging. Cheng et al. (2014) compared the sorption capacity of newly produced BC and historical BC that was left in the soil in the mid-nineteenth century toward organics and heavy metals. In this chapter, we mainly focus on organics. Results showed that compared with newly produced BC, sorption of diuron and atrazine were reduced by 56–91% in aged BC. This difference can be attributed to the formation of a greater number of O-containing groups on the BC surface during aging in soil, thereby hindering the sorption sites for diuron and atrazine. Furthermore, we believe that adsorption of DOCs and other substances that exist in the soil solution may compete for the available sorption sites on the BC surface, block pore structures, or both.

8.3 Conclusions

BC is a promising sorbent for water and wastewater treatment due to its relatively high surface area, developed pore structure, and abundant surface functional groups, all of which can facilitate the sorption of organics and inorganics. BC may serve as a good substitute for AC due to its cost-effectiveness, because BC needs no pre- or posttreatment or activation. BC production is also a good way to dispose of biowaste that is produced in agriculture and horticulture. However, detailed research before application of BC to water and wastewater treatment is necessary. For example, there is the possibility for the release of polycyclic aromatic hydrocarbons and inorganic nutrients (K^+, phosphate ions) from BC, as these chemicals are generated during BC production.

References

Ali, I., and V.K. Gupta. 2007. Advances in water treatment by adsorption technology. *Nature Protocols* 1: 2661–2667.

Ahmad, M., A.U. Rajapaksha, J.E. Lim, et al. 2014. Biochar as a sorbent for contaminant management in soil and water: A review. *Chemosphere* 99: 19–33.

Azanova, V.V., and J. Hradil. 1999. Sorption properties of macroporous and hypercrosslinked copolymers. *React Funct Polym* 41: 163–175.

Bosso, S.T., and J. Enzweiler. 2002. Evaluation of heavy metal removal from aqueous solution onto scolecite. *Water Res* 36: 4795–4800.

Cao, X., L. Ma, B. Gao, and W. Harris. 2009. Dairy-manure derived bio-char effectively sorbs lead and atrazine. *Environ Sci Technol* 43: 3285–3291.

Chen, B., D. Zhou, and L. Zhu. 2008. Transitional adsorption and partition on nonpolar and polar aromatic contaminants by biochars of pine needles with different pyrolytic temperatures. *Environ Sci Technol* 42: 5137–5143.

Chen, J., D. Zhu, and C. Sun. 2007. Effects of heavy metals on the sorption of hydrophobic organic compounds to wood charcoal. *Environ Sci Technol* 41: 2536–2541.

Chen, Z., B. Chen, D. Zhou, et al. 2012. Bisolute Sorption and Thermodynamic Behavior of Organic Pollutants to Biomass-derived Biochars at Two Pyrolytic Temperatures. *Environ Sci Technol* 46: 12476–12483.

Cheng, C.H., T.P. Lin, J. Lehmann, L.J. Fang, Y.W. Yang, O.V. Menyailo, K.H. Chang, and J.S. Lai. 2014. Sorption properties for black carbon (wood char) after long term exposure in soils. *Organic Geochem* 70: 53–61.

Chern, J.M., and Y.W. Chien. 2003. Competitive adsorption of benzoic acid and p-nitrophenol onto activated carbon: Isotherm and breakthrough curves. *Water Res* 37: 2347–2356.

Chiou, C.T. 2002. *Partition and Adsorption of Organic Contaminants in Environmental Systems*. Wiley, New York.

Chiou, C.T., L.J. Peters, and H. Freed. 1979. A physical concept of soil-water equilibria for nonionic organic compounds. *Science* 206: 831–832.

Chiou, C.T., P.E.D. Porter, and W. Schmedding. 1983. Partition equilibria of nonionic organic compounds between soil organic matter and water. *Environ Sci Technol* 17: 227–231.

Chun, Y., G.Y. Sheng, C.T. Chiou, and B.S. Xing. 2004. Compositions and sorptive properties of crop residue-derived chars. *Environ Sci Technol* 38: 4649–4655.

Faust, S.D., and O.M. Aly. 1983. *Chemistry of Water Treatment*. Butterworth, Stoneham, MA.

Freundlich, H.M.F. 1906. Über die adsorption in lösungen. *Z Phys Chem* 57(A): 385–470.

Ji, L.L., W. Chen, S.R. Zheng, Z.Y. Xu, and D. Zhu. 2009. Adsorption of sulfonamide antibiotics to multiwalled carbon nanotubes. *Langmuir* 25: 11608–11613.

Ji, L.L., Y.Q. Wan, S.R. Zheng, et al. 2011. Adsorption of tetracycline and sulfamethoxazole on crop residue-derived ashes: Implication for the relative importance of black carbon to soil sorption. *Environ Sci Technol* 45: 5580–5586.

Jia, M.Y., F. Wang, Y.R. Bian, et al. 2013. Effects of pH and metal ions onoxytetracycline sorption to maize-straw-derived biochar. *Bioresour Tech* 136: 87–93.

Jung, M.W., K.H. Ahn, Y. Lee, et al. 2001. Adsorption characteristics of phenol and chlorophenols on granular activated carbons (CAC). *Microchem J* 70: 123–131.

Katsoyiannis, A., and C. Samara. 2005. Persistent organic pollutants (POPs) in the conventional activated sludge treatment process: Fate and mass balance. *Environ Res* 97: 245–257.

Langmuir, I. 1916. The constitution and fundamental properties of solids and liquids. *J Am Chem Soc* 38(11): 2221–2295.

Lian, F., B.B. Sun, Z.G. Song, L.Y. Zhu, X.H. Qi, and B.S. Xing. 2014. Physicochemical properties of herb-residue biochar and its sorption to ionizable antibiotic sulfamethoxazole. *Chem Eng J* 248: 128–134.

Mahmoud, D.K., M.A.M. Salleh, W.A.W.A. Karim, A. Idris, and Z.Z. Abidin. 2012. Batch adsorption of basic dye using acid treated kenaf fibre char: Equilibrium, kinetic and thermodynamic studies. *Chem Eng J* 181–182: 449–457.

Nguyen, T.H., H.H. Cho, D.L. Poster, and W.P. Ball. 2007. Evidence for a pore-filling mechanism in the adsorption of aromatic hydrocarbons to a natural wood char. *Environ Sci Technol* 41: 1212–1217.

Petrasek, A.C., I.J. Kugelman, B.M. Austern, et al. 1983. Fate of toxic organic compounds in wastewater treatment plants. *J Water Pollut Contr Fed* 55: 1286–1296.

Pham, T.T., and S. Proulx. 1997. PCBs and PAHs in the Montreal urban community (Quebec, Canada) wastewater treatment plant and in the effluent plume in the St Lawrence River. *Water Res* 31: 1887–1896.

Pignatello, J.J., S. Kwon, and Y.F. Lu. 2006. Effect of natural organic substances on the surface and adsorptive properties of environmental black carbon (char): Attenuation of surface activity by humic and fulvic acids. *Environ Sci Technol* 40: 7757–7763.

Qiu, Y., Z. Zheng, Z. Zhou, and G.D. Sheng. 2009. Effectiveness and mechanisms of dye adsorption on a straw-based bio-char. *Bioresour Technol* 100: 5348–5351.

Robinson, T., B. Chandran, and P. Nigam. 2002. Studies on desorption of individual textile dyes and a synthetic dye effluent from dye adsorbed agricultural residues using solvents. *Bioresour Technol* 84: 299–301.

Shen, Y.H. 2002. Removal of phenol from water by adsorption flocculation using organobentonite. *Water Res* 36: 1107–1114.

Site, A.D. 2001. Factors affecting sorption of organic compounds in natural sorbent/water systems and sorption coefficients for selected pollutants. A review. *J Phys Chem Ref Data* 30: 187–439.

Sun, K., M. Keiluweit, M. Kleber, et al. 2011. Sorption of fluorinated herbicides to plant biomass-derived biochars as a function of molecular structure. *Bioresour Technol* 102: 9897–9903.

Sun, K., J. Jin, M. Keiluweit, M. Kleber, et al. 2012. Polar and aliphatic domains regulate sorption of phthalic acid esters (PAEs) to biochars. *Bioresour Technol* 118: 120–127.

Sun, L., S. Wan, and W. Luo. 2013. Biochars prepared from anaerobic digestion residue, palm bark, and eucalyptus for adsorption of cationic methylene blue dye: Characterization, equilibrium, and kinetic studies. *Bioresour Technol* 140: 406–413.

Teixidó, M., J.J. Pignatello, J.L. Beltrán, et al. 2011. Speciation of the ionizable antibiotic sulfamethazine on black carbon (biochar). *Environ Sci Technol* 45: 10020–10027.

Weber, W.J., P. M. McGinley, and L.E. Katz. 1992. A distributed reactivity model for sorption by soils and sediments. 1. Conceptual basis and equilibrium assessments. *Environ Sci Technol* 26: 1955–1962.

Xiao, L.W., E.P. Bi, B.B. Du, et al. 2014. Surface characterization of maize-straw-derived biochars and their sorption performance for MTBE and benzene. *Environ Earth Sci* 71: 5195–5205.

Xie, M.X., W. Chen, Z.Y. Xu, et al. 2014. Adsorption of sulfonamides to demineralized pine wood biochars prepared under different thermochemical conditions. *Environ Pollut* 186: 187–194.

Xing, B.S., and J.J. Pignatello. 1997. Dual-mode sorption of low-polarity compounds in glassy poly(vinylchloride) and soil organic matter. *Environ Sci Technol* 31: 792–796.

Xing, B.S., J.J. Pignatello, and B. Gigliotti. 1996. Competitive sorption between atrazine and other organic compounds in soils and model sorbents. *Environ Sci Technol* 30: 2432–2440.

Xu, R.K., S.C. Xiao, J.H. Yuan, and A.Z. Zhao. 2011. Adsorption of methyl violet from aqueous solutions by the biochars derived from crop residues. *Bioresour Technol* 102: 10293–10298.

Yao, Y., B. Gao, H. Chen, et al. 2012. Adsorption of sulfamethoxazole on biochar and its impact on reclaimed water irrigation. *J hazard Mater* 209-210: 408–413.

Zhang, G.X., X.T. Liu, K. Sun, F. He, Y. Zhao, and C.Y. Lin. 2012. Competitive sorption of metsulfuron-methyl and tetracycline on corn straw biochars. *J Environ Qual* 41(6): 1906–1915.

Zheng, W., M. Guo, T. Chow, et al. 2010. Sorption properties of greenwaste biochar for two triazine pesticides. *J Hazard Mater* 181: 121–126.

Zheng, H., Z. Wang, J. Zhao, et al. 2013. Sorption of antibiotic sulfamethoxazole varies with biochars produced at different temperatures. *Environ Pollut* 181: 60–67.

Zhu, D.Q., S. Kwon, and J.J. Pignatello. 2005. Adsorption of single-ring organic compounds to wood charcoals prepared under different thermochemical conditions. *Environ Sci Technol* 39: 3990–3998.

SECTION IV

Agronomic Applications and Climate Change Mitigation

9

Biochar Effects on Soil Fertility and Nutrient Cycling

Chapter 9

Biochar Effects on Soil Fertility and Nutrient Cycling

Yanjiang Cai and Scott X. Chang

Chapter Outline

9.1 Introduction

Good soil fertility is associated with sufficient availability of nutrients and favorable environmental conditions for plant and crop growth. Soil nutrients required by plants are called essential nutrients and include macronutrients (C, H, O, N, P, K, S, Ca, and Mg) and micronutrients (B, Mn, Cu, Zn, Fe, Mo, and Cl). Having sufficient availability of essential nutrients is critical for supporting soil fertility and productivity. In agricultural production systems, incorporation of crop residues into the soil and application of manure, chemical fertilizers, both a combination are general measures to enhance soil fertility through improving nutrient availability, retention, and cycling (Whitbread et al. 2003; Moyin-Jesu 2007).

Recent research has demonstrated that biochar application to soils can improve soil fertility (Figure 9.1), in addition to biochar's main role of climate change mitigation by storing carbon in the soil in a stable form (Lehmann 2007; Atkinson et al. 2010; Alburquerque et al. 2013; Gomez et al. 2014). Biochar, or biomass-derived charcoal, is a carbonaceous material produced during thermal decomposition of biomass under low-oxic conditions (Lehmann 2007; Ding et al. 2010). Many organic material types, such as wood, plant litter, crop residue, animal manure, and waste products (Figure 9.1), can be used for biochar production (Ding et al. 2010; Ippolito et al. 2012; Kloss et al. 2014a). Most of the Ca, Mg, K, P, and plant micronutrients and about half of the N and S in the biomass can be partitioned into the biochar fraction in the biochar production process (Laird et al. 2010a). Biochar application is therefore expected to directly improve soil fertility by adding nutrients to the soil that

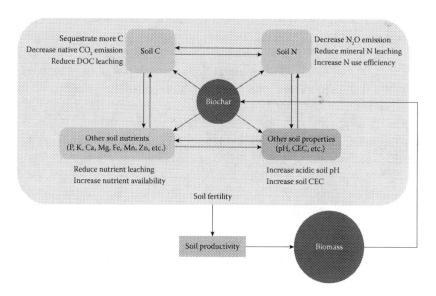

Figure 9.1 Role of biochar application in increasing soil C storage and improving soil fertility.

previously exist in the biomass used for biochar production (Glaser et al. 2002; Lehmann et al. 2003; Chan et al. 2008; Kloss et al. 2014a). Biochar application can also significantly increase soil cation exchange capacity (CEC), water-holding capacity, and bioavailability of nutrients, such as P, Ca, S, N, Zn, and Mn, due to its porous nature and high surface area (Liang et al. 2006; Novak et al. 2009; Hass et al. 2012; Chintala et al. 2014a). In addition, biochar application to cropland may reduce the N demand by plants through enhanced N use efficiency (Zhang et al. 2012) that, in turn, reduces the emission of greenhouse gases from the N fertilizer industry, if the amount of fertilizer N to be applied can be reduced due to increased N use efficiency. Thus, biochar can be considered as an ideal amendment for improving soil fertility.

The improvement in soil fertility after biochar addition may be due to beneficial effects of biochar on soil chemical, physical, and biological properties by altering pH, CEC, moisture content, and microbial compositions (Glaser et al. 2002; Asai et al. 2009; Atkinson et al. 2010; Sohi et al. 2010). However, there is still considerable controversy about the effect of biochar on soil fertility. Some studies showed that biochar increased the adsorption capability of plants for soil nutrients and thereby reduced the leaching losses of nutrients (Cheng et al. 2008; Laird et al. 2010a). In contrast, other studies pointed out that biochar addition might limit soil N availability in N-deficient soils due to the high C/N ratio of biochar and therefore might reduce crop productivity, at least temporarily (Lehmann et al. 2003; Asai et al. 2009). The exact effect of biochar application on soil fertility depends on many factors, such as the feedstock type used for biochar production, pyrolysis condition, application patterns (e.g., application rate, size of the biochar, field management), soil properties, and environmental conditions (Joseph et al. 2010; Alburquerque et al. 2013). This chapter provides an overview of our current understanding of the effect of biochar application on soil fertility and nutrient cycling, mainly focusing on the dynamics of soil nutrients and chemical characteristics that are closely related to soil fertility.

9.2 Biochar Effects on Soil Organic Matter

Soil organic C (SOC) content is one of the key indicators of soil fertility. Increasing SOC content is one of the main means to decrease the carbon dioxide (CO_2) concentration in the atmosphere (Andrews et al. 2004). Converting labile organic materials to a more recalcitrant form renders the organic C more resistant to biodegradation. As such, converting biomass to biochar and applying biochar to the soil has been used to sequester C, improve soil fertility, and offset anthropogenic CO_2 emissions (Lehmann 2007; Sohi et al. 2010; Hass et al. 2012; Akhtar et al. 2014). Biochar was responsible for the high soil organic matter content and enhanced soil fertility of anthropogenic soils (terra preta) found in the central Amazon (Glaser et al. 2002). Biochar addition could increase the formation and stabilization of soil microaggregates (Brodowski et al. 2006) that often play a greater role in providing physical protection to SOC than macroaggregates (Cheng et al. 2008; Hua et al. 2014). Adding biochar to the soil may therefore help trap SOC within aggregates and reduce the decomposition of SOC, and C occlusion within aggregates might be one of the main reasons for the stabilization of SOC in biochar-amended soils. For example, in an incubation experiment, Hua et al. (2014) found that the amendment of biochar significantly increased SOC content and promoted the formation of soil aggregates after 7 months of incubation. Other research similarly reported that SOC increased with increasing biochar application rate (Streubel et al. 2011; Uzoma et al. 2011a). In addition, biochar addition could result in much greater increase in SOC in soils with a low initial C content (Table 9.1).

Manure-derived biochars, having higher labile C concentrations than those of plant biomass–derived biochars, could be mineralized to a greater extent and show higher positive priming effects on native SOC mineralization (Singh and Cowie 2014), whereas biochars produced from grasses can exert greater effects on native SOC mineralization than those from hardwoods (Zimmerman et al. 2011). Moreover, greater positive priming effects on native SOC mineralization were

TABLE 9.1

Effects of Biochar Application on SOC and DOC Concentrations and Soil CO_2 Emissions

Background C Content (%)	Biochar				Dose	Effects (% Change)[a]			Reference
	Feedstock	PT[b] (°C)	Act.[b]	Size (mm)		SOC	DOC	CO_2	
1.60	Wood chip	420	N	NA	0.5%	/	/	NS	Marchetti and Castelli (2013)
	Swine solids	420	N	NA	0.5%	/	/	NS	
0.55	Oak wood	600	N	≤0.25	0.5%	+56	−3.6	/	Demisie et al. (2014)
					1%	+181	−10	/	
					2%	+342	NS	/	
	Bamboo	600	N	≤0.25	0.5%	+37	NS	/	
					1%	+98	−11	/	
					2%	+281	−9.0	/	
1.21	Corn straw	600	N	<1	8%	+41	/	−29	Hua et al. (2014)
0.49	Corn straw	600	N	<1	8%	+75	/	−39	
1.60	Wood chips	525	N	<2	3%	+119	/	/	Kloss et al. (2014a)
	Wheat straw	525	N	<2	3%	+188	/	/	
	Vineyard pruning	400	N	<2	3%	+106	/	/	
	Vineyard pruning	525	N	<2	3%	+169	/	/	

(Continued)

TABLE 9.1 *(Continued)*
Effects of Biochar Application on SOC and DOC Concentrations and Soil CO_2 Emissions

Background C Content (%)	Biochar					Effects (% Change)[a]			Reference
	Feedstock	PT[b] (°C)	Act.[b]	Size (mm)	Dose	SOC	DOC	CO_2	
2.87	Willow chips	470	N	>2	0.5%	/	/	NS	Prayogo et al. (2014)
					2%	/	/	−10	
1.97	Poultry litter	450	N	NA	10 Mg ha^{-1}	+19	/	/	Chan et al. (2008)
					25 Mg ha^{-1}	+40	/	/	
					50 Mg ha^{-1}	+80	/	/	
		550	Y	NA	10 Mg ha^{-1}	+16	/	/	
					25 Mg ha^{-1}	+27	/	/	
					50 Mg ha^{-1}	+64	/	/	
2.27	Wood mixtures	450	N	NA	50 Mg ha^{-1}	+22	NS	NS	Jones et al. (2012)

CO_2, carbon dioxide; DOC, dissolved organic carbon; SOC, soil organic carbon.

[a] / indicates that the subject was not studied and +, −, and NS indicate stimulative, inhibitory, and lack of significant effect, respectively.

[b] PT, pyrolysis temperature; Act., activation.

observed for biochars produced under low temperatures (250–400°C) compared with those produced under high temperatures (500–650°C) (Zimmerman et al. 2011; Ippolito et al. 2012; Singh and Cowie 2014).

However, there is conflicting evidence in the literature about the effect of biochar on native SOC minerlization (Table 9.1); positive, negative, or no priming effects are all reported in previous studies (Zimmerman et al. 2011; Ippolito et al. 2012; Prayogo et al. 2014; Singh and Cowie 2014). Nevertheless, even though positive priming effects may occur, the amount of native C lost by positive priming is very small compared to the amount of C added in the form of biochar, and soil C storage can also be enhanced over the long term (over many years) (Table 9.1). In addition, soil dissolved organic carbon (DOC), one of the most important forms of soil labile organic C, can be influenced by biochar addition (Table 9.1). Some studies showed that biochar application decreased DOC, and this reduction may be due to the increase in microbial activity induced by the biochar or to the marked adsorption of DOC to soil particles in the presence of biochar (Yuan and Xu 2011; Hass et al. 2012; Guerena et al. 2013; Yin et al. 2014). In contrast, other studies did not find any effect of biochar application on soil DOC (Jones et al. 2012; Demisie et al. 2014). Therefore, further research should be conducted to understand the mechanisms involved in biochar regulating soil native SOC minerlization and the dynamics of labile organic C, which are closely related to the activities of soil microorganisms, plant growth, and the efficiency of C sequestration.

9.3 Biochar Effects on Soil Nitrogen

Nitrogen loss to the environment from intensively managed agricultural production systems cause negative feedbacks to environmental and human health, such as water eutrophication, ozone depletion, and contamination of drinking water (Guerena et al. 2013). Biochar addition to the soil can be beneficial both to the environment and to the crop productivity

(Lehmann 2007). Some studies showed that biochar application markedly increased soil N retention (Zheng et al. 2012; Zhao et al. 2013), which was due to decreased leaching and increased recovery of applied fertilizer N (Guerena et al. 2013). Moreover, the growth of soil microorganisms that promote the complete denitrification or dissimilatory reduction of nitrate (NO_3^-) to ammonium (NH_4^+) (DNRA) may potentially be enhanced by the addition of biochar, thereby reducing N losses to leaching or gaseous fluxes (Lehmann et al. 2003; Novak et al. 2009; Laird et al. 2010b; Anderson et al. 2011; Dempster et al. 2012a). However, in a 3-year field experiment conducted on a sandy clay loam soil in Wales, UK, Jones et al. (2012) reported that biochar had no notable short- or long-term effect on soil N cycling (Table 9.2). Even though soil N

TABLE 9.2
Effects of Biochar Application on TN, and AP and AK Concentrations

Biochar			Effects (% Change)[a]			
Feedstock	PT[b] (°C)	Dose	TN	AP	AK	Reference
Poultry litter	450	10 Mg ha^{-1}	+13	+343	+96	Chan et al. (2008)
		25 Mg ha^{-1}	+22	+700	+221	
		50 Mg ha^{-1}	+43	+1129	+500	
	550	10 Mg ha^{-1}	NS	+63	+48	
		25 Mg ha^{-1}	NS	+71	+113	
		50 Mg ha^{-1}	+14	+113	+263	
Peanut hull	400	11 Mg ha^{-1}	NS	+39	+88	Gaskin et al. (2010)
		22 Mg ha^{-1}	+24	+76	+162	
Pine chip	400	11 Mg ha^{-1}	+20	NS	NS	
		22 Mg ha^{-1}	+17	NS	NS	
Wood mixtures	450	50 Mg ha^{-1}	NS	NS	/	Jones et al. (2012)
Wood chips	525	1%	NS	+25	+105	Kloss et al. (2014a)
		3%	NS	+70	+296	

AK, available K; AP, available P; TN, total N.

[a] / indicates that the subject was not studied and + and NS indicate stimulative and lack of significant effect, respectively.

[b] PT, pyrolysis temperature.

transformations or microbial community structure under ruminant urine patches was not affected, biochar addition increased the relative abundance of nitrifiers and denitrifiers (Anderson et al. 2014). Compared with the control soil, biochar application increased the presence of N-fixing and denitrification genes in an eroded calcareous soil (Ducey et al. 2013). Further research is needed to better understand the biogeochemical processes regulating N cycling, N retention in particular, and changes of associated microbiological properties that occur in biochar-amended soils.

Manure-derived biochars had higher N contents than those produced from plant biomass, but plant biomass–derived biochars had higher C/N ratios (Sigua et al. 2014). Soil C/N ratio generally increases with increasing biochar application rates, whereas rates of net N mineralization decrease and rates of N immobilization increase with the increase in soil C/N ratio (Lehmann et al. 2003; Chan et al. 2008; Uzoma et al. 2011a). Net N mineralization was therefore found significantly decreased with the increasing rate of addition of plant biomass–derived biochar (Streubel et al. 2011; Dempster et al. 2012b). Ameloot et al. (2013) found higher net N mineralization rates in a soil applied with swine manure digestate–derived biochar than that applied with willow wood–derived biochar, which had greater C/N ratios, further demonstrating the potential impact of the C/N ratio of the applied biochar on N cycling. However, although an increase in soil N mineralization was expected in response to chicken manure–derived biochar addition, biochar application decreased the NO_3^--N level in a loamy soil, probably due to enhanced N immobilization (Hass et al. 2012). Increased N immobilization by biochar addition has also been reported in other studies (Lehmann et al. 2003; Novak et al. 2010; Wang et al. 2011). Therefore, soil N retention could be increased through the addition of manure-derived biochar, whereas addition of plant biomass–derived biochar had little potential effect on soil N retention, probably more so for grass biomass–derived biochar than wood biomass–derived biochar, which generally has a higher C/N ratio (Schomberg et al. 2012).

In addition to the feedstock type used, conditions used for pyrolysis can influence soil N cycling as well. The application of biochars produced at high temperatures ($\geq 500°C$) induced greater net N mineralization and lower N immobilization than those produced at low temperatures ($\leq 400°C$); as a result, biochars produced at low temperatures may induce less mineral N leaching loss and greater soil N retention (Ippolito et al. 2012; Kameyama et al. 2012; Nelissen et al. 2012; Yao et al. 2012). In addition, application of fresh biochar produced from wheat straw by slow pyrolysis increased net N mineralization, whereas application of biochar produced through fast pyrolysis led to N immobilization (Bruun et al. 2012); but until recently, few studies have examined the effect of rate of biochar pyrolysis (slow versus fast) on soil N dynamics.

The effect of biochar on soil N cycling may also be related to biochar application rate or the particle size of biochar as well as field management. Soil mineral N concentrations as well as N concentrations in leachates decrease after biochar application, probably due to increased microbial immobilization and reduced nitrification activity (Ippolito et al. 2014; Sika and Hardie 2014). In contrast, Zhao et al. (2014a) reported that the application of rice straw–derived biochar stimulated nitrification in two subtropical acidic arable soils. However, biochar application either inhibited or enhanced soil nitrification, depending on the type of N fertilizer used (Zheng et al. 2013). Another study showed that biochar application did not affect total N leaching at low N fertilization rate, but decreased total N leaching at high fertilization rate (Guerena et al. 2013). Similarly, biochar did not affect the loss of NO_3^-, NO_4^-, or nitrous oxide (N_2O) in unfertilized soils, but decreased the cumulative NO_3^- leaching and N_2O emission and had inconsistent effects on NH_4^+ leaching in N fertilized soils (Zheng et al. 2012). It should be noted that the effect of biochar application on soil NH_4^+ leaching was also dependent on the particle size of the biochar, with greater NH_4^+ leaching with the small- (<250-μm) than the large-sized (>250-μm) biochar treatment (Zheng et al. 2012). Those differences in the effect of biochar on soil N leaching were probably due to the different biochar effects on N transformation (e.g., mineralization, immobilization, and nitrification) potentials and the large discrepancy in the absorption capacity of different biochars (Zhao et al. 2013). Therefore, whether biochar application to soils results in a reduction in N leaching or not depends on the feedstock type, pyrolysis condition, biochar application rate, biochar particle size, fertilizer type, and fertilization rate.

9.4 Biochar Effects on Soil Phosphorus and Potassium

The concentration of available P in the soil is lower than that of N and K, and sorption of P by soil constituents is a dominant mechanism for P retention in the soil (Chintala et al. 2014a). The application of biochar has been found to increase the retention of P and the availability of P and K in soils; the increase is positively related to the biochar application rate (Table 9.2). There are two possible explanations for the increase of soil available P content after biochar application: (1) the high P content in the biochar applied and (2) the increased CEC and decreased soluble Al in acidic soils as a result of biochar application would increase P availability in the soil (Hass et al. 2012; Guo et al. 2014). However, biochar application effects on soil P availability are not universal (Gaskin et al. 2010; Liang et al. 2014). In addition, the incorporation of biochar into agricultural soils can increase the leaching of P and K, at least in the short term (Angst et al. 2014; Guo et al. 2014). For the leachates, ~30–70% of the dissolved P occurred as phosphate (PO_4^{3-})-P after the application of chicken manure–derived biochar, and the increase in total P levels in leachate was more pronounced with the application of activated biochars (with steam activation at a flow rate of 3 mL·min^{-1} for 45 min at 800°C) than in their nonactivated counterparts, probably because steam activation increased P solubility in the activated biochars (Hass et al. 2012).

The impact of biochar application on soil dissolved P varies with biochar feedstock type and soil property. For example, the incorporation of poultry litter–derived biochar generally increased PO_4^{3-}-P, whereas pine chip–derived biochar had a minimal or no effect on PO_4^{3-}-P in soil solution (Abit et al. 2012).

However, Laird et al. (2010a) observed a significant decrease in the total amount of P leached from manure-amended columns as mixed hardwood-derived biochar application rates increased, in spite of the biochar itself added a substantial amount of P to the columns. When wood-derived biochars applied to a low-fertility, acidic soil, leaching of K significantly increased and leaching of P significantly decreased at the 0.6-m soil depth, whereas leaching of both P and K significantly decreased at the 1.2-m soil depth, indicating that the effect of biochar on P and K also depends on the depth of leachate collection (Major et al. 2012). In addition, the impact of biochar on P retention and release depends on soil pH. The incorporation of biochars significantly decreased P sorption and increased its availability in acidic soils, but it increased P sorption and decreased its availability in calcareous soils (Chintala et al. 2014b). Therefore, the effect of biochar application on soil P and K concentrations is closely related to the P and K contents in the biochar and their availabilities in the soil, with increased sorption and reduced leaching leading to high P and K retention and benefiting soil fertility.

9.5 Biochar Effects on Other Soil Nutrients

Biochar application may affect some other soil nutrients, such as Ca, Mg, S, Zn, Fe, and Mn, through changed physical or chemical processes, with little connection with changed microbiological processes in the soil. Given that most biochars have a large surface area, microporous structure, and high pH and soluble salts, biochar application can potentially reduce the solubility of some heavy metals in soils through adsorption and precipitation (Zhang et al. 2013a; Lu et al. 2014). Therefore, biochar may reduce the bioavailability of plant essential nutrients or toxic trace elements in soils (Beesley et al. 2011; Kloss et al. 2014b). A pot experiment showed that the concentrations of toxicity characteristic leaching procedure–extractable Cu and Zn in the soil significantly decreased with increasing rates of biochar application (Lu et al. 2014). Similarly, Beesley et al. (2010) reported that a hardwood-derived biochar addition significantly decreased the water-extractable Zn concentration in the soil. However, there are many inconsistent results in the literature. For example, Novak et al. (2009) found that pecan shell–derived biochar applied to a loamy sand soil reduced soil S and Zn availabilities but increased Ca and Mn availabilities. In contrast, the amendment of mixed hardwood-derived biochar significantly increased soil extractable Mg and Ca, but it did not affect extractable S, Cu, and Zn in a repacked soil column experiment (Laird et al. 2010a). Another study found that the application of filtercake-derived biochar strongly altered nutrient availability in a Ferralsol; the extractability of Fe, Mn, and Zn was significantly increased only under a high application rate (5%), the Ca availability markedly increased even under a low application rate (1.25%), and soil S and B availabilities were not influenced by biochar addition (Eykelbosh et al. 2014). Moreover, Lehmann et al. (2003) and Zhao et al. (2014b) found that biochar addition had little effect on exchangeable Ca and Mg in soils.

Soil nutrient status after biochar application may be not only linked to the feedstock type but also the pyrolysis conditions and application rates. A field study showed that soil Ca and Mg, but not S, in the surface soil linearly increased with increasing rate of application of peanut hull–derived biochar; in contrast, pine chip–derived biochar did not affect soil S, Ca, and Mg concentrations (Gaskin et al. 2010). Another study by Ippolito et al. (2014) reported that increasing hardwood-derived biochar application rate to alkaline soils increased soil Fe and Mn availabilities, but it had no effect on soil Zn and Cu availabilities. With regard to the effect of pyrolysis temperature, the availabilities of soil Cu, Mg, and Zn increased, whereas that of Fe, Mn, and S decreased with the application of biochar produced with an increasing pyrolysis temperature between 350°C and 700°C (Hass et al. 2012). Similarly, Ippolito et al. (2012) found that Ca and Mg leaching was lower, whereas extractable Mn and Ni concentrations were higher in two Aridisols after the application of switchgrass biomass–derived biochar produced at 250°C than those produced at 500°C, probably due to the greater total negative surface charge favoring divalent cation sorption in the biochar produced at low than at high temperature. Therefore, the effect of biochar application on the availability of other soil nutrients is mainly related to the characteristics of the soil and the nutrient of concern, the adsorption or precipitation characteristics of the biochar and the biochar-amended soil.

9.6 Biochar Effects on Soil pH and CEC

The effects of biochar application on soil pH are closely related to the biochar feedstock type and the pyrolysis process (Table 9.3). Streubel et al. (2011) found that the average pH of the herbaceous biomass–derived biochars (9.4) was 2 units higher than woody biomass–derived biochars (7.4), and the increases in soil pH were 0.4–0.8 and 0.1–0.4 units for soils amended with herbaceous- and woody biomass–derived biochars, respectively. In addition, biochars produced at high pyrolysis temperatures (≥500°C) generally had greater pH than those produced at low temperatures (≤400°C) (Table 9.3). Many studies showed that the application of high-pH biochars led to much higher soil pH compared with the application of low-pH biochars (Chan et al. 2008; Abit et al. 2012; Chintala et al. 2014b). There is a significant positive linear relationship between biochar pH and pH of biochar-treated soil, and biochar alkalinity might be a key factor in controlling the liming effect on acid soils (Yuan and Xu 2011).

The background soil pH value should influence the effect of biochar application on soil pH as well. When biochars are added to acidic soils, soil pH can significantly increase with increasing application rate of biochar (Uzoma et al. 2011b; Jien and Wang 2013). However, it is unlikely to be beneficial to plant growth if biochar application resulted in soil pH > 8.0, which may temporarily adversely affect the availability of some elements by lowering their water-soluble fractions (Artiola et al. 2012). But it should be noted that the addition of biochar at high rates could result in overliming of the soil; addition of biochar to an acidic soil (pH 5.14) significantly increased soil pH to 6.80, 7.34, and 8.42 at application levels of 0.5, 2.5, and 10%, respectively (Sika and Hardie 2014). In addition, compared to silt loam soils (pH < 5.0), sand soils (pH 7.1) exhibited the greatest

TABLE 9.3
Effects of Biochar Application on Soil pH and CEC

Soil Type or Textural Class	Biochar					Effects (% Change)[a]		Reference
	Feedstock	PT[b] (°C)	Act.[b]	Size (mm)	Dose	pH	CEC	
Alfisol (Chromosol)	Poultry litter	450	N	NA	10 Mg ha^{-1}	+21	+19	Chan et al. (2008)
					25 Mg ha^{-1}	+32	+35	
					50 Mg ha^{-1}	+41	+65	
		550	Y	NA	10 Mg ha^{-1}	+38	+19	
					25 Mg ha^{-1}	+51	+56	
					50 Mg ha^{-1}	+61	+76	
Loamy sand	Peanut hull	400	Y	NA	11 Mg ha^{-1}	NS	/	Gaskin et al. (2010)
					22 Mg ha^{-1}	+3.9	/	
	Pine chip	400	Y	NA	11 Mg ha^{-1}	+1.6	/	
					22 Mg ha^{-1}	−1.1	/	
Sandy clay loam	Wood mixtures	450	N	NA	50 Mg ha^{-1}	+4.7	/	Jones et al. (2012)

(Continued)

TABLE 9.3 (Continued)
Effects of Biochar Application on Soil pH and CEC

| Soil Type or Textural Class | Biochar | | | | | Effects (% Change)[a] | | Reference |
	Feedstock	PT[b] (°C)	Act.[b]	Size (mm)	Dose	pH	CEC	
Sandy loam	Wood chip	620	N	NA	8%	NS	NS	Cely et al. (2014)
	Paper sludge and wheat husks	500	N	NA	8%	NS	NS	
	Sewage sludge	600	N	NA	8%	NS	NS	
Clay loam	Corn stover	650	N	<0.2	2–6%	+	+	Chintala et al. (2014b)
	Switchgrass	650	N	<0.2	2–6%	+	+	
Sandy loam	Wood chips	525	N	<2	3%	+30	+35	Kloss et al. (2014a)
	Wheat straw	525	N	<2	3%	+23	+25	
	Vineyard pruning	400	N	<2	3%	+25	+29	
	Vineyard pruning	525	N	<2	3%	+25	+29	
Clay loam	Willow chips	470	N	>2	0.5%	NS	/	Prayogo et al. (2014)

CEC, cation exchange capacity.

[a] /indicates that the subject was not studied and +, −, and NS indicate stimulative, inhibitory, and lack of significant effect, respectively.

[b] PT, pyrolysis temperature; Act., activation.

and most rapid increase in pH after biochar application, due to the lower buffering capacity of sand soils than silt loam and clayey soils (Streubel et al. 2011). Therefore, the pH increase after biochar application was greater in sandy and loamy soils than in clayey soils (Glaser et al. 2002). The average increase in pH after biochar application was 1.69 and 1.15 units in paddy (pH 5.41) and forest (pH 3.84) soils, respectively, implying that the soil pH buffering capacity was higher in the forest soil (Yu et al. 2013). Interestingly, a 2-year field study showed that application of alkaline biochars to acidic soils decreased soil pH in the second year (Gaskin et al. 2010). Prommer et al. (2014) reported that the effect of hardwood-derived biochar on soil pH was minimal, with soil pH decreased from 7.5 to 7.4 in biochar-treated soils, possibly because more carboxylic functional groups were created by oxidation reactions on the surface of the biochar over time (Cheng et al. 2006), but this phenomenon might be related to the weakened liming effect of biochar, when biochar incorporation into acidic soils accelerated nitrification (Zhao et al. 2014a).

The CEC and specific surface area are indirect measures of the capacity of soils to retain water, nutrients, and various contaminants (Laird et al. 2010a). Biochar additions to the soil have been reported to increase CEC. For example, in a loamy Mollisol in Iowa, USA, biochar addition resulted in a 20% increase in soil CEC (Laird et al. 2010a). Chintala et al. (2014b) found that the increase in soil CEC values was significantly higher in corn stover–derived biochar treatments at all application rates than that in switchgrass biomass–derived biochar treatments, and soil CEC values increased with increasing application rates of both biochars. However, biochar application did not affect CEC in an alkaline soil (Zhang et al. 2013b). In contrast, Prommer et al. (2014) reported that soil CEC decreased slightly from an initial 22.5 to 20.8 cmolc·kg^{-1} in hardwood-derived biochar-treated soils. In addition, it should be noted that effective CEC was significantly and positively correlated with soil pH (Cheng et al. 2008), and similar to soil pH, changes in CEC have also been shown to be a function of biochar age (time interval between biochar application to the soil and at the time measurement was made), as surface oxidation over time increases the number of negatively charged sites as the age of biochar increases (Cheng et al. 2008; Zimmerman et al. 2011). The CEC of a biochar-amended soil would thus likely change over time and the effect of biochar addition on the CEC would change over time, with the aged biochar having less influence on soil CEC.

9.7 Summary

In this chapter, we reviewed the current literature on the effects of biochar application on soil fertility and nutrient cycling. The information contained in this chapter will be useful for researchers and biochar users in understanding the agronomic implications of biochar use in agricultural production systems. The literature is clear that biochar characteristics are greatly influenced by the nature of the feedstock type and pyrolysis conditions used for biochar production. Therefore, the effect of biochar on soil fertility is dependent on the properties of the biochars applied and other factors such as soil properties, biochar particle sizes, and management practices.

The benificial effects of biochar addition on increasing SOC are consistent across a range of soil and biochar types, and applying C-rich biochars to C-poor and acidic soils will lead to greater benefits in increasing SOC than applying to C-rich and basic soils. However, contradictory results have been reported about the effect of biochar application on native SOC minerlization and the dynamics of soil labile organic C. Much more research is needed to fully understand the mechanisms involved in biochar effects on SOC dynamics.

Biochar addition is expected to increase soil N retention, which may be greatly increased through applying manure-derived biochars, whereas plant biomass–derived biochars have less potential to supply soil N. Increases in soil C are greater than that in soil N with increasing biochar application rate, which can result in high soil C/N ratio and low soil N availability. Soils amended with biochars produced at a high temperature can cause greater net N mineralization and NO_3^--N leaching loss compared with those produced at a low temperature. The effect of biochar on soil N leaching is dependent on the characteristics of N transformation processes in the soil and the N absorption capacity of different biochars.

The P and K contents in the biochar and the availabilities of native P and K in the soil influence the effect of biochar application on P and K availabilities in soils. Biochar application would most likely increase and decrease P availability in acidic and calcareous soils, respectively. The effect of biochar application on the availability of other soil nutrients, such as Ca, Mg, Fe, Mn, Zn, and S, is linked to the feedstock type, the pyrolysis condition, and the biochar application rate; the effect of biochar application on those nutrients is more complicated than that on soil C and N.

Overliming should be prevented when applying biochar to soils, with high-pH biochars having a greater effect on increasing soil pH. The pH of acidic soils with a low buffering capacity can be greatly influenced by biochar application, and aging of biochar substantially reduces the liming effect of biochar on acidic soils. Similarly, effects of biochar addition on soil CEC range from positive to negative, and the CEC of a biochar-amended soil will also likely change over time, with aged biochar (after long-term oxidation) having little effect on soil CEC.

Our mechanistic understanding on the effects of biochar on soil fertility and nutrient dynamics is still quite limited; therefore, further research is urgently needed in improving the mechanistic understanding of biochar effects on soil properties and processes to help design biochars for specific applications. Interactions among biochar, soil, microorganisms, and plants in biochar-amended soils can be very complex, and mechanistic understanding of those interactions is still lacking. Moreover, most of the literature on biochar application is based on short-term laboratory, column, or pot studies; thus, long-term field experiments studying the processes involved in soil nutrient cycling and the relevant microbiological properties are urgently needed (Joseph et al. 2010; Sohi et al. 2010; Jones et al. 2012; Guerena et al. 2013). Furthermore, the longevity of biochar benefits on soil fertility and nutrient cycling should be addressed in future long-term field studies.

References

Abit, S.M., C.H. Bolster, P. Cai, and S.L. Walker. 2012. Influence of feedstock and pyrolysis temperature of biochar amendments on transport of *Escherichia coli* in saturated and unsaturated soil. *Environ. Sci. Technol.* 46: 8097–8105.

Akhtar, S.S., G.T. Li, M.N. Andersen, and F.L. Liu. 2014. Biochar enhances yield and quality of tomato under reduced irrigation. *Agric. Water Manage.* 138: 37–44.

Alburquerque, J.A., P. Salazar, V. Barron, J. Torrent, M.D. del Campillo, A. Gallardo, and R. Villar. 2013. Enhanced wheat yield by biochar addition under different mineral fertilization levels. *Agron. Sustain. Dev.* 33: 475–484.

Ameloot, N., S. Neve, K. Jegajeevagan, G. Yildiz, D. Buchan, Y.N. Funkuin, W. Prins, L. Bouckaert, and S. Sleutel. 2013. Short-term CO_2 and N_2O emissions and microbial properties of biochar amended sandy loam soils. *Soil Biol. Biochem.* 57: 401–410.

Anderson, C.R., L.M. Condron, T.J. Clough, M. Fiers, A. Stewart, R.A. Hill, and R.R. Sherlock. 2011. Biochar induced soil microbial community change: Implications for biogeochemical cycling of carbon, nitrogen and phosphorus. *Pedobiologia.* 54: 309–320.

Anderson, C.R., K. Hamonts, T.J. Clough, and L.M. Condron. 2014. Biochar does not affect soil N-transformations or microbial community structure under ruminant urine patches but does alter relative proportions of nitrogen cycling bacteria. *Agr. Ecosyst. Environ.* 191: 63–72.

Andrews, S.S., D.L. Karlen, and C.A. Cambardella. 2004. The soil management assessment framework: A quantitative soil quality evaluation method. *Soil Sci. Soc. Am. J.* 68: 1945–1962.

Angst, T.E., J. Six, D.S. Reay, and S.P. Sohi. 2014. Impact of pine chip biochar on trace greenhouse gas emissions and soil nutrient dynamics in an annual ryegrass system in California. *Agr. Ecosyst. Environ.* 191: 17–26.

Artiola, J.F., C. Rasmussen, and R. Freitas. 2012. Effects of a biochar-amended alkaline soil on the growth of romaine lettuce and bermudagrass. *Soil Sci.* 177: 561–570.

Asai, H., B.K. Samson, H.M. Stephan, K. Songyikhangsuthor, K. Homma, Y. Kiyono, Y. Inoue, T. Shiraiwa, and T. Horie. 2009. Biochar amendment techniques for upland rice production in Northern Laos 1. Soil physical properties, leaf SPAD and grain yield. *Field Crop. Res.* 111: 81–84.

Atkinson, C.J., J.D. Fitzgerald, and N.A. Hipps. 2010. Potential mechanisms for achieving agricultural benefits from biochar application to temperate soils: A review. *Plant Soil.* 337: 1–18.

Beesley, L., E. Moreno-Jiménez, and J.L. Gomez-Eyles. 2010. Effects of biochar and greenwaste compost amendments on mobility, bio-availability and toxicity of inorganic and organic contaminants in a multi-element polluted soil. *Environ. Pollut.* 158: 2282–2287.

Beesley, L., E. Moreno-Jiménez, J.L. Gomez-Eyles, E. Harris, B. Robinson, and T. Sizmur. 2011. A review of biochars' potential role in the remediation, revegetation and restoration of contaminated soils. *Environ. Pollut.* 159: 3269–3282.

Brodowski, S., B. John, H. Flessa, and W. Amelung. 2006. Aggregate-occluded black carbon in soil. *Eur. J. Soil Sci.* 57: 539–546.

Bruun, E.W., P. Ambus, H. Egsgaard, and H. Hauggaard-Nielsen. 2012. Effects of slow and fast pyrolysis biochar on soil C and N turnover dynamics. *Soil Biol. Biochem.* 46: 73–79.

Cely, P., A.M. Tarquis, J. Paz-Ferreiro, A. Mendez, and G. Gasco. 2014. Factors driving the carbon mineralization priming effect in a sandy loam soil amended with different types of biochar. *Solid Earth* 5: 585–594.

Chan, K.Y., L. Van Zwieten, I. Meszaros, A. Downie, and S. Joseph. 2008. Using poultry litter biochars as soil amendments. *Soil Res.* 46: 437–444.

Cheng, C.H., J. Lehmann, and M.H. Engelhard. 2008. Natural oxidation of black carbon in soils: Changes in molecular form and surface charge along a climosequence. *Geochim. Cosmochim. Acta.* 72: 1598–1610.

Cheng, C.H., J. Lehmann, J.E. Thies, S.D. Burton, and M.H. Engelhard. 2006. Oxidation of black carbon by biotic and abiotic processes. *Org. Geochem.* 37: 1477–1488.

Chintala, R., J. Mollinedo, T.E. Schumacher, D.D. Malo, and J.L. Julson. 2014b. Effect of biochar on chemical properties of acidic soil. *Arch. Agron. Soil Sci.* 60: 393–404.

Chintala, R., T.E. Schumacher, L.M. McDonald, D.E. Clay, D.D. Malo, S.K. Papiernik, S.A. Clay, and J.L. Julson. 2014a. Phosphorus sorption and availability from biochars and soil/biochar mixtures. *Clean-Soil Air Water.* 42: 626–634.

Demisie, W., Z. Liu, and M. Zhang. 2014. Effect of biochar on carbon fractions and enzyme activity of red soil. *Catena* 121: 214–221.

Dempster, D.N., D.B. Gleeson, Z.M. Solaiman, D.L. Jones, and D.V. Murphy. 2012a. Decreased soil microbial biomass and nitrogen mineralisation with Eucalyptus biochar addition to a coarse textured soil. *Plant Soil.* 354: 311–324.

Dempster, D.N., D.L. Jones, and D.V. Murphy. 2012b. Clay and biochar amendments decreased inorganic but not dissolved organic nitrogen leaching in soil. *Soil Res.* 50: 216–221.

Ding, Y., Y.X. Liu, W.X. Wu, D.Z. Shi, M. Yang, and Z.K. Zhong. 2010. Evaluation of biochar effects on nitrogen retention and leaching in multi-layered soil columns. *Water Air Soil Pollut.* 213: 47–55.

Ducey, T.F., J.A. Ippolito, K.B. Cantrell, J.M. Novak, and R.D. Lentz. 2013. Addition of activated switchgrass biochar to an aridic subsoil increases microbial nitrogen cycling gene abundances. *Appl. Soil Ecol.* 65: 65–72.

Eykelbosh, A.J., M.S. Johnson, E.S. de Queiroz, H.J. Dalmagro, and E.G. Couto. 2014. Biochar from sugarcane filtercake reduces soil CO_2 emissions relative to raw residue and improves water retention and nutrient availability in a highly-weathered tropical soil. *PLOS One.* 9: e98523.

Gaskin, J.W., R.A. Speir, K. Harris, K.C. Das, R.D. Lee, L.A. Morris, and D.S. Fisher. 2010. Effect of peanut hull and pine chip biochar on soil nutrients, corn nutrient status, and yield. *Agron. J.* 102: 623–633.

Glaser, B., J. Lehmann, and W. Zech. 2002. Ameliorating physical and chemical properties of highly weathered soils in the tropics with charcoal—A review. *Biol. Fert. Soils.* 35: 219–230.

Gomez, J.D., K. Denef, C.E. Stewart, J. Zheng, and M.F. Cotrufo. 2014. Biochar addition rate influences soil microbial abundance and activity in temperate soils. *Eur. J. Soil Sci.* 65: 28–39.

Guerena, D., J. Lehmann, K. Hanley, A. Enders, C. Hyland, and S. Riha. 2013. Nitrogen dynamics following field application of biochar in a temperate North American maize-based production system. *Plant Soil.* 365: 239–254.

Guo, Y.J., H. Tang, G.D. Li, and D.T. Xie. 2014. Effects of cow dung biochar amendment on adsorption and leaching of nutrient from an acid yellow soil irrigated with biogas slurry. *Water Air Soil Pollut.* 225: 1820.

Hass, A., J.M. Gonzalez, I.M. Lima, H.W. Godwin, J.J. Halvorson, and D.G. Boyer. 2012. Chicken manure biochar as liming and nutrient source for acid Appalachian soil. *J. Environ. Qual.* 41: 1096–1106.

Hua, L., Z. Lu, H. Ma, and S. Jin. 2014. Effect of biochar on carbon dioxide release, organic carbon accumulation, and aggregation of soil. *Environ. Prog. Sustain. Energy.* 33: 941–946.

Ippolito, J.A., J.M. Novak, W.J. Busscher, B.M. Ahmedna, D. Rehrah, and D.W. Watts. 2012. Switchgrass biochar affects two aridisols. *J. Environ. Qual.* 41: 1123–1130.

Ippolito, J.A., M.E. Stromberger, R.D. Lentz, and R.S. Dungan. 2014. Hardwood biochar influences calcareous soil physicochemical and microbiological status. *J. Environ. Qual.* 43: 681–689.

Jien, S.H., and C.S. Wang. 2013. Effects of biochar on soil properties and erosion potential in a highly weathered soil. *Catena.* 110: 225–233.

Jones, D.L., J. Rousk, G. Edwards-Jones, T.H. DeLuca, and D.V. Murphy. 2012. Biochar-mediated changes in soil quality and plant growth in a three year field trial. *Soil Biol. Biochem.* 45: 113–124.

Joseph, S.D., M. Camps-Arbestain, Y. Lin, P. Munroe, C.H. Chia, J. Hook, L. van Zwieten, et al. 2010. An investigation into the reactions of biochar in soil. *Aust. J. Soil Res.* 48: 501–515.

Kameyama, K., T. Miyamoto, T. Shiono, and Y. Shinogi. 2012. Influence of sugarcane bagasse-derived biochar application on nitrate leaching in calcaric dark red soil. *J. Environ. Qual.* 41: 1131–1137.

Kloss, S., F. Zehetner, E. Oburger, J. Buecker, B. Kitzler, W.W. Wenzel, B. Wimmer, and G. Soja. 2014b. Trace element concentrations in leachates and mustard plant tissue (*Sinapis alba* L.) after biochar application to temperate soils. *Sci. Total Environ.* 481: 498–508.

Kloss, S., F. Zehetner, B. Wimmer, J. Buecker, F. Rempt, and G. Soja. 2014a. Biochar application to temperate soils: Effects on soil fertility and crop growth under greenhouse conditions. *J. Plant Nutr. Soil Sci.* 177: 3–15.

Laird, D., P. Fleming, B.Q. Wang, R. Horton, and D. Karlen. 2010b. Biochar impact on nutrient leaching from a Midwestern agricultural soil. *Geoderma.* 158: 436–442.

Laird, D.A., P. Fleming, D.D. Davis, R. Horton, B.Q. Wang, and D.L. Karlen. 2010a. Impact of biochar amendments on the quality of a typical Midwestern agricultural soil. *Geoderma.* 158: 443–449.

Lehmann, J. 2007. Bio-energy in the black. *Front. Ecol. Environ.* 5: 381–387.

Lehmann, J., J.P. da Silva, C. Steiner, T. Nehls, W. Zech, and B. Glaser. 2003. Nutrient availability and leaching in an archaeological Anthrosol and a Ferralsol of the Central Amazon basin: Fertilizer, manure and charcoal amendments. *Plant Soil.* 249: 343–357.

Liang, B., J. Lehmann, D. Solomon, J. Kinyangi, J. Grossman, B. O'Neill, J.O. Skjemstad, et al. 2006. Black carbon increases cation exchange capacity in soils. *Soil Sci. Soc. Am. J.* 70: 1719–1730.

Liang, F., G.T. Li, Q.M. Lin, and X.R. Zhao. 2014. Crop yield and soil properties in the first 3 years after biochar application to a calcareous soil. *J. Integr. Agric.* 13: 525–532.

Lu, K.P., X. Yang, J.J. Shen, B. Robinson, H.G. Huang, D. Liu, N. Bolan, J.C. Pei, and H.L. Wang. 2014. Effect of bamboo and rice straw biochars on the bioavailability of Cd, Cu, Pb and Zn to Sedum plumbizincicola. *Agr. Ecosyst. Environ.* 191: 124–132.

Major, J., M. Rondon, D. Molina, S.J. Riha, and J. Lehmann. 2012. Nutrient leaching in a Colombian savanna Oxisol amended with biochar. *J. Environ. Qual.* 41: 1076–1086.

Marchetti, R., and F. Castelli. 2013. Biochar from swine solids and digestate influence nutrient dynamics and carbon dioxide release in soil. *J. Environ. Qual.* 42: 893–901.

Moyin-Jesu, E.I. 2007. Use of plant residues for improving soil fertility, pod nutrients, root growth and pod weight of okra (*Abelmoschus esculentum L*). *Biores. Tech.* 98: 2057–2064.

Nelissen, V., T. Rutting, D. Huygens, J. Staelens, G. Ruysschaert, and P. Boeckx. 2012. Maize biochars accelerate short-term soil nitrogen dynamics in a loamy sand soil. *Soil Biol. Biochem.* 55: 20–27.

Novak, J.M., W.J. Busscher, D.L. Laird, M. Ahmedna, D.W. Watts, and M.A.S. Niandou. 2009. Impact of biochar amendment on fertility of a Southeastern Coastal Plain soil. *Soil Sci.* 174: 105–112.

Novak, J.M., W.J. Busscher, D.W. Watts, D.A. Laird, M.A. Ahmedna, and M.A.S. Niandou. 2010. Short-term CO_2 mineralization after additions of biochar and switchgrass to a Typic Kandiudult. *Geoderma.* 154: 281–288.

Prayogo, C., J.E. Jones, J. Baeyens, and G.D. Bending. 2014. Impact of biochar on mineralisation of C and N from soil and willow litter and its relationship with microbial community biomass and structure. *Biol. Fert. Soils.* 50: 695–702.

Prommer, J., W. Wanek, F. Hofhansl, D. Trojan, P. Offre, T. Urich, C. Schleper, S. Sassmann, B. Kitzler, G. Soja, and R.C. Hood-Nowotny. 2014. Biochar decelerates soil organic nitrogen cycling but stimulates soil nitrification in a temperate arable field trial. *PLoS One.* 9: e86388.

Schomberg, H.H., J.W. Gaskin, K. Harris, K.C. Das, J.M. Novak, W.J. Busscher, D.W. Watts, et al. 2012. Influence of biochar on nitrogen fractions in a coastal plain soil. *J. Environ. Qual.* 41: 1087–1095.

Sika, M.P., and A.G. Hardie. 2014. Effect of pine wood biochar on ammonium nitrate leaching and availability in a South African sandy soil. *Eur. J. Soil Sci.* 65: 113–119.

Sigua, G.C., J.M. Novak, D.W. Watts, K.B.Cantrell, P.D. Shumaker, A.A. Szogi, and M.G. Johnson. 2014. Carbon mineralization in two ultisols amended with different sources and particle sizes of pyrolyzed biochar. *Chemosphere* 103: 313–321.

Singh, B.P., and A.L. Cowie. 2014. Long-term influence of biochar on native organic carbon mineralisation in a low-carbon clayey soil. *Sci. Rep.* 4: 3687.

Sohi, S.P., E. Krull, E. Lopez-Capel, and R. Bol. 2010. A review of biochar and its use and function in soil. *Adv. Agron.* 105: 47–82.

Streubel, J.D., H.P. Collins, M. Garcia-Perez, J. Tarara, D. Granatstein, and C.E. Kruger. 2011. Influence of contrasting biochar types on five soils at increasing rates of application. *Soil Sci. Soc. Am. J.* 75: 1402–1413.

Uzoma, K.C., M. Inoue, H. Andry, H. Fujimaki, A. Zahoor, and E. Nishihara. 2011a. Effect of cow manure biochar on maize productivity under sandy soil condition. *Soil Use Manag.* 27: 205–212.

Uzoma, K.C., M. Inoue, H. Andry, A. Zahoor, and E. Nishihara. 2011b. Influence of biochar application on sandy soil hydraulic properties and nutrient retention. *J. Food Agric. Environ.* 9: 1137–1143.

Wang, J., M. Zhang, Z. Xiong, P. Liu, and G. Pan. 2011. Effects of biochar addition on N_2O and CO_2 emissions from two paddy soils. *Biol. Fert. Soils.* 47: 887–896.

Whitbread, A., G. Blair, Y. Konboon, R. Lefroy, and K. Naklang. 2003. Managing crop residues, fertilizers and leaf litters to improve soil C, nutrient balances, and the grain yield of rice and wheat cropping systems in Thailand and Australia. *Agric. Ecosyst. Environ.* 100: 251–263.

Yao, Y., B. Gao, M. Zhang, M. Inyang, and A.R. Zimmerman. 2012. Effect of biochar amendment on sorption and leaching of nitrate, ammonium, and phosphate in a sandy soil. *Chemosphere.* 89: 1467–1471.

Yin, Y.F., X.H. He, R. Gao, H.L. Ma, and Y.S. Yang. 2014. Effects of rice straw and its biochar addition on soil labile carbon and soil organic carbon. *J. Integr. Agric.* 13: 491–498.

Yu, L.Q., J. Tang, R.D. Zhang, Q.H. Wu, and M.M. Gong. 2013. Effects of biochar application on soil methane emission at different soil moisture levels. *Biol. Fert. Soils.* 49: 119–128.

Yuan, J.H., and R.K. Xu. 2011. The amelioration effects of low temperature biochar generated from nine crop residues on an acidic Ultisol. *Soil Use Manag.* 27: 110–115.

Zhang, A.F., Y.M. Liu, G.X. Pan, Q. Hussain, L.Q. Li, J.W. Zheng, and X.H. Zhang. 2012. Effect of biochar amendment on maize yield and greenhouse gas emissions from a soil organic carbon poor calcareous loamy soil from Central China Plain. *Plant Soil*. 351: 263–275.

Zhang, Q.Z., X.H. Wang, Z.L. Du, X.R. Liu, and Y.D. Wang. 2013a. Impact of biochar on nitrate accumulation in an alkaline soil. *Soil Res.* 51: 521–528.

Zhang, X., H. Wang, L. He, K. Lu, A. Sarmah, J. Li, N. Bolan, J. Pei, and H. Huang. 2013b. Using biochar for remediation of soils contaminated with heavy metals and organic pollutants. *Environ. Sci. Poll. Res.* 20: 8472–8483.

Zhao, X., S.Q. Wang, and G.X. Xing. 2014a. Nitrification, acidification, and nitrogen leaching from subtropical cropland soils as affected by rice straw-based biochar: Laboratory incubation and column leaching studies. *J. Soils Sedim.* 14: 471–482.

Zhao, X., X.Y. Yan, S.Q. Wang, G.X. Xing, and Y. Zhou. 2013. Effects of the addition of rice-straw-based biochar on leaching and retention of fertilizer N in highly fertilized cropland soils. *Soil Sci. Plant Nutr.* 59: 771–782.

Zhao, X.R., D. Li, J. Kong, and Q.M. Lin. 2014b. Does biochar addition influence the change points of soil phosphorus leaching? *J. Integr. Agric.* 13: 499–506.

Zheng, H., Z.Y. Wang, X. Deng, S. Herbert, and B.S. Xing. 2013. Impacts of adding biochar on nitrogen retention and bioavailability in agricultural soil. *Geoderma*. 206: 32–39.

Zheng, J.Y., C.E. Stewart, and M.F. Cotrufo. 2012. Biochar and nitrogen fertilizer alters soil nitrogen dynamics and greenhouse gas fluxes from two temperate soils. *J. Environ. Qual.* 41: 1361–1370.

Zimmerman, A.R., B. Gao, and M.Y. Ahn. 2011. Positive and negative carbon mineralization priming effects among a variety of biochar-amended soils. *Soil Biol. Biochem.* 43: 1169–1179.

10

Biochar Substrate for Hydroponic Vegetable Production

Chapter 10

Biochar Substrate for Hydroponic Vegetable Production

Ataullah Khan, Nick Savidov,
John Zhang, Jian Yang, Tim Anderson,
Don Harfield, and Anthony O. Anyia

Chapter Outline

10.1 Introduction

Biochar is a carbon-rich solid produced by pyrolysis of biomass under partial or complete exclusion of oxygen. The process converts labile carbon in biomass into recalcitrant carbon, which resists degradation and can be sequestered in the soil for centuries. Biochar is generally used for soil amendments, reclamation and remediation, oil sand tailing treatment, lake water de-eutrophication, and carbon sequestration [1–7]. An emerging global new use of biochar is as a growth medium for greenhouse vegetable production [1] and as a carrier for microbial inoculants [8] to enhance soil fertility and crop productivity. This chapter focuses on biochar substrate use in hydroponic vegetable production, as the knowledge in this emergent field is very limited.

10.2 Biochar in Hydroponic Production Systems

10.2.1 Hydroponics

Hydroponics is a method of growing plants using fertilizer solutions in an inert media [9]. A wide variety of substrates can be used, such as rockwool, sawdust, coconut coir, peat moss, sand, perlite, vermiculite, biochar, and zeolite. The desired properties of an ideal hydroponics growth media, such as adequate water-holding capacity, air (void) space, and cation exchange capacities, are expected to remain consistent throughout a growing cycle. Hydroponics is one of the fastest developing methods to produce food in protected agriculture. It is far more effective than traditional methods in terms of productivity per square meter. Ecologically, only soilless culture methods can meet the ever stricter demands to drastically reduce water consumption, groundwater eutrophication, and nitrate reduction [10,11]. Soilless substrates also reduce plant disease problems in greenhouse production, especially for soilborne diseases. Hydroponics is a product of the scientific advances made in understanding the physiology of plant mineral nutrition in the first half of 20th century [12]. Plant growth rate is greatly dependent on environmental conditions in the root zone, such as availability of water, nutrients, and oxygen. Therefore, selection of substrates for soilless greenhouse production is often a key factor in the success of the greenhouse operation as vegetable yields depend on the properties of the substrate material and on its stability over extended use [1].

The main roles of hydroponic substrate are providing air and water to the root zone and protecting the root zone [9]. Optimal particle size distribution is critical for the growing medium to promote fast seed germination, strong root growth, and adequate water drainage. Both physical and chemical characteristics of the substrate are critical in hydroponic applications. The desirable physical characteristics of a substrate and growing medium are bulk density, particle size distribution, porosity, pore size distribution, hydraulic conductivity, and stability, whereas desirable chemical characteristics of the substrate are, for example, surface charge, nutrient retention, and pH buffer capacity [9]. The physical characteristics of substrate govern the availability of adequate water and air in the root zone and thus the irrigation regime. Irrigation regimes have to be tailored to the specific substrate. In contrast, the chemical characteristics of the substrate govern the nutrient retention, ion exchange, and pH buffering capacities, and ultimately the plant mineral nutrition.

Mineral, organic, and synthetic substrates are the three main kinds of substrates used for soilless production [9]. Mineral substrates include sand, rockwool, glasswool, pumice stone, perlite, vermiculite, zeolite [13], pyroclastic (volcanic) material, gravel, and expanded clay. Organic substrates include peat, compost, sawdust, bark, rice husk, and coconut coir. Synthetic materials such as polyurethane foam are also used as a greenhouse growth substrate. In Canada, the hydroponic substrate market is dominated by organic substrates such as sawdust and coconut coir and to a lesser extent by inorganic substrates. Biochar is an emerging hydroponic substrate with high compatibility, and it offers unique advantages for sustainable production operations.

10.2.2 Biochar: An Emerging Substrate

The stability issues associated with organic growth substrates such as sawdust and coir can be overcome by conversion to biochar, a recalcitrant (stable) material with superior physicochemical characteristics that make it a suitable candidate for hydroponic substrate use. Being carbonaceous, biochar is highly inert (recalcitrant) and consistent during its entire life cycle. The carbonization process increases substrate porosity favorably by imparting high air porosity and water retention capacity.

Better physical characteristics and buffering capacity imply decreased volume of the substrate per grow bag, thereby translating to cost saving for growers. Algae and fungal growth can be significantly suppressed in biochar, which implies biochar is better for disease control. Biochar can be tailored (porosity, particle size, bulk density and ion exchange, nutrient retention, buffer capacity) for specialty crops by selecting suitable feedstocks and carbonization process conditions. Biochar's high adsorption capacity toward organic compounds could mitigate accumulation of phytotoxic organic compounds, such as polyphenols that are naturally produced by the plants. Furthermore, biochar has been shown to immobilize toxic compounds, including herbicides, as a result of its high adsorption capacity, thereby offering an additional advantage to growers who use dugout waters with potential herbicide contamination [1,14,15]. Moreover, it can be reused for multiple growing seasons, thus providing cost savings to the growers.

10.2.3 Biochar Performance in Hydroponic Vegetable Production

In a pioneering study, Savidov [1] of Alberta Agriculture and Rural Development in collaboration with Alberta Research Council (ARC) staff conducted extensive feasibility and precommercial trials on biochar obtained from diverse feedstocks, including wheat straw, spent coconut coir, and forestry residues [4]. The growth trials were performed on the three common greenhouse vegetable crops, cucumber, tomato, and pepper, from 2008 to 2014. Various aspects related to biochar growing media (Figure 10.1) were studied: technical feasibility, media reusability, comparative performance, and plant stress alleviation. The data collected in the course of the 7-year study are presented in Table 10.1. The 10 trials included 6 with cucumber, 1 with bell pepper, and 3 with tomato and were conducted under a typical commercial greenhouse setup. The overall plant performance, fruit yield, and plant stress results demonstrated that biochar was a viable substrate for hydroponics production. The growth performance of the test crops on biochar was comparable or superior to the growth on other growth media tested, especially after repeated use (Figure 10.2).

Figure 10.1 (**See color insert.**) Cucumber plant grown in wheat-straw biochar substrate at Crop Diversification Centre South (Brooks, AB, Canada).

As seen in Figure 10.3, a considerable compaction was observed in coconut coir, but not in biochar, after two consecutive years of use, which might have contributed to better results produced with reused biochar compared to reused coconut coir. As shown in Figure 10.4, algal growth was noted in both sawdust and coir growth media, but was absent in biochar growth medium, another positive aspect of the use of a carbonized substrate [1].

In a follow-up study, researchers at Alberta Innovates—Technology Futures (AITF) (formerly ARC) evaluated the performance of vegetable crops on biochar blend produced from the thermomechanical pulp sludge (TMPS) and pin chips from Alberta Newsprint Company (ANC) [14,15]. Four greenhouse trials (three on cucumber and one on tomato) were conducted in the research greenhouses of AITF, Vegreville facility. The pulp sludge and pin chips biochar were produced in separate batches by using similar conditions. The pin chip and TMPS biochars were then blended in different ratios, packaged in 25-L hydroponics grow bags (as shown in Figure 10.1), and evaluated for various performance parameters (physical, chemical, and biological tests) by using coconut coir and wheat-straw biochar as benchmarks.

TABLE 10.1
Biochar Performance Data Derived from Greenhouse Trials Conducted at Crop Diversification Centres South and North (AB, Canada)

Crop	Cultivar	No. of Plants in Each Treatment	Trial No.	Period	Design	Averaged Fruit Yield (kg/plant)						
						Carbonized Sawdust	Carbonized Straw	Carbonized Coir	Rice Husk Biochar	Sawdust	Coco	Carbonized Sawdust + Zeolite
Cucumber	24-146	36 Total 18 Std. 18 Low	1[a]	July–Nov. 2008	RBD	7.53	**8.67**	7.72	N/A	7.34	8.27	N/A
	24-146	36 Total 18 Std. 18 Low	2[b]	Nov. 2008–Jan. 2009	RBD	**1.65**	N/A	N/A	N/A	1.63	1.57	N/A
	24-146	20 Total 10 Std. 10 Low	3[b]	Jan.–May 2009	CRD	11.14	11.85	12.36	12.6	11.0	**13.05**	N/A
	24-146	20 Total 10 Std. 10 Low	4[b]	June–Sept. 2009	CRD	16.37	16.1	17.05	17.76	15.76	**18.01**	N/A
	24-146	72 Total 36 Std. 36 Low	5[b]	Sept.–Nov. 2009	RBD	2.79	N/A	N/A	N/A	N/A	3.26	**3.26**

(Continued)

TABLE 10.1 (Continued)
Biochar Performance Data Derived from Greenhouse Trials Conducted at Crop Diversification Centres South and North (AB, Canada)

Crop	Cultivar	No. of Plants in Each Treatment	Trial No.	Period	Design	Averaged Fruit Yield (kg/plant)						
						Carbonized Sawdust	Carbonized Straw	Carbonized Coir	Rice Husk Biochar	Sawdust	Coco	Carbonized Sawdust + Zeolite
Cucumber	Kasja	16 Total	6[b]	Dec. 2013–May 2014	CRD	9.3	N/A	N/A	N/A	N/A	7.87	N/A
Bell pepper	cv. Bossanova	108 Total 54 Std. 54 Low	1[a]	Oct. 2009–Nov. 2010	RBD	9.65	N/A	N/A	N/A	N/A	8.47	10.4
Tomato	cv. Endeavor cv. Komeett	108 Total 54 Endeav 54 Komeet	1[a]	Nov. 2010–June 2011	RBD	8.17	N/A	N/A	N/A	N/A	8.91	8.77
	cv. Tradiro	48 Total	2[b]	Mar.–Nov. 2012	CRD	7.37	N/A	N/A	N/A	N/A	6.18	6.74
	cv. Torero	64 Total	3[b]	Dec. 2013–May 2014	RBD	4.88[c]	N/A	N/A	N/A	N/A	4.525	N/A

Std, standard irrigation; low, less frequent irrigation; N/A, not attempted; RBD, randomized block design; CRD, completely randomized design.

[a] Fresh substrate.
[b] Used substrate from previous trial.
[c] Fresh coconut coir and used biochar substrates (fourth year).

Figure 10.1 Cucumber plant grown in wheat-straw biochar substrate at Crop Diversification Centre South (Brooks, AB, Canada).

Figure 10.2 Tomato plants grown on 4-year old biochar (left) and fresh coconut coir (right) at Crop Diversification Centre North (Edmonton, AB, Canada).

Figure 10.3 Compaction of greenhouse substrates after two consecutive years of repeated use: coconut coir (left) and biochar (right) in greenhouse trials at Crop Diversification Centre North (Edmonton, AB, Canada).

Figure 10.4 Algal control in cucumber plants grown in carbonized sawdust (left) and raw sawdust (right) at Crop Diversification Centre South (Brooks, AB, Canada).

Figure 10.5 Cucumber plants inoculated with *Pythium* grown in biochar (left) [T1] and coconut coir (right) [T3] at Alberta Innovates—Technology Futures, Vegreville facility (AB, Canada).

Figure 10.2 **(See color insert.)** Tomato plants grown on 4-year old bio-char (left) and fresh coconut coir (right) at Crop Diversification Centre North (Edmonton, AB, Canada).

Figure 10.3 **(See color insert.)** Compaction of greenhouse substrates after two consecutive years of repeated use: coconut coir (left) and bio-char (right) in greenhouse trials at Crop Diversification Centre North (Edmonton, AB, Canada).

Figure 10.4 **(See color insert.)** Algal control in cucumber plants grown in carbonized sawdust (left) and raw sawdust (right) at Crop Diversification Centre South (Brooks, AB, Canada).

The results of these trials are summarized in Table 10.2 [15] and indicate comparable performance between biochar blends and coconut coir (industry standard). The biochar blend of 25% wood chips and 75% TMPS produced higher cucumber yield over coconut coir, which was statistically significant in the first cucumber experiment. In trial 2, cucumber plants encountered a salt toxicity problem that was subsequently overcome by flashing the bags and adjusting (acidifying) the pH of the nutrient solution. In the first and second trials, 5–7 fruit pickings took place, and in the third trial 10 fruit pickings were achieved. As a result, the third trial on cucumber had better growth and higher yield compared to the previous two trials. In the tomato trial, the plants grown in biochar blend (25% wood chip biochar +75% sludge biochar) performed better. The nutrient solution strength, pH, and EC were adjusted accordingly to overcome the nutrient deficiency or surplus. Quality characterization and chemical analysis of cucumber fruit with regard to heavy metals and polycyclic aromatic hydrocarbons (PAHs) indicated that the fruit was safe for human consumption. In addition to supporting vigorous and healthy plant growth that was comparable or superior to coconut coir, biochar displayed unique properties, including stability, porosity, and noncompactability. With a better rate of drying, the water content of biochar blends was lower than that of coconut coir. Visual observations of growth media indicated fewer incidences of algal and fungal growth in biochar blends than in coconut coir. Thus, this study was successful in establishing that the wood chip– and pulp sludge–derived biochars and their blends can potentially replace coconut coir as a viable greenhouse growth media through carefully designed nutrient and pH management programs [15].

Additional biochar hydroponics trials were performed in a commercial greenhouse by John Zhang and coworkers at Alberta Agriculture's Greenhouse Research and Production Complex (Brooks, AB, Canada) in association with the Red Hat Cooperative Ltd. (Redcliff, AB, Canada) between December 2012 and November 2013. To make the trials practical and convincing to growers, the trials were conducted year-round on three common greenhouse vegetable crops, cucumber, pepper, and tomato, to account for varying climatic conditions and standard practices of irrigation and operation. Three different

TABLE 10.2
Biochar Performance Data Derived from Greenhouse Trials Conducted at Alberta Innovates—Technology Futures, Vegreville Facility (AB, Canada)

Crop	Cultivar	Design	No. of Plants in Each Treatment	Trial No.	Period	Fruit Yield (kg/plant)					
						Wood Chip Biochar	Wheat-Straw Biochar	75% Wood Chip Biochar + 25% Pulp Sludge Compost	25% Wood Chip Biochar + 75% Pulp Sludge Biochar	75% Wood Chip Biochar + 25% Pulp Sludge Biochar	Coco
Cucumber	cv. Verdon	CRD	12	1[a]	July 26–Oct. 22 2010	2.0	1.63	1.30	2.50	1.41	1.48
		CRD	12	2[b]	Oct. 14 2010–Jan. 4 2011	1.60	1.43	1.32	1.78	1.68	1.28
		CRD	12	3[b]	Jan. 5–Mar. 25 2011	4.79	4.31	3.70	4.60	4.46	4.30
Tomato	cv. Endeavor	CRD	12	1	July 26–Dec. 16 2010	3.70	3.35	3.10	4.10	3.82	3.73

% Volume basis; pulp sludge compost sourced from ANC.
[a] Fresh substrate.
[b] Used substrate from previous trial.

blends of wood chip biochar and sludge biochar in varying proportions were used for growing tomato, cucumber, and pepper crops. The crop performance and fruit yield data obtained in this trial are documented in Table 10.3. The overall performance of biochar growth media was statistically

TABLE 10.3
Biochar Performance Data Derived from Greenhouse Trials Conducted at Greenhouse Research Production Complex (Brooks, AB, Canada)

				Fruit Yield (kg/m²)	
Crop	**Cultivar**	**Trial**	**Harvest Period**	**Biochar[a]**	**Coir[b]**
Tomato[c]	Komeett	1[d]	Dec. 5 2012–	34.69	39.94
	Brioso		Aug. 16 2013	28.16	28.17
Pepper	Morraine	1[d]	Jan. 25–	15.39	15.31
	Baselga		Aug. 16 2013	16.67	20.62
	Sympathy			15.29	17.36
	Healey			15.75	18.06
	Bentley			15.63	18.44
	DRS0711			16.54	19.43
	Blocky pepper			15.88	18.20
	Mini Red			10.95	10.82
	Mini Yellow			12.67	13.15
	Mini Orange			14.97	14.78
	Mini pepper			12.86	12.92
Mini-cucumber	Picowell	1[d]	Jan. 21–April 16 2013	6.5	6.6
		2[e]	May 13–July 25 2013	12.5	12.0
		3[e]	Sept. 6–Nov. 4 2013	6.4	6.2

[a] 25 vol % wood chip biochar + 75 vol % pulp sludge biochar for tomato trials; 40 vol % wood chip biochar + 60 vol % pulp sludge biochar for pepper trials; 50 vol % wood chip biochar + 50 vol % pulp sludge biochar for mini-cucumber trials.
[b] 'Millenium' for tomato and cucumber trials and 'Forteco' for pepper trials.
[c] Two-week lag for biochar crop due to delayed transplant.
[d] Fresh substrate.
[e] Used substrate.

comparable to the performance of coconut coir in terms of fruit yield per plant (Table 10.3) [16].

As all of the above-reported greenhouse studies were confined to Albertan climatic conditions, so there was a need to test biochar under different climatic conditions. Accordingly, Deborah Henderson and coworkers conducted a precommercial greenhouse trial on biochar and coconut coir growth media at Kwantlen Polytechnic University (Surrey, BC, Canada) to replicate the most common greenhouse growing conditions and practices prevalent in the lower mainland. The biochar blend of 25% wood chips and 75% sludge was evaluated against the current industry standard growth medium, coconut coir sourced from Millennium Coir. Four trials, with two on mini-cucumber, one on tomato, and one on pepper, were conducted between January and August 2013. The pertinent information collected during the course of these trials is presented in Table 10.4, indicating that biochar performance was similar to or slightly better than that of coconut coir [17].

The collective knowledge gathered from >20 greenhouse hydroponic trials between 2008 and 2014 on common greenhouse vegetables at various locations unequivocally proves the superiority of biochar in terms of enhanced substrate stability, sterility, water holding, nutrient retention, pH buffering capacities, and disease suppression, with biochar imparting systemic disease tolerance and alleviating plant stress and ultimately achieving greater yield [1,15–17].

10.2.4 Biochar for Disease Suppression

The impact of biochar on promoting plant growth and stimulating beneficial microflora is well established in the scientific literature. However, the ability of biochar to impart systemic resistance to plant pathogens is not fully understood, and only a few studies have addressed this topic. As an initially sterile material due to the production conditions (high temperatures), biochar is expected to be free of any indigenous microbial populations [18,19] but large porous network of biochar can potentially provide habitat to beneficial microorganisms. In addition, the chemical compounds present in the volatile residue of biochar may prove toxic or even lethal to the pathogens. Several chemical

TABLE 10.4
Biochar Performance Data Derived from Greenhouse Trials Conducted at Kwantlen Polytechnic University (Surrey, BC, Canada)

Crop	Cultivar	Trial No.	Plants/Slab (No.) and Plant Density (plants/m²)	No. of Plants in Each Treatment	Harvest Period	Fruit Yield (kg/plant)		No. Fruit/Plant	
						25 Vol % Wood Chip Biocha + 75 Vol % Pulp Sludge Biochar	Coconut Coir	25% Wood Chip Biochar + 75% Pulp Sludge Biochar	Coconut Coir
Mini-cucumber	Picowell	1[a]	2 and 0.9	18	Jan. 14–April 5 2013	0.57	0.54	97.4	93.2
		2[b]	3 and 1.3	27	June 10–Aug. 31 2013	0.52	0.52	134.7	136.0
Tomato	Endeavor	1[a]	2 and 0.9	18	Mar. 15–July 19 2013	1.26	1.23	149.1	153.1
Pepper	Red line	1[a]	2 and 0.9	18	April 12–July 19 2013	0.52	0.54	56.1	59.3

[a] Fresh substrate.
[b] Used substrate from trial 1.

compounds associated with biochar, such as ethylene glycol, propylene glycol, hydroxyl-propionic, butyric acids, benzoic acid, *o*-cresol, quinones, and 2-phenoxyethanol, have been identified to cause acute toxicity to microbiota and, in turn, result in the proliferation of resistant microbial communities [18,19]. The complexity involved in the biochar–soil–plant–water environment makes the isolation of biochar effects difficult. To understand the true impact of biochar on plant systemic disease resistance, a series of experiments using pepper and tomato plants were conducted that eliminated the nutritional and physical benefits from biochar [18]. The study found lower disease severity toward necrotrophic and biotrophic foliar pathogens (*Botrytis cinerea* and *Leveillula taurica*) in pepper and tomato plants and toward a common mite pest (*Polyphagotarsonemus latus*) in pepper plants grown in biochar-amended media; this effect was attributed to the presence of low-level phytotoxic compounds or to the formation of chemical elicitors [19]. The authors concluded that the biochar impact goes beyond the obvious nutritional benefit and improves soil physical properties. Accordingly, they put forth two hypotheses. The first hypothesis related to the physical and chemical attributes of biochar that preferentially promote the population of beneficial plant growth rhizobacteria or fungi, many species of which are known to induce systemic resistance to plant diseases. The second hypothesis related to the chemical facilitation termed *hormesis* that involves stimulation of the plant immune mechanism through low doses of biochar-borne chemicals, most of which have known phytotoxicity (biocidality) at high doses [18]. Unlike potted plants where biochar is used in a blended soil matrix, in hydroponics the plant roots are in direct contact with the biochar substrate; therefore, the requirement for biochar quality in hydroponics is extremely stringent.

The effects of growth media on disease severity and pathogen population were investigated using cucumber and tomato plants grown in biochar and coconut coir (control) growth media in a greenhouse hydroponic study conducted at AITF, Vegreville facility. *Pythium* root rot was studied in cucumber, and *Fusarium* root rot was studied in tomato. The cucumber trial was terminated 16 days after inoculation, and the tomato trial was terminated 4 weeks after inoculation (Table 10.5). Plant roots were rated for infection based on a 0–5 scale, where 0 = healthy,

TABLE 10.5
Impact of Biochar on Plant Disease Suppression

	Root Disease Rating	% Root Discoloration	Fusarium CFU/g (Wet)	Total Bacteria CFU/g (Wet)
Tomato				
Biochar + *Fusarium*	0.50b	10.0b	3.5×10^4	2.1×10^8
Biochar	0.00c	0.0c	<100	1.6×10^8
Coconut coir + *Fusarium*	1.00a	15.0a	8.5×10^4	2.6×10^9
Coconut coir	0.17c	5.4c	<100	2.1×10^9
Cucumber				
Biochar + *Pythium*	0.83a	16.7b	ND	ND
Biochar	0.00a	0.0c	ND	ND
Coconut coir + *Pythium*	3.00b	60.0a	ND	ND
Coconut coir	0.17a	2.5c	ND	ND

CFU, colony-forming unit; ND, not determined.
Lowercase letters denote Fisher's least significant different test ($p = 0.05$).

1 = <20% root discoloration, 2 = 21–40% root discoloration, 3 = 41–60% root discoloration, 4 = 61–80% root discoloration and infection, and 5 = 81–100% root infection. Both trials were rated using the same scale for root infection as given in Table 10.5. As seen from Figure 10.5, the disease rating in the cucumber crop grown in *Pythium*-inoculated coconut coir was 3.6 times more severe than that grown in *Pythium*-inoculated biochar. Disease rating in the tomato crop grown in *Fusarium*-inoculated coconut coir and *Fusarium*-inoculated biochar showed similar trends. Furthermore, higher root mass was noted for tomato plants grown in biochar compared to root mass of plants grown in coconut coir. The microbial analysis of growth media from the tomato trial showed that the total *Fusarium* population was higher in inoculated coconut coir than that in inoculated biochar (Table 10.5). These results indicated that biochar does not support plant pathogen growth and development in the rhizosphere. Therefore, biochar can delay and reduce plant root diseases, thereby minimizing crop yield losses and disease control costs in commercial greenhouse hydroponic operations [20].

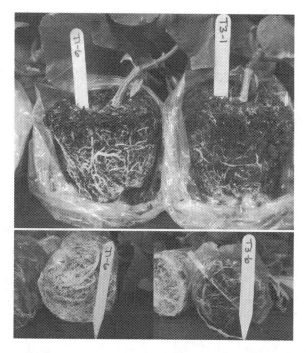

Figure 10.5 **(See color insert.)** Cucumber plants inoculated with *Pythium* grown in biochar (left) [T1] and coconut coir (right) [T3] at Alberta Innovates—Technology Futures, Vegreville facility (AB, Canada).

10.2.5 Food Safety Implications of Biochar in Hydroponic Production Systems

The prospect of the existence and bioavailability of PAHs, pyrolytic compounds produced by the incomplete pyrolysis of biochar, is a valid concern that could potentially hinder biochar's acceptance with greenhouse growers and cooperators. Khan et al. [21] conducted a series of slow pyrolysis experiments on ANC TMPS at different temperatures and residence times to generate nine biochars that were evaluated for proximate and elemental compositions of toxic compounds (PAHs). The sum of PAHs varied from 0.4 to 236 µg/g biochar. A significant correlation between the content of PAHs and toxicity was noted and was matched to the production conditions. Toxicological estimation of the biochars was performed on the basis of a battery of biotests, such as radish germination bioassay and lettuce growth trials. Except for one biochar (generated at 450°C, 60 min), all the TMPS-derived biochars and three controls (coconut coir, quartz sand, and commercial potting mix) passed the germination bioassay. Similar inference could be derived from the lettuce growth trails. The stunted growth noted in lettuce plants grown in biochar generated at 450°C, 60 min, can be attributed to the high concentrations of PAHs detected and their plausible bioavailability and translocatability. Interesting correlations between the PAH content, toxicity, and process conditions were obtained in the form of process toxicity relationships. Thus, based on the result of this study, it is recommended that TMPS feedstock will yield benign-quality biochar when processed at a minimum 500°C for an optimum residence time of 30 min. This study strengthens the fact that not all biochars are created equal, and with proper feedstock selection and process optimization, it is possible to create a benign-quality biochar suitable for safe greenhouse-based food production [21].

10.3 Concluding Remarks

Biochar use in hydroponic production systems is an emerging opportunity with potential economic and environmental benefits if managed carefully. The research reviewed in this chapter shows that biochar can be a potentially high-value material for hydroponic production systems. In addition to providing comparable to or better plant growth performance than coconut coir, biochar-based media also provide unique advantages, such as greater stability (important for repeated use), biological inertness (important for suppressing algal and fungal growth), and carbon sequestration potential. Unlike other growth media that may pose recycling or disposal issues, spent biochar will be a valuable material that can be repackaged and sold as amendments for degraded soils with potential for carbon credits. Due to its outstanding stability and filtration capacity, biochar represents an ideal substrate in biologically active (self-sustaining and self-regulating) integrated multitrophic systems such as aquaponics, that has three major components: livestock (fish) production, waste-water treatment, and soilless plant culture.

References

1. Savidov, N. *Supergrow—Greenhouse trials and commercial applications.* Final Report to ACIDF, Alberta Crop Industry Development Fund Ltd, 2013, Alberta, Canada.
2. Savidov, N. Evaluation of aquaponics technology in Alberta Canada. *Aquaponics J.* Issue#37, 2nd Quarter, 20–25 (2005).
3. Savidov, N. Use of carbonized organic material for production of greenhouse crops. *Presentation at the Canadian Biochar Initiative*, December 12, 2008, Montreal, Quebec, 2008.
4. Graber, E.R., and Elad, Y. Biochar Impact on Plant Resistance to Disease, in *Biochar and Soil Biota*, edited by N. Ladygina and F. Rineau, CRC Press, Taylor & Francis Group, FL, 40–68 (2013).
5. Iranmanesh, S., Harding, T., Abedi, J., Seyedeyn-Azad, F., and Layzell, D.B. Adsorption of naphthenic acids on high surface area activated carbons. *J Environ Sci Health A Tox Hazard Subst Environ Eng.* 49 (2014): 913–22.
6. Zeng, Z., Zhang, S.-d., Li, T.-q., Zhao, F.-l., He, Z.-l. Zhao, H.-p., Yang, X.-e., Wang, H.-l., Zhao, J., and Rafiq, M.T. Sorption of ammonium and phosphate from aqueous solution by biochar derived from phytoremediation plants. *J Zhejiang Univ Sci B.* 14 (2013): 1152–61.
7. Cowie, A. Woolf, D. Gaunt, J. Brandao, M. Anaya del la Rosa, R. Biochar, Carbon Accounting and Climate Change in Biochar for Environmental Management, edited by J. Lehmann and S. Joseph, Earthscan from Routledge (2nd ed.), Oxfordshire, 763–794 (2015).
8. Crowley, D., McGiffen, M., and Hale, L. *Biochar as a carrier for microbial inoculants*, University of California, Riverside, CA, 2011.
9. Jones Jr., J.B. Hydroponics: A Practical Guide for the Soilless Grower, CRC Press (2nd ed), FL, 2005.
10. van Os, E.A. New developments in recirculation systems and disinfection methods for greenhouse crops. *Horticult Eng.* 16 (2001): 2.

11. Savidov, N.A., Hutchings, E. and Rakocy, J.E. Fish and Plant Production in a Recirculating Aquaponic System: A New Approach to Sustainable Agriculture in Canada. *Acta Hort. (ISHS).* 724 (2007): 209–221.

12. Marschner, H. *Mineral nutrition of higher plants,* 2nd ed., Academic Press, New York, 1995.

13. Savidov, N. *Evaluation of greenhouse substrates containing zeolite and secondary use of spent substrate,* Report 2005, Alberta Agricultural Funding Consortium, Alberta, Canada.

14. Anyia, A., Gibson, R., and Wang, S. *Biochar—Potential technology to sequester carbon and sustainably improve agricultural and horticultural productivity in Alberta,* AITF Report, 2010, Alberta, Canada.

15. Anderson, T., Dobson, J., Yang, J., Drozdowski, B., and Anyia, A. *High-value biochar from forestry wastes,* FP Innovations— Innovation Fund Report, 2011, Alberta, Canada.

16. Zhang, J., Mohammed, N., Cote, P., Dalpe, S., and Dufresne, G. *Greenhouse trials on biochar as the growth media for cucumber, tomato, and pepper hydroponic vegetable production,* Alberta Agriculture Report, 2013, Alberta, Canada.

17. Henderson, D., Torres, A., and Vechter, D. *BC pre-commercial demonstration of 3 greenhouse vegetable crops grown in biochar media compared to industry standard coco coir media,* Kwantlen Polytechnic University Report, AITF JDI Report, ABI Report# ABI2014.003, Alberta, Canada.

18. Graber, E.R., Y. Meller-Harel, M. Kolton, E. Cytryn, A. Silber, D. Rav David, L. Tsechansky, M. Borenshtein, and Y. Elad. Biochar impact on development and productivity of pepper and tomato grown in fertigated soilless media. *Plant Soil.* 337 (2010): 481–96.

19. Elad, Y., David, D.R., and Harel, Y.M. Induction of systemic resistance in plants by biochar. *Am Phytopathol Soc.* 100 (2010): 913–21.

20. Yang, J. *Impact of biochar derived from pulp sludge waste on microbial population and the plant diseases of greenhouse vegetables,* ABI Report# ABI2014.007, Alberta, Canada.

21. Khan, A., Mirza, M., Fahlman, B., Rybchuk, R., Yang, Y., Harfield, D., and Anyia, A. Mapping TMPS biochar characteristics for greenhouse produce safety. *J Agric Food Chem.* 63 (2015): 1648–57.

11

Effects of Straw-Derived Biochar on Rice Paddy

Chapter 11

Effects of Straw-Derived Biochar on Rice Paddy

Da Dong, Weixiang Wu, and Ting Zhong

Chapter Outline

11.1 Introduction

Rice (*Oryza sativa*) is the world's most important food crop, raised on ~22% of the total grain-cultivated area and representing 28% of the total grain production. Most rice is grown in Asia, followed by Africa and the Americas (South America, Northern America, Central America and Caribbean), whereas only relatively small rice production areas are situated in Oceania and Europe (Haefele et al. 2014). To meet the demand of the rapidly increasing human population, it is expected that the world's annual rice production must increase from the present 520 million to at least 880 million tonnes by 2025 (Harada et al. 2005). However, with the increasing production of rice from paddy fields, some severe environmental challenges need to be addressed. It is well known that a rice paddy is an important human-made ecosystem and emission source for global greenhouse gases (GHGs), including methane (CH_4) and nitrous oxide (N_2O). Conventional burning and direct incorporation of rice residues into the soil have resulted in increasing soot and CH_4 emissions (Gustafsson et al. 2009; Yuan et al. 2014). In addition, the overuse of conventional chemical fertilizers results in serious nutrient leaching and losses from paddy fields through ammonia volatilization, denitrification, surface runoff, and leaching (Lin et al. 2007; Tian et al. 2007; Xing et al. 2002), resulting not only in severe eutrophication in waterways and lakes of irrigated areas but also in unsafe drinking water (Conley et al. 2009). Furthermore, soil degradation resulting from very intensive crop rotation is becoming increasingly serious worldwide. One-third of the total rice in the world is grown on very poor soils, with very limited indigenous nutrient supplies, low nutrient retention capacity, and very acidic soil reaction (Haefele et al. 2014). Therefore, a main challenge for the sustainable development of modern agriculture is to increase rice productivity while maintaining paddy soil health and reducing various nonpoint pollutions from the paddy field ecosystem.

Biochar has received increasing attention by scientists because of its beneficial effects on carbon sink and land improvement. A wide range of biomass sources, such as straw, wood, manure, and other organic wastes, can be used to produce biochar. In consideration of the richness of raw materials, straw is a potentially attractive feedstock in making biochar for a soil amendment. More than 3000 million tonnes of cereal straw is produced in the world each year (Spiridon 2005), and a large part of this cereal straw is directly burnt or returned to the field. In recent years, research has indicated that straw-derived biochar (SDB) has excellent physical and chemical properties, including high pH, excellent porosity, large specific surface area, and extensive nutrients (Wang et al. 2013b; Xiao et al. 2014). Incorporation of SDB into paddy soil will influence soil physical and chemical characteristics that may then lead to agronomic benefits, such as improving soil quality, increasing rice yield, and reducing GHG emissions.

As an important human-made ecosystem, a rice paddy has many characteristics that are different from those of other ecosystems. Flooded rice fields are predominantly anaerobic, except for some distinct zones including the surface of floodwater and rhizosphere of rice. Periodic wetting and drying cycles of the water regime would significantly affect the redox potential and soil aeration of paddy soil. Moreover, paddy soil is rich in nitrate (NO_3^-), Mn^{4+}, Fe^{2+}, and sulfate that could act as electron acceptors in carbon dioxide (CO_2) reduction to change the flow of electron direction in microbial metabolism (Kögel-Knabner et al. 2010). Therefore, this chapter focuses on the nutrient and surface chemical characteristics of SDB, relating how these characteristics are influenced by the pyrolysis temperature and residence time. In addition, the effects of SDB on nitrogen cycling, rice yield, and GHG emissions in paddy soil are reviewed.

11.2 Properties of SDB

11.2.1 Nutrient Characteristics of SDB

Biochar can be produced from a diverse range of crop residues, including herbs and woody plants (Clough et al. 2010; Inyang et al. 2010; Regmi et al. 2012; Tong et al. 2012; Xu et al. 2012). SDB has special elemental characteristics compared with bamboo-, wood-, rice husk–, and coconut shell–derived biochars (Butnan et al. 2015; Yao et al. 2012; Zhao et al. 2013a; Wang et al. 2013b). The elemental contents of C, N, P, Ca, K, and Mg in different types of biochars are shown in Table 11.1. The N content of SDB is 1.18%, higher than that of wood-based biochar, and results in a very low C/N ratio.

The ash content of biochars mainly depends on their feedstock (Crombie et al. 2013). Usually, a higher ash content of feedstock results in an equally high ash content of the corresponding biochar, because the basic inorganic mineral components are retained in the pyrolysis process. As a result, the ash content of SDB is much higher than that of wood- and bamboo-derived biochars. This observation was confirmed by Dong et al. (2013),

TABLE 11.1
Total Elemental Content of Biochars Produced from Different Feedstocks

Biochar Feedstock	C (%)	N (%)	C/N	Ash (%)	P (%)	K (%)	Ca (%)	Mg (%)	Reference
Rice straw	47.2	1.18	40.1	27.9	5.63	9.94	1.05	0.47	Wang et al. (2013b)
Rice husk	68.2	1.23	55.4	42.3	0.12	0.68	0.13	0.06	Zhao et al. (2013a)
Peanut hull	81.5	1	81.5	—	0.09	0.94	0.33	0.13	Yao et al. (2012)
Bamboo	76.9	0.20	384.5	—	0.36	0.35	0.29	0.19	Yao et al. (2012)
Eucalyptus wood	—	—	—	2.35	0.12	1.20	1.27	0.10	Butnan et al. (2015)
Brazilian pepperwood	75.6	0.3	252	—	0.07	0.25	1.32	0.23	Yao et al. (2012)

who found that the ash content of rice straw–derived biochar (RSDB), 381.3 g·kg⁻¹, was much higher than that of bamboo-derived biochar. In addition, as shown in Table 11.1, SDB has greater contents of P and K than wood-derived biochar. Although SDB is richer in nutrients than wood-derived biochar, studies on the availability of these nutrients to the rice plant in a paddy field are not extensive; there are only a few studies focusing on the nutrient availability of biochar (Dong et al. 2013; Major et al. 2010; Uzoma et al. 2011). As shown in Table 11.2, significant differences in available nutrients can be observed among various types of biochar. In comparison, the available nutrients in RSDB are much higher than those of bamboo-derived biochar, especially available K and P (Dong et al. 2013).

Other than the raw material, pyrolysis temperature and residence time are also key factors affecting the ash, volatile, and available nutrient contents in SDB. The ash and volatile contents of RSDB over a range of temperatures and residence times are shown in Figure 11.1. Volatile matter content of the biochar decreased with increasing temperature from 49 to 13%. In contrast, ash content increased with the increasing temperature.

The effects of pyrolysis temperature on the availability of P, K, Ca, Na, and Mg are shown in Figures 11.2 and 11.3. The contents of available K, Ca, Na, Mg, and P increased from 300°C and reached their maximum values at 400°C. However, the residence time did not show any significant impact on the available nutrient contents in RSDB in comparison to the pyrolysis temperature.

Cation exchange capacity (CEC) is also one of the important characteristics of biochar and plays an essential role in nutrient leaching. SDB has a relatively higher CEC value than other biochars. Dong et al. (2013) found that the CEC in RSDB and bamboo-derived biochar was 44.7 and 15.3 cmol·kg⁻¹, respectively. Similar results were reported by other studies. Wang et al. (2013b) reported that CEC ranged from 20.1 to 34.8 cmol·kg⁻¹ in SDB pyrolyzed at 500°C for 4–16 h and only from 1.3 to 3.0 cmol·kg⁻¹ in wood-derived biochars. The CEC value is also lower in cow manure biochar, ~4.84 cmol·kg⁻¹ (Uzoma et al. 2011). In contrast to the residence time, the CEC value of RSDB is more significantly affected by the pyrolysis temperature (Figure 11.4), with higher CEC values below 400°C than above 400°C.

TABLE 11.2
Available Nutrients of Biochars Produced with Different Feedstocks

Biochar Feedstock	Production (°C)	pH	Ash (%)	P (mg·kg⁻¹)	K (mg·kg⁻¹)	Ca (mg·kg⁻¹)	Mg (mg·kg⁻¹)	CEC (cmol·kg⁻¹)	Reference
Bamboo	600	9.8	6.23	77.6	2614.2	128.9	194.8	15.3	Dong et al. (2013)
Wood	—	9.2	—	30	464	331	49	11.2	Major et al. (2010)
Rice straw	600	10.2	38.1	1672.3	3758.2	3604.2	1216.0	44.7	Dong et al. (2013)
Dry cow manure	500	9.2	—	—	0.14[a]	2.12[a]	1.40[a]	4.84	Uzoma et al. (2011)

[a] Exchangeable nutrients (cmol·kg⁻¹).

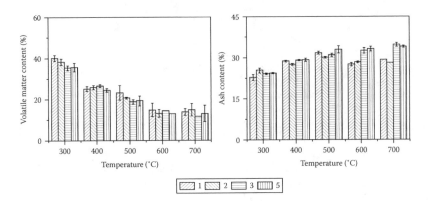

Figure 11.1 Ash and volatile matter of rice straw–derived biochar, including ash and volatile contents over a range of temperatures and residence times (1, 2, 3, and 5 refer to residence time in hours). (From Wu W, et al. 2012. *Biomass and Bioenergy* 47: 268–276.)

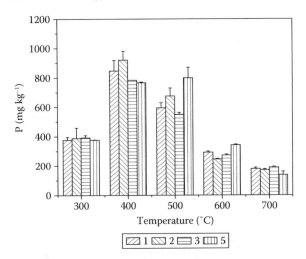

Figure 11.2 Available phosphorus (P) of rice straw–derived biochar produced at different pyrolysis temperatures and residence times (1, 2, 3, and 5 refer to residence time in hours). (From Wu W, et al. 2012. *Biomass and Bioenergy* 47: 268–276.)

11.2.2 Surface Chemical Characteristics of SDB

The surface of biochar can be slowly oxidized after being applied into the soil through biotic and abiotic oxidation beginning from the surface of the biochar particles and progressing inward. The formation of surface oxygen groups will possibly

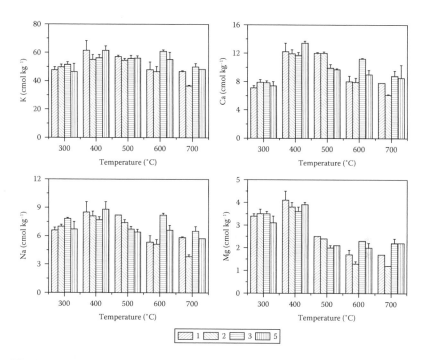

Figure 11.3 Extractable cations (K^+, Ca^{2+}, Na^+, and Mg^{2+}) of rice straw–derived biochar produced at different pyrolysis temperatures and residence times (1, 2, 3, and 5 refer to residence time in hours). (From Wu W, et al. 2012. *Biomass and Bioenergy* 47: 268–276.)

weaken the stability of biochar and enhance its oxidation. Therefore, understanding the surface chemical characteristics of SDB is essential to understanding the environmental behavior and effects of SDB in a paddy field.

Advanced techniques such as Fourier transform infrared (FTIR) spectroscopy, x-ray diffraction (XRD), and solid-state ^{13}C nuclear magnetic resonance have been widely used to determine the chemical compositions of compounds and to deduce their structures. The effects of pyrolysis temperature and residence time on the surface chemical characteristics of RSDB determined by FTIR spectroscopy are shown in Figure 11.5. With increasing pyrolysis temperature, functional groups of RSDB experienced the following changes. At 300°C, the strength of aliphatic C–H stretching (2950 and 2850 cm^{-1}) and C–O stretching (1110–1030 cm^{-1}) clearly declined due to the progressive dehydration and depolymerization. The loss of

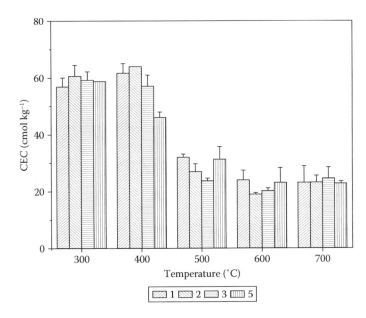

Figure 11.4 Cation exchange capacity of rice straw–derived biochar produced at different pyrolysis temperatures and residence times (1, 2, 3, and 5 refer to residence time in hours). (From Wu W, et al. 2012. *Biomass and Bioenergy* 47: 268–276.)

the functional groups in RSDB was much easier than that in wood-derived biochar because hemicellulose and cellulose are the main components of rice straw (Yang et al. 2007). It could also explain that the decrease of H/C and O/C of SDB was more significant than those of wood- or bamboo-derived biochar under the same temperature and residence time. With the pyrolysis temperature extended from 400 to 500°C, stretching vibration of aliphatic C–H disappeared, whereas bending vibration of aromatic C–H became more apparent. This showed that greater dehydration and increased aromatization occurred in this temperature range, which was in accordance with the result of decreasing of H/C and O/C. With the further increase in the pyrolysis temperature, the absorption intensity of all peaks significantly decreased, except for stretching vibration of C–H, which was observed in the biochar-derived corn stover (Lee et al. 2010) and explained by the presence of pyranose rings and guaiacyl monomers of cellulose and hemicellulose (Figure 11.5).

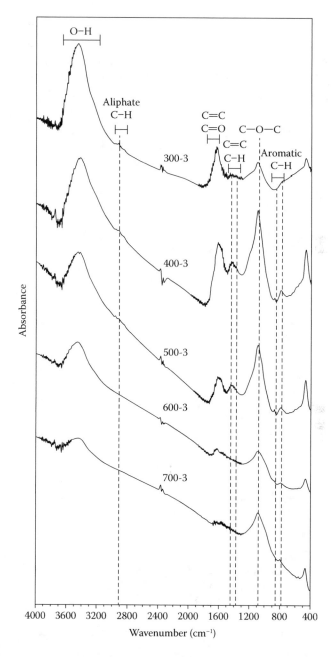

Figure 11.5 Fourier transform infrared spectra of rice straw–derived biochar generated at temperatures ranging from 300 to 700°C for 3 h. (From Wu W, et al. 2012. *Biomass and Bioenergy* 47: 268–276.)

XRD was also used to determine the presence of crystalline compounds in biochar. The sharp peaks in the spectra are mainly inorganic crystals in RSDB, including silicon dioxide (peaks at 4.27 and 3.35 Å), potassium chloride (peaks at 3.14 and 2.21 Å), and calcium carbonate (peaks at 3.04 Å) (Figure 11.6). It differed from biochars derived from other feedstocks, such as wood, grass, corn straw, and peanut straw, as rice straw is low in lignin and high in Si and K (Bourke et al. 2007; Keiluweit et al. 2010).

Table 11.3 shows the degree of aromaticity of RSDB determined by solid-state ^{13}C NMR. The degree of aromaticity of RSDB was higher when pyrolyzed at 500°C (98.5%) than at 400°C (82.9%).

Compared with the properties of other kinds of biochar, SDB, especially RSDB, contains considerably more nutrients

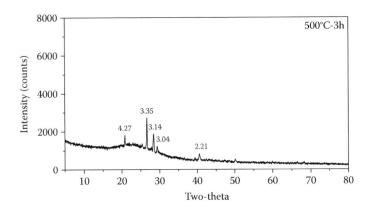

Figure 11.6 XRD spectra of rice straw–derived biochar produced at 500°C for 3 h.

TABLE 11.3
Quantitative Direct Polarization ^{13}C Nuclear Magnetic Resonance Analysis of Rice Straw–Derived Biochar

Sample (°C)	Carbonyl Group C=O	Aromatic Carbon		Alkyl Group CH
		$C–O_{0.75}H_{0.5}$	C/C–H	
400	9.6	12.8	70.1	7.5
500	0	24.7	73.8	1.5

that can be directly taken up by a rice plant, e.g., the available K and P of RSDB were higher than those of other biochars. Moreover, RSDB generated at 500°C not only contains highly aromatic structures but also is rich in available nutrients. Therefore, application RSDB produced at 500°C might be a better option for carbon sequestration and soil amendment.

11.3 Effects of SDB on Nitrogen Cycling in Paddy Soil

11.3.1 Effects of SDB on Nitrogen Leaching and Retention

Nitrogen is an essential nutrient for rice growth, and urea fertilizer amendment is a key factor in the maintenance of high rice yield in a paddy field. Farmers often rely on convenient and quick-acting chemical fertilizers, especially nitrogen fertilizers, to increase rice yield in their paddy fields. However, overuse of chemical fertilizers has resulted in severe nitrogen losses from paddy fields through ammonia volatilization, denitrification, and leaching. The ability of biochar to reduce nitrogen leaching and increase nitrogen in paddy soil has been reported in several studies. Ding et al. (2010) reported that addition of 0.5% bamboo-derived biochar to the surface paddy soil layer could retard the vertical movement of NH_4^+-N into the deeper layers, and cumulative losses of NH_4^+-N via leaching at 20 cm could be decreased by 15.2% over the experimental period. Another soil column experiment with isotope labeling showed that nitrogen retention could be increased by 5.6–26% and leaching decreased by 26–49% after amendment with RSDB (Zhao et al. 2013b). The nitrogen retention capacity of biochar is mainly attributed to its great surface area and high CEC (Bruun et al. 2012; Lehmann et al. 2003). The Brunauer–Emmett–Teller surface area of SDB with 700°C pyrolysis temperature could be as high as 112–378 $m^2 \cdot g^{-1}$, which is a comparable level with that of wood-derived biochars (Wang et al. 2013b). Compared with other wood-derived biochars, SDB has higher CEC (Table 11.2), suggesting that SDB may have stronger adsorption capability than wood-derived biochar. Dong et al. (2013) reported that RSDB amendment resulted in a significant increase in CEC in the rice rhizosphere soil compared with the control treatment in two successive years.

11.3.2 Effects of SDB on Nitrogen Nitrification and Denitrification in Paddy Soil

Nitrification is an important nitrogen transformation process in paddy field because urea- and ammonium (NH_4^+)-based fertilizers need to be transformed to NO_3^-, which could increase nitrogen uptake by rice. However, only a few studies have assessed the impacts of SDB on this process in paddy soil. Zhao et al. (2014b) found that the amendment of wheat straw–derived biochar stimulated nitrification in Alfisols and Oxisols in a laboratory incubation experiment. The surface of biochar contains high concentration of metal oxides, including calcium oxide, magnesium oxide, iron(III) oxide, and titanium dioxide, which might catalyze the photooxidation of NH_4^+ (Koukouzas et al. 2007; Lee et al. 2005). As the first rate-limiting step of nitrification, ammonia oxidation had attracted much attention. However, with regard to the link between biochar and its effect on ammonia oxidation, there is disagreement as to its mechanism. One mechanism is microbial ammonia oxidation that may occur in different ways, depending on the environmental conditions. Ammonia-oxidizing archaea (AOA) and ammonia-oxidizing bacteria (AOB), e.g., are quite sensitive to the changes in soil conditions. Existing studies showed that the amendment of biochar had the potential to stimulate ammonia oxidation in forest soil (Ball et al. 2010; Taketani and Tsai 2010). In particular, Ball et al. (2010) found that the abundance of AOB in all three soil horizons samples from recent-fire forest soils was significantly greater than that in the corresponding control soils. Furthermore, soil pH is an important factor that affects ammonia oxidation in soil. Song et al. (2014) found that application of biochar to a coastal alkaline soil stimulated the ammonia oxidation rate in the first 4 weeks of incubation, whereas significantly inhibited the ammonia oxidation rate in the next 2 weeks. In contrast nitrification can be accelerated by RSDB in acidic arable soils of subtropical regions in the presence of NH_4^+ fertilizer, and nitrification-induced acidification can be offset by the overall amelioration effects of biochar (Zhao et al. 2014b). Long-term effects of RSDB amendment on abundance of AOA and AOB in acidic paddy soil were also analyzed by quantitative polymerase chain reaction (qPCR). Results showed that RSDB amendment could significantly inhibit the abundance of AOA and that this effect would gradually decrease over time. However, RSDB application without urea would not significantly affect the abundance of AOB in paddy soil (Wu, unpublished data).

Denitrification is the reduction of NO_3^- to nitrite, and then to nitric oxide (NO), N_2O, and nitrogen gas (N_2), an important process that leads to the GHG emissions and nitrogen loss from the paddy soil. Some publications indicated that biochar amendment in soil could stimulate denitrification (Anderson et al. 2011; Luz Cayuela et al. 2013; Troy et al. 2014). Luz Cayuela et al. (2013) pointed that biochar can facilitate the transfer of electrons to denitrifying microorganisms in soil, thereby promoting denitrification. However, others suggested that the decreasing emission of N_2O in biochar-amended soil resulted from the inhibition of denitrification (Felber et al. 2014; Nelissen et al. 2014). Characterized by a high nitrogen level and the alternation of drying and wetting as well as severe anthropogenic disturbance, the rice paddy ecosystem is completely different from natural ecosystems. Therefore, the effects of SDB amendment on denitrification in a paddy field need to be further studied.

11.4 Effects of SDB on Rice Productivity

Rice is a staple food as it feeds more than half the global population. Its planting areas had reached 11% of the world's cultivated land in 2005. There is the potential for a growing demand of rice with the increasing world population. However, a large amount of global agricultural land is reported to be low in crop productivity, including 21% of the paddy fields in China (Chen et al. 2014). Natural forces and intensive cropping systems, severe and destructive soil erosion, and progressive loss of plant nutrients have become serious problems in these areas. In particular, rice growth is limited by available nutrients, especially available K and P in the paddy field (Chen et al. 2011; Nakamura et al. 2013; Zorb et al. 2014). In addition, Al and Mn toxicity in acid paddy soils seriously threaten rice productivity.

An important renewable resource and rich in nutrients, straw becomes potentially attractive as a feedstock in making biochar for soil amendment. Evidence thus far shows that amendment of SDB to paddy soil is an effective way to achieve agronomic benefits, particularly an increase in rice yield (Table 11.4). Zhang et al. (2012) reported that applying biochar derived from wheat straw exerted a significant positive role in enhancing rice yield by 9.2–11.8% in the first cycle and by 9.2–27.6% in the subsequent cycle. In a field experiment, Dong et al. (2013) found that rice yield was significantly increased by 13.5% in 2010 and 6.1% in 2011 under RSDB treatment compared with the control. In addition, an unpublished study has pointed to a possible long-term effect of RSDB on rice yield in a paddy field with low productivity. A single application of 1% (w/w) RSDB into a paddy field with low productivity in 2009 has successively increased rice yield for 5 years (Wu, unpublished data).

The quick positive response of rice yield to the SDB amendment can be attributed to the direct supply of nutrients from the biochar. Biochar derived from the low-temperature pyrolysis of rice straw had notably high available nutrients, with a P content of 1.6 $g \cdot kg^{-1}$, a K content of 3.7 $g \cdot kg^{-1}$, a Ca content of 3.6 $g \cdot kg^{-1}$, and a Mg content of 1.2 $g \cdot kg^{-1}$ (Dong et al. 2013). Nutrient input by RSDB amendment results in a significant increase in available K and P of rhizosphere soil that can enhance nutrient uptake and benefit rice growth. Apart from the nutrient release from biochar, it has been well demonstrated that SDB also plays a significant role in reducing nutrient leaching due to its large surface area, negative surface charge, and change in soil density (Ding et al. 2010; Liang et al. 2006).

However, the initial supplement of nutrient elements cannot sustain long-term benefits in rice productivity. When the available nutrients in biochar decline through soil adsorption, water leaching, and rice uptake, the increase in rice yield may

TABLE 11.4
Effects of Straw-Derived Biochar on Rice Yield in Paddy Soil

Type of Experiment	Pyrolysis Temperature (°C)	Pyrolysis Time (h)	Type of Straw-Derived Biochar	Biochar Rate (t·ha⁻¹)	Water Management[a]	Result	Reference
Field	350–550	—	Wheat straw	10, 20, and 40	Conventional practices	No significant differences	Cui et al. (2011)
Field	300–500	—	Wheat straw	10 and 40	F-D-F-M	In the first cycle (by 9.2–11.8%) and in the second cycle (by 9.2–27.6%)	Zhang et al. (2012)
Cross-site field	450–550	—	Wheat straw	20 and 40	F-D-M	40 t·ha⁻¹ significantly increased with nitrogen fertilizer in Changsha	Liu et al. (2012)
Field	—	—	Wheat straw	20 and 40	Conventional practices	Late rice at the Jinxin experiment site, increasing by 5.18% (20 t·ha⁻¹) and 7.95% (40 t·ha⁻¹)	Qu et al. (2012)
Field	350–500	—	Wheat straw	5, 10, and 20	F-D-F-M	No significant effects	Meng et al. (2013)
Pot	500	8	Rice straw	4.5 and 9	—	Grain yield increased by 11.4–60.5%	Wang et al. (2013a)
Field	600	2	Rice straw	22.5	F-D-F-M	Significantly increase by 13.5% in 2010 and 6.1% in 2011	Dong et al. (2013)

[a] F, flooding; D, drainage; M, moist.

then be closely related to the improvement of soil health, especially soil pH and CEC. Furthermore, the bioavailability of some elements toxic to rice growth, such as Al and Mn, might be reduced with the increase in soil pH due to the SDB amendment, whereas the bioavailability of K, P, Ca, and Mg is enhanced. In addition, SDB has a high CEC and usually increases with age due to its surface oxidation, leading to a greater nutrient retention potential. In our paddy field trial, soil CEC significantly increased from 7.44 to 8.09 $cmol \cdot kg^{-1}$ with the RSDB amendment (Dong et al. 2013). In addition, many studies have indicated that other indirect effects resulted from RSDB amendment, such as an increase of soil moisture and a decrease of soil density, effects that may also enhance rice yield (Dong et al. 2015; Zhang et al. 2012). Therefore, the increase in rice yield in a paddy field is a combination direct nutrient supplementation from SDB, improvement of soil health, and indirect effects of the biochar amendment.

11.5 Effects of SDB on GHG Emissions in Paddy Soil

11.5.1 Effects of SDB on CH_4 Emission

Rice fields are regarded as an important source of atmospheric CH_4, with global warming potential 25 times higher than that of CO_2 over a 100-year horizon (IPCC 2007). As a sustainable means to mitigate GHG emissions, the application of SDB to agricultural land has received increasing attention. As shown in Table 11.5, net emission of CH_4 in paddy soil was found to decrease by the addition of SDB in small-scale field and short-term incubation studies. In a laboratory incubation experiment, CH_4 emission from paddy soil, after amendment with RSDB at a high rate (2.5% of the soil, w/w), could almost be completely inhibited compared to the nonamendment control soil with or without the addition of rice straw (Liu et al. 2011). It has also been reported that the CH_4 emissions from paddy soil amended with wheat straw biochar generated at 300 and 500°C were significantly less than those in the corresponding

TABLE 11.5
Effects of Straw-Derived Biochar on CH$_4$ Emission in Paddy Soil

Type of Experiment	Pyrolysis Temperature (°C)	Pyrolysis Time (h)	Type of Biochar	Biochar Rate	Water Management[a]	Result	Reference
Field	300–500	—	Wheat straw	10 and 40 t·ha^{-1}	F-D-F-M	Significantly increased	Zhang et al. (2010)
Serum bottle	600	—	Rice straw	0.5, 1.5, and 2.5% w/w	F	Significantly decreased	Liu et al. (2011)
Pot	Set at 200°C initially and then elevated stepwise to 250, 300, 350, and 400°C	1.5 for each temperature step	Cornstalk	12 t·ha^{-1}	—	No significant effects	Xie et al. (2013)
Pot	300, 400, and 500	1.5	Cornstalk	24 t·ha^{-1}	—	300 and 500°C significantly decreased; 400°C no significant effect	Feng et al. (2012)
Field	350–550	—	Wheat straw	10, 20, and 40 t·ha^{-1}	F-D-F-M	20 and 40 t·ha^{-1} significantly increased; 10 t·ha^{-1} no significantly effect	Zhang et al. (2012)

(Continued)

TABLE 11.5 (*Continued*)
Effects of Straw-Derived Biochar on CH$_4$ Emission in Paddy Soil

Type of Experiment	Pyrolysis Temperature (°C)	Pyrolysis Time (h)	Type of Biochar	Biochar Rate	Water Management[a]	Result	Reference
Field	600	2	Rice straw	22.5 t·ha^{-1}	—	Significantly decreased at heading stage	Dong et al. (2013)
Incubation	350 and 500	4	Rice straw	5% w/w	40% water-holding capacity	350°C significantly decreased; 500°C increased	Li et al. (2013)
Pot	500	8	Rice straw	4.5 and 9 t·ha^{-1}	Remained flooded	No obvious differences in CH$_4$ emission between control and biochar treatments	Zhao et al. (2014b)

[a] F, flooding; D, drainage; M, moist.

controls (Feng et al. 2012). In addition, Dong et al. (2013) reported that a single addition of RSDB (1% w/w) could result in significant reduction in CH_4 emissions from the paddy field in the successive 2 years. However, some studies observed the opposite results. For example, Zhang et al. (2012) found that under wheat straw–biochar amendment at 20 and 40 $t \cdot ha^{-1}$, CH_4 emission from paddy soil increased by 153 and 51% in 2009 and by 54 and 38% in 2010, respectively, compared to the corresponding controls. Other researchers reported that SDB amendment had no significant effects on CH_4 emissions from paddy soil (Xie et al. 2013; Zhao et al. 2014a). It seems that soil CH_4 emission in response to the biochar amendment may vary with the characteristics and application rate of SDB, type of paddy soil, and method of water management.

CH_4 emission from paddy soil encompasses production, oxidation, and transportation. Its net emission mainly depends on the relative rate of CH_4 production and oxidation. The potential pathways for the impacts of SDB amendment on CH_4 emission from a paddy field can be summarized as follows. First, the addition of SDB can decrease soil bulk density and compaction, and increase soil aeration, all of which lead to lower CH_4 production and higher CH_4 oxidation. Second, the large amount of specific surface area of SDB facilitates CH_4 uptake and oxidation in soil. Third, the increasing rice plant biomass enhanced by the SDB application may inhibit methane emission by increasing CH_4 oxidation in the rhizosphere through the enlargement of aerenchyma conduits (Cheng et al. 2000), where O_2 is transferred from the atmosphere through the rice aerenchyma system. Fourth, chemical substances such as ethylene that can be easily generated from the pyrolysis of biomass and have been widely known to reduce microbial ammonium nitrification (Porter 1992) and soil methanotrophic activity (Jackel et al. 2004) may increase in paddy soil due to the amendment of SDB, leading to increased CH_4 emission. Finally, the emission of CH_4 in paddy soil amended with SDB may have a close relationship with the content of easily decomposable organic carbon in the biochar. It is reported that this carbon can act as the preferential substrate for soil methanogens, resulting in increased CH_4 emission, particularly at the beginning of the experiment. In addition, soil pH is an important parameter that governs CH_4 emission rate in paddy soil. Most of the methanogenic archaea grow at a pH near neutral, with a range of 6.5–7.5 (Wang et al. 1993; Yang and Chang 1998). Therefore, the liming effects of SDB may play an important role in CH_4 emission, especially in acidic paddy soil.

However, the amount of CH_4 emission from paddy soils should eventually be directly influenced by the abundance, activity, and community structure of methanogenic archaea and methanotrophic proteobacteria. Compared with the control treatment, CH_4 oxidation activity in the rhizosphere soil at the seeding and maturing stages could be significantly stimulated by RSDB amendment, whereas methanogenic activity in the rhizosphere could only be increased at the seedling stage (Yang et al. 2013). However, in another incubation experiment, Liu et al. (2011) found that there were no significant differences in the community composition of methanogenic archaea and methanotrophic proteobacteria in the paddy soils between the RSDB amendment and the nonamendment control. Nevertheless, by using qPCR and denaturing gradient gel electrophoresis, Feng et al. (2012) observed that the amendment of SDB could stimulate a more significant effect on the abundance of methanotrophic proteobacteria than that of methanogenic archaea in paddy soil. The above-mentioned controversial results suggest that further studies are needed to elucidate the microbial mechanisms of SDB amendment on CH_4 production and oxidation in paddy soil.

11.5.2 Effects of SDB on N₂O Emission

N_2O is the single most important ozone-depleting GHG, with global warming potential 298 times higher than that of CO_2 over a 100-year horizon (IPCC 2007; Ravishankara et al. 2009). An agricultural field is one of the main N_2O emission sources and is estimated to contribute to 42% of anthropogenic emissions worldwide (Akiyama et al. 2010; Cayuela et al. 2013). In addition, N_2O emission is estimated to increase 35–60% by 2030 due to the extensive application of nitrogen fertilizer (Nelissen et al. 2014). SDB has been shown in several field and laboratory studies to mitigate N_2O emission from paddy fields. As shown in Table 11.6, Zhang et al. (2010) reported that N_2O

TABLE 11.6
Effects of Straw-Derived Biochar on N_2O Emissions in Paddy Soil

Type of Experiment	Pyrolysis Temperature (°C)	Pyrolysis Time (h)	Type of Biochar	Biochar Rate	Water Management[a]	Result	Reference
Field	300–500	—	Wheat straw	10 and 40 $t \cdot ha^{-1}$	F-D-F-M	Sharply decreased by 40–51 and 21–28%	Zhang et al. (2010)
Field	300–500	—	Wheat straw	10, 20, and 40 $t \cdot ha^{-1}$	F-D-F-M	Decreased by 40–51 and 31–56% in the first and second cycle, respectively	Zhang et al. (2012)
Field	450–550	—	Wheat straw	20 and 40 $t \cdot ha^{-1}$	F-D-M	A great reduction in total N_2O emission by 40 and by 60% under biochar amendment rate of 20 and 40 $t \cdot ha^{-1}$	Liu et al. (2012)
Incubation	450	Fast pyrolysis	Wheat straw	0.29% w/w	60% water-holding capacity	Similar N_2O production rate	Cheng et al. (2012)

(Continued)

TABLE 11.6 (*Continued*)
Effects of Straw-Derived Biochar on N$_2$O Emissions in Paddy Soil

Type of Experiment	Pyrolysis Temperature (°C)	Pyrolysis Time (h)	Type of Biochar	Biochar Rate	Water Management[a]	Result	Reference
Field	500	—	Wheat straw	24 and 48 t·ha^{-1}	Local farming practice	Increased by 12.6 to 80.1%, 618.8 to 594.1%, and 24.3 to19.6% in the early rice, late rice and fallow seasons, respectively	Liu et al. (2014)
Microcosm	Step to 400	1.5 h	Cornstalk	12 t·ha^{-1}	F-D-F-M	Significantly increased N$_2$O emissions by the Inceptisol	Xie et al. (2013)
Incubation	350 and 500	4 h	Rice straw	5% w/w	40% water-holding capacity	350°C inhibited or decreased N$_2$O emission; 500°C significantly decreased	Li et al. (2013)
Pot	500	2 h	Rice straw	2.5% w/w	F	No effects, inhibited or decreased	Unpublished

[a] F, flooding; D, drainage; M, moist.

emission from a paddy field could be decreased by 40 and 51% with N fertilization when the soil was amended with wheat straw–derived biochar at the rate of 10 and 40 $t \cdot ha^{-1}$, respectively. Liu et al. (2012) found that the 40–60% of the cumulative N_2O emission reduction from rice paddy could be obtained in three sites across southern China due to the amendment of wheat straw–derived biochar. However, it seems that response of N_2O emissions in paddy soil amended with SDB depends on the soil characteristics and types as well as the properties of the biochar. Xie et al. (2013) observed that, in contrast to the significantly higher N_2O emission in the slightly alkaline Inceptisol, the N_2O emission in the acidic Ultisol amended with SDB was similar to the control. Li et al. (2013) discovered that the addition of SDB pyrolyzed at 500°C could significantly reduce N_2O emission ($p < 0.05$) in all five types of soils, but only in three soils with amended with SDB pyrolyzed at 350°C. Moreover, the influence of SDB on the N_2O emission in soil might be affected by the dosage of fertilizer. Zhang et al. (2010) found the amendment of SDB resulted in 40–51% of total N_2O emission reduction in paddy soil with N fertilization in comparison to only 21–28% without N fertilization.

The reasons for the reduction of N_2O emission in paddy soil after biochar amendment could be attributed to four aspects. First, SDB, with its high surface area and porosity, can decrease available NH_4^+ and NO_3^- in paddy soil due to the adsorption by biochar after fertilizer application, thereby decreasing substrate availability for N_2O production (Lee et al. 2005). Second, SDB amendment might be able to promote denitrification (change N_2O to N_2) through the liming effect (Yanai et al. 2007). The alkaline property of SDB can stimulate the activity of N_2O reductase, thereby leading to a decrease in the N_2O/N_2 ratio. Third, ethylene, a product from the pyrolysis of biomass, may inhibit N_2O production by affecting denitrifying microorganisms (Spokas et al. 2010). Finally, the increased rice growth in a paddy field should also be considered as this growth alone can significantly reduce the total amount of N in soil through plant uptake.

Agricultural N_2O emission primarily stems from two microbially mediated processes, nitrification and denitrification (Henault et al. 2012). It appears that SDB amendment alters the functional gene copy number involved in soil denitrification. Liu et al. (2014) found a significant inhibition in *nirS* gene copy number with the amendment of SDB (2% w/w) in paddy soil. In a pot experiment, we also detected an increase in *nirK* and *nirS* gene copy numbers in paddy soil with the addition of RSDB (2.5% w/w) (Wu, unpublished data). In addition, it has been reported that biochar amendment could significantly increase *nosZ* gene copy number, which will mitigate N_2O emission by further reducing it to N_2 (Harter et al. 2014; Xu et al. 2014). Although it is generally assumed that denitrification is responsible for N_2O emission from paddy soil, nitrification may also play a role in the production of N_2O (Chapuis-Lardy et al. 2007). The SDB amendment may potentially alter N_2O emission through the impact of abundance and diversity of AOB and AOA in the paddy. Liu et al. (2014) observed that the abundance of AOB in SDB-amended paddy soil was significantly lower than that in the control treatment. These results indicate that it is of great importance to consider the biochar properties, soil types, and environmental factors together with the microbial analysis when assessing the amendment of SDB on N_2O emissions in paddy soil.

11.6 Future Research

The application of SDB to paddy soil has the potential to greatly change soil physical, chemical, and biological characteristics. This chapter has provided studies showing that SDB as a soil amendment can, under certain conditions, reduce nutrient leaching and GHG emissions and increase rice yield in paddy fields. However, to date, it appears that the effects may depend on the properties of the biochar, the soil structure, and environmental factors. There is still limited knowledge on the potential mechanisms for carbon and nitrogen cycling over agriculturally relevant time frames. More research should be conducted in the following areas:

1. It is well established that the amendment of paddy soil with SDB is able to significantly increase rice growth and productivity. However, the increase depends on the

properties of SDB. Thus, it is essential to standardize the quality of SDB for its application in paddy soil.
In addition, it is still unclear how this function can be
kept for several rice cultivation cycles, especially in the
interaction with N fertilizer, with respect to soil quality, nutrient cycling, and the aging of SDB.

2. Most studies on the stability of SDB in paddy soil
   are carried out in the laboratory without plants.
   Therefore, it is of great importance to perform equivalent studies in the paddy field over a long-time scale.
   Interactions with other environmental factors, such
   as soil erosion, runoff, and transport, also needed for
   the assessment of the carbon sequestration role of
   SDB in paddy fields.

3. Although most studies have indicated that SDB amendment can significantly reduce CH_4 and N_2O emissions
   from paddy soils, several unresolved issues still exist. It
   is well known that CH_4 and N_2O emissions in paddy fields
   are influenced by many factors, such as soil characteristics; soil microorganisms involved in C, N, and nutrient cycling; and environmental conditions. However, in
   a certain soil type, CH_4 and N_2O emission rates will be
   directly related to the diversity and activity of microorganisms involved in C and N cycling. Therefore, it is of
   great importance to analyze the changes in microbial
   diversity and activities of the functional microbes to
   elucidate the potential mechanisms of SDB and its role
   in the mitigation of GHG emissions from paddy fields.

References

Akiyama H, Yan X, Yagi K. 2010. Evaluation of effectiveness of
enhanced-efficiency fertilizers as mitigation options for N_2O
and NO emissions from agricultural soils: Meta-analysis.
Global Change Biology 16: 1837–1846.

Anderson CR, Condron LM, Clough TJ, Fiers M, Stewart A, Hill
RA, Sherlock RR. 2011. Biochar induced soil microbial community change: Implications for biogeochemical cycling of carbon,
nitrogen and phosphorus. *Pedobiologia* 54: 309–320.

Ball PN, MacKenzie MD, DeLuca TH, Holben WE. 2010. Wildfire and charcoal enhance nitrification and ammonium-oxidizing bacterial abundance in dry montane forest soils. *Journal of Environmental Quality* 39: 1243–1253.

Bourke J, Manley-Harris M, Fushimi C, Dowaki K, Nunoura T, Antal MJ. 2007. Do all carbonized charcoals have the same chemical structure? 2. A model of the chemical structure of carbonized charcoal. *Industrial & Engineering Chemistry Research* 46: 5954–5967.

Bruun EW, Petersen C, Strobel BW, Hauggaard-Nielsen H. 2012. Nitrogen and carbon leaching in repacked sandy soil with added fine particulate biochar. *Soil Science Society of America Journal* 76: 1142–1148.

Butnan S, Deenik JL, Toomsan B, Antal MJ, Vityakon P. 2015. Biochar characteristics and application rates affecting corn growth and properties of soils contrasting in texture and mineralogy. *Geoderma* 237: 105–116.

Cayuela M, van Zwieten L, Singh B, Jeffery S, Roig A, Sánchez-Monedero M. 2013. Biochar's role in mitigating soil nitrous oxide emissions: A review and meta-analysis. *Agriculture, Ecosystems & Environment* 191: 5–16.

Chapuis-Lardy L, Wrage N, Metay A, Chotte JL, Bernoux M. 2007. Soils, a sink for N_2O? A review. *Global Change Biology* 13: 1–17.

Chen J, Zhang Y, Zeng X, Tan Z, Zhou Q. 2011. Ecological effects of balanced fertilization on red earth paddy soil with P-deficiency. *Acta Ecologica Sinica* 31: 1877–1887.

Chen Z, Xu C, Zhu Q, Cui H, Lu S, Li J, Zhang D, Gu Y. 2014. Effects of different types of cold water paddy field Fe^{2+} on physiological activity of rice. *Chinese Agricultural Science Bulletin* 30: 63–70.

Cheng W, Chander K, Inubushi K. 2000. Effects of elevated CO_2 and temperature on methane production and emission from submerged soil microcosm. *Nutrient Cycling in Agroecosystems* 58: 339–347.

Cheng Y, Cai ZC, Chang SX, Wang J, Zhang JB. 2012. Wheat straw and its biochar have contrasting effects on inorganic N retention and N_2O production in a cultivated Black Chernozem. *Biology and Fertility of Soils* 48: 941–946.

Clough TJ, Bertram JE, Ray J, Condron LM, O'Callaghan M, Sherlock RR, Wells N. 2010. Unweathered wood biochar impact on nitrous oxide emissions from a bovine-urine-amended pasture soil. *Soil Science Society of America Journal* 74: 852–860.

Conley DJ, Paerl HW, Howarth RW, Boesch DF, Seitzinger SP, Havens KE, Lancelot C, Likens GE. 2009. ECOLOGY controlling eutrophication: Nitrogen and phosphorus. *Science* 323: 1014–1015.

Crombie K, Masek O, Sohi SP, Brownsort P, Cross A. 2013. The effect of pyrolysis conditions on biochar stability as determined by three methods. *Global Change Biology Bioenergy* 5: 122–131.

Cui L, Li L, Zhang A, Pan G, Bao D, Chang A. 2011. Biochar amendment greatly reduces rice Cd uptake in a contaminated paddy soil: A two-year field experiment. *Bioresources* 6: 2605–2618.

Ding Y, Liu YX, Wu WX, Shi DZ, Yang M, Zhong ZK. 2010. Evaluation of biochar effects on nitrogen retention and leaching in multi-layered soil columns. *Water, Air, & Soil Pollution* 213: 47–55.

Dong D, Feng QB, McGrouther K, Yang M, Wang H, Wu W. 2015. Effects of biochar amendment on rice growth and nitrogen retention in a waterlogged paddy field. *Journal of Soils and Sediments* 15: 153–162. doi: 10.1007/s11368-014-0983-3.

Dong D, Yang M, Wang C, Wang H, Li Y, Luo J, Wu W. 2013. Responses of methane emissions and rice yield to applications of biochar and straw in a paddy field. *Journal of Soils and Sediments* 13: 1450–1460.

Felber R, Leifeld J, Horak J, Neftel A. 2014. Nitrous oxide emission reduction with greenwaste biochar: Comparison of laboratory and field experiments. *European Journal of Soil Science* 65: 128–138.

Feng Y, Xu Y, Yu Y, Xie Z, Lin X. 2012. Mechanisms of biochar decreasing methane emission from Chinese paddy soils. *Soil Biology and Biochemistry* 46: 80–88.

Gustafsson O, Krusa M, Zencak Z, Sheesley RJ, Granat L, Engstrom E, Praveen PS, Rao PSP, Leck C, Rodhe H. 2009. Brown clouds over South Asia: Biomass or fossil fuel combustion? *Science* 323: 495–498.

Haefele SM, Nelson A, Hijmans RJ. 2014. Soil quality and constraints in global rice production. *Geoderma* 235–236: 250–259.

Harada N, Otsuka S, Nishiyama M, Matsumoto S. 2005. Influences of indigenous phototrophs on methane emissions from a straw-amended paddy soil. *Biology and Fertility of Soils* 41: 46–51.

Harter J, Krause H-M, Schuettler S, Ruser R, Fromme M, Scholten T, Kappler A, Behrens S. 2014. Linking N_2O emissions from biochar-amended soil to the structure and function of the N-cycling microbial community. *ISME Journal* 8: 660–674.

Henault C, Grossel A, Mary B, Roussel M, Léonard J. 2012. Nitrous oxide emission by agricultural soils: A review of spatial and temporal variability for mitigation. *Pedosphere* 22: 426–433.

Inyang M, Gao B, Pullammanappallil P, Ding W, Zimmerman AR. 2010. Biochar from anaerobically digested sugarcane bagasse. *Bioresource Technology* 101: 8868–8872.

Intergovernmental Panel on Climate Change (IPCC). 2007. *Climate change 2007*: The physical science basis. Contribution of Working Group I to the fourth assessment report of the Intergovernmental Panel on Climate Change. Cambridge, United Kingdom and New York. Cambridge University Press.

Jackel U, Schnell S, Conrad R. 2004. Microbial ethylene production and inhibition of methanotrophic activity in a deciduous forest soil. *Soil Biology & Biochemistry* 36: 835–840.

Kögel-Knabner I, Amelung W, Cao Z, Fiedler S, Frenzel P, Jahn R, Kalbitz K, Kölbl A, Schloter M. 2010. Biogeochemistry of paddy soils. *Geoderma* 157: 1–14.

Keiluweit M, Nico PS, Johnson MG, Kleber M. 2010. Dynamic molecular structure of plant biomass-derived black carbon (biochar). *Environmental Science & Technology* 44: 1247–1253.

Koukouzas N, Hämäläinen J, Papanikolaou D, Tourunen A, Jäntti T. 2007. Mineralogical and elemental composition of fly ash from pilot scale fluidised bed combustion of lignite, bituminous coal, wood chips and their blends. *Fuel* 86: 2186–2193.

Lee DK, Cho JS, Yoon WL. 2005. Catalytic wet oxidation of ammonia: Why is N_2 formed preferentially against? *Chemosphere* 61: 573–578.

Lee JW, Kidder M, Evans BR, Paik S, Buchanan AC, III, Garten CT, Brown RC. 2010. Characterization of biochars produced from cornstovers for soil amendment. *Environmental Science & Technology* 44: 7970–7974.

Lehmann J, Pereira da Silva J, Steiner C, Nehls T, Zech W, Glaser B. 2003. Nutrient availability and leaching in an archaeological anthrosol and a ferralsol of the central amazon basin: Fertilizer, manure and charcoal amendments. *Plant and Soil* 249: 343–357.

Li FY, Cao XD, Zhao L, Yang F, Wang JF, Wang SW. 2013. Short-term effects of raw rice straw and its derived biochar on greenhouse gas emission in five typical soils in China. *Soil Science and Plant Nutrition* 59: 800–811.

Liang B, Lehmann J, Solomon D, Kinyangi J, Grossman J, O'Neill B, et al. 2006. Black carbon increases cation exchange capacity in soils. *Soil Science Society of America Journal* 70: 1719–1730.

Lin DX, Fan XH, Hu F, Zhao HT, Luo JF. 2007. Ammonia volatilization and nitrogen utilization efficiency in response to urea application in rice fields of the Taihu Lake region, China. *Pedosphere* 17: 639–645.

Liu L, Shen G, Sun M, Cao X, Shang G, Chen P. 2014. Effect of biochar on nitrous oxide emission and its potential mechanisms. *Journal of the Air & Waste Management Association* 64: 894–902.

Liu XY, Qu JJ, Li LQ, Zhang AF, Jufeng Z, Zheng JW, Pan GX. 2012. Can biochar amendment be an ecological engineering technology to depress N_2O emission in rice paddies?—A cross site field experiment from South China. *Ecological Engineering* 42: 168–173.

Liu Y, Yang M, Wu Y, Wang H, Chen Y, Wu W. 2011. Reducing CH_4 and CO_2 emissions from waterlogged paddy soil with biochar. *Journal of Soils and Sediments* 11: 930–939.

Liu J, Shen J, Li Y, Su Y, Ge T, Jones DL, Wu J. 2014. Effects of biochar amendment on the net greenhouse gas emission and greenhouse gas intensity in a Chinese double rice cropping system. *European Journal of Soil Biology* 65: 30–39.

Luz Cayuela M, Angel Sanchez-Monedero M, Roig A, Hanley K, Enders A, Lehmann J. 2013. Biochar and denitrification in soils: When, how much and why does biochar reduce N_2O emissions? *Scientific Reports* 3: 1–7.

Major J, Rondon M, Molina D, Riha SJ, Lehmann J. 2010. Maize yield and nutrition during 4 years after biochar application to a Colombian savanna oxisol. *Plant and Soil* 333: 117–128.

Meng M, Lv C, Li Y, Qin X, Wan Y, Gao Q. 2013. Effect of biochar on CH_4 and N_2O emissions from early rice field in South China. *Chinese Journal of Agrometeorology* 34: 396–402.

Nakamura S, Fukuda M, Nagumo F, Tobita S. 2013. Potential utilization of local phosphate rocks to enhance rice production in Sub-Saharan Africa. *Japan Agricultural Research Quarterly* 47: 353–363.

Nelissen V, Saha BK, Ruysschaert G, Boeckx P. 2014. Effect of different biochar and fertilizer types on N_2O and NO emissions. *Soil Biology & Biochemistry* 70: 244–255.

Porter LK. 1992. Ethylene inhibition of ammonium oxidation in soil. *Soil Science Society of America Journal* 56: 102–105.

Qu JJ, Zheng JW, Zheng JF, Zhang XH, Li LQ, Pan GX, Ji XH, Yu XC. 2012. Effects of wheat-straw-based biochar on yield of rice and nitrogen use efficiency of late rice. *Journal of Ecology and Rural Environment* 28: 288–293.

Ravishankara AR, Daniel JS, Portmann RW. 2009. Nitrous Oxide (N_2O): The dominant ozone-depleting substance emitted in the 21st century. *Science* 326: 123–125.

Regmi P, Garcia Moscoso JL, Kumar S, Cao X, Mao J, Schafran G. 2012. Removal of copper and cadmium from aqueous solution using switchgrass biochar produced via hydrothermal carbonization process. *Journal of Environmental Management* 109: 61–69.

Song YJ, Zhang XL, Ma B, Chang SX, Gong J. 2014. Biochar addition affected the dynamics of ammonia oxidizers and nitrification in microcosms of a coastal alkaline soil. *Biology and Fertility of Soils* 50: 321–332.

Spiridon I. 2005. Hydrolytic enzymes effects on straw cellulosic pulp. *Revue Roumaine de Chimie* 50: 541–545.

Spokas KA, Baker JM, Reicosky DC. 2010. Ethylene: Potential key for biochar amendment impacts. *Plant and Soil* 333: 443–452.

Taketani RG, Tsai SM. 2010. The influence of different land uses on the structure of archaeal communities in Amazonian anthrosols based on 16S rRNA and *amoA* genes. *Microbial Ecology* 59: 734–743.

Tian YH, Yin B, Yang LZ, Yin SX, Zhu ZL. 2007. Nitrogen runoff and leaching losses during rice-wheat rotations in Taihu Lake Region, China. *Pedosphere* 17: 445–456.

Tong X, Li J, Yuan J, Xu R, Zhou L. 2012. Adsorption of Cu(II) on rice straw char from acidic aqueous solutions. *Environmental Chemistry* 31: 64–68.

Troy SM, Lawlor PG, Flynn CJO, Healy MG. 2014. The impact of biochar addition on nutrient leaching and soil properties from tillage soil amended with pig manure. *Water Air and Soil Pollution* 225: 1–15.

Uzoma KC, Inoue M, Andry H, Fujimaki H, Zahoor A, Nishihara E. 2011. Effect of cow manure biochar on maize productivity under sandy soil condition. *Soil Use and Management* 27: 205–212.

Wang J, Zhou C, Zhao X, Xu H, Wang S, Xing G. 2013a. Effects of crop-straw biochar on paddy soil productivity and carbon sequestration. *Research of Environmental Sciences* 26: 1325–1332.

Wang Y, Hu Y, Zhao X, Wang S, Xing G. 2013b. Comparisons of biochar properties from wood material and crop residues at different temperatures and residence times. *Energy & Fuels* 27: 5890–5899.

Wang ZP, DeLaune RD, Masscheleyn PH, Patrick WH, Jr. 1993. Soil redox and pH effects on methane production in a flooded rice soil. *Soil Science Society of America Journal* 57: 382–385.

Wu W, Yang M, Feng Q, McGrouther K, Wang H, Lu H, Chen Y. 2012. Chemical characterization of rice straw-derived biochar for soil amendment. *Biomass and Bioenergy* 47: 268–276.

Xiao L, Bi E, Du B, Zhao X, Xing C. 2014. Surface characterization of maize-straw-derived biochars and their sorption performance for MTBE and benzene. *Environmental Earth Sciences* 71: 5195–5205.

Xie ZB, Xu YP, Liu G, Liu Q, Zhu JG, Tu C, Amonette JE, Cadisch G, Yong JWH, Hu SJ. 2013. Impact of biochar application on nitrogen nutrition of rice, greenhouse-gas emissions and soil organic carbon dynamics in two paddy soils of China. *Plant and Soil* 370: 527–540.

Xing GX, Cao YC, Shi SL, Sun GQ, Du LJ, Zhu JG. 2002. Denitrification in underground saturated soil in a rice paddy region. *Soil Biology & Biochemistry* 34: 1593–1598.

Xu HJ, Wang XH, Li H, Yao HY, Su JQ, Zhu YG. 2014. Biochar impacts soil microbial community composition and nitrogen cycling in an acidic soil planted with rape. *Environmental Science & Technology* 48: 9391–9399.

Xu T, Lou L, Luo L, Cao R, Duan D, Chen Y. 2012. Effect of bamboo biochar on pentachlorophenol leachability and bioavailability in agricultural soil. *Science of the Total Environment* 414: 727–731.

Yanai Y, Toyota K, Okazaki M. 2007. Effects of charcoal addition on N_2O emissions from soil resulting from rewetting air-dried soil in short-term laboratory experiments. *Soil Science and Plant Nutrition* 53: 181–188.

Yang H, Yan R, Chen H, Lee DH, Zheng C. 2007. Characteristics of hemicellulose, cellulose and lignin pyrolysis. *Fuel* 86: 1781–1788.

Yang M, Liu Y, Sun X, Dong D, Wu W. 2013. Biochar improves methane oxidation activity in rice paddy soil. *Transactions of the Chinese Society of Agricultural Engineering* 29: 145–151.

Yang SS, Chang HL. 1998. Effect of environmental conditions on methane production and emission from paddy soil. *Agriculture, Ecosystems & Environment* 69: 69–80.

Yao Y, Gao B, Zhang M, Inyang M, Zimmerman AR. 2012. Effect of biochar amendment on sorption and leaching of nitrate, ammonium, and phosphate in a sandy soil. *Chemosphere* 89: 1467–1471.

Yuan Q, Pump J, Conrad R. 2014. Straw application in paddy soil enhances methane production also from other carbon sources. *Biogeosciences* 11: 237–246.

Zhang A, Bian R, Pan G, Cui L, Hussain Q, Li L, Zheng J, Zheng J, Zhang X, Han X, Yu X. 2012. Effects of biochar amendment on soil quality, crop yield and greenhouse gas emission in a Chinese rice paddy: A field study of 2 consecutive rice growing cycles. *Field Crops Research* 127: 153–160.

Zhang A, Cui L, Pan G, Li L, Hussain Q, Zhang X, Zheng J, Crowley D. 2010. Effect of biochar amendment on yield and methane and nitrous oxide emissions from a rice paddy from Tai Lake plain, China. *Agriculture, Ecosystems & Environment* 139: 469–475.

Zhao L, Cao X, Wang Q, Yang F, Xu S. 2013a Mineral constituents profile of biochar derived from diversified waste biomasses: Implications for agricultural applications. *Journal of Environal Quaitylity* 42: 545–552.

Zhao X, Wang J, Wang S, Xing G. 2014a. Successive straw biochar application as a strategy to sequester carbon and improve fertility: A pot experiment with two rice/wheat rotations in paddy soil. *Plant and Soil* 378: 279–294.

Zhao X, Wang S, Xing G. 2014b. Nitrification, acidification, and nitrogen leaching from subtropical cropland soils as affected by rice straw-based biochar: Laboratory incubation and column leaching studies. *Journal of Soils and Sediments* 14: 471–482.

Zhao X, Yan XY, Wang SQ, Xing GX, Zhou Y. 2013b. Effects of the addition of rice-straw-based biochar on leaching and retention of fertilizer N in highly fertilized cropland soils. *Soil Science and Plant Nutrition* 59: 771–782.

Zorb C, Senbayram M, Peiter E. 2014. Potassium in agriculture—Status and perspectives. *Journal of Plant Physiology* 171: 656–669.

12

Biochar Effects on Soil Organic Carbon Storage

Chapter 12
Biochar Effects on Soil Organic Carbon Storage

Hongjie Zhang, R. Paul Voroney,
and Gordon W. Price

Chapter Outline

12.1 Introduction

One of the most significant challenges to managing agroecosystems sustainably is to increase soil organic carbon (SOC) storage so as to maintain soil quality and food production (Lal 2004). Agricultural farming practices during the past 150 years have resulted in a global loss of 78 gigatons of SOC, accounting for

66% of historic C storage; causes for SOC loss include clearing of native vegetation, cultivation of soils for annual cropping, and removal of crop residues for alternative uses (Lal 2004, 2009, 2010). Within Canada, studies have shown that agriculture has led to SOC losses of ~30% within the first 20 years of cultivation, specifically in Ontario, and losses up to 75% in some agricultural soils after a century (Oelbermann and Voroney 2011; Stockmann et al. 2013; Congreves et al. 2014). Thus, restoring the SOC level in cultivated agricultural soils is fundamental to maintaining their long-term soil productivity.

Current management strategies recommended for increasing SOC storage include implementing reduced soil tillage, growing cover crops, returning crop residues, adding animal manures, and applying sewage sludge as soil amendments (Lal 2004). However, these practices for managing organic amendments in agricultural systems should be reconsidered in light of promoting C storage in soil because the organic C contained in these amendments is rapidly decomposed in soils, returning C as carbon dioxide (CO_2) to the atmosphere and resulting in a C storage potential near neutral, especially for plant residue amendments (Lehmann 2007). In this regard, production of biochar for use as a soil amendment has been proposed as a technology to restore SOC levels (Lehmann 2007). Biochar, a C-rich solid residue of thermal degradation of organic matter at temperatures <700°C under oxygen (O_2)-limited conditions (Lehmann and Joseph 2009), is an ideal product for increasing SOC when used as a soil amendment. The recalcitrance of biochar-C in soil, coupled with its large surface area, highly porous structure, and high cation exchange capacity (Keiluweit et al. 2010; Zhang et al. 2015), make it a strong candidate for increasing soil C storage.

The capacity for C storage in biochar-amended soils has been predicted to be up to 130 Pg over a century at a global level, a capacity based on the sustainable production of biochar and application rates of at least 50 Mg·C ha^{-1} (Woolf et al. 2010). Biochar amendments to agricultural soils would serve to both increase storage of organic C in the long term (Lehmann 2007; Lal 2009) and contribute to climate change mitigation by decreasing greenhouse gas emissions from soils (Woolf et al. 2010). Biochar has also been shown to improve other soil properties, including increased retention of plant nutrients and available water (Glaser et al. 2002; Sohi et al. 2010), and to improve plant nutrient-use efficiency (Van Zwieten et al. 2010; Ippolito et al. 2012).

The objectives of this chapter are to highlight the impact of biochar amendments on SOC storage and to examine biochar production methods, feedstock properties, and the relationship between biochar-C chemical structures and SOC storage. The focus is on biochar-C stability as it relates to SOC storage potential in soils under different management systems.

12.2 Production Conditions and Chemical Property Effects on Biochar-C Stability

12.2.1 Production Methods

Biochar can be produced from pyrolysis (slow and fast pyrolysis), gasification, and flash carbonization of plant biomass and organic wastes (Brewer et al. 2011; Meyer et al. 2011). Based on the definition of Meyer et al. (2011), pyrolysis can be divided into slow and fast pyrolysis with variable residence times that is conducted in the absence of oxygen. Gasification is carried out in closed chambers or bubbling fluidized bed reactors at temperatures <800°C under conditions of low air influx. For flash carbonization, organic materials are flash ignited at elevated pressure of air (1–2 MPa). The technologies used for biochar production are shown in Table 12.1, and it is these differences in production methods that result in differences in biochar-C stability.

Brewer et al. (2011) investigated 17 biochars, produced from corn stover, switchgrass, and woody material, and used fast pyrolysis, slow pyrolysis, and gasification processes. Also, quantitative ^{13}C solid-state nuclear magnetic resonance (NMR) spectroscopy was applied to identify biochars with potential to increase SOC storage. One of the key properties identified by Brewer et al. (2011) to affect biochar-C stability was their aromatic-C content (Figure 12.1). Brewer et al. (2011) suggested that biochar-C stability was greater in biochars produced through pyrolysis (slow and fast pyrolysis) than through gasification.

TABLE 12.1
Biochar Production Technologies and Corresponding Biochar Yield and Carbon Content

Process Type	Process Temperature (°C)	Residence Time	Solid Yield Based on a Dry Woodstock (in Mass %)	Carbon Content of Solid Product (in Mass %)	Carbon Yield[a]
Slow pyrolysis	~400	Minutes to days	~30	95	~0.58
Fast pyrolysis	~500	~1 s	12–26	74	0.2–0.26
Gasification	~800	~10–20 s	~10	NA	NA
Flash carbonization	~300–600	<30 min	37	~85	~65

From Meyer, S. et al. 2011. *Environmental Science & Technology* 45 (22): 9473–9483.
[a] Carbon yield = $mass_{carbon\ product}/mass_{carbon\ feedstock}$.
NA, not available.

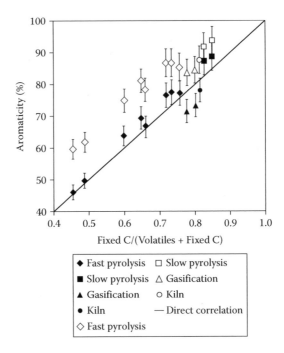

Figure 12.1 Relationship between biochar aromaticity and fixed-C content. Unshaded points represent aromaticity calculated on a molar basis, and shaded points represent aromaticity calculated on a mass basis. (From Brewer, C. E. et al. 2011. *Bioenergy Research* 4 (4): 312–323.)

12.2.2 Biochar-C Recalcitrance

Chemical properties of biochars are highly variable, depending on the composition of the feedstock and the thermal conditions used for biochar production (Brewer et al. 2011; Zimmerman et al. 2011). Biochars produced from switchgrass and corn stover residues, containing small proportions of lignin, have lower aromatic C contents than those produced from woody materials with high lignin contents (Brewer et al. 2011). Zhang et al. (2015), using ^{13}C solid-state NMR techniques, showed how cellulose and lignin in woody materials were transformed with increasing pyrolysis temperature (Figure 12.2). The NMR spectra of untreated woody material was dominated by high peaks of aliphatic C compounds indicative of cellulose and hemicelluloses, and represented 87% of O-alkyl C; biochars produced at 200°C had a similar spectrum to the original woody material.

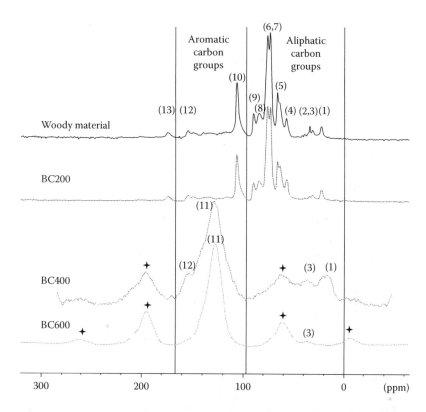

Figure 12.2 Solid-state ^{13}C cross-polarization magic angle spinning carbon spectra for woody material and corresponding biochar produced at 200, 400, and 600°C. Signal assignments and spinning sidebands are marked by numbers and asterisks. (From Zhang, H. et al. 2015. *Soil Biology and Biochemistry* 83: 19–28.)

By contrast, the spectra of biochars produced at 400 and 600°C had dominantly fused aromatic C ring structures, containing 82 and 98% aromatic C, respectively. Other studies also have shown that pyrolysis converts labile C fractions in feedstocks into aromatic C structures characteristic of biochars (typically >80% aromatic C content), and this contributes to the low decomposition rates of biochar-C in soils (Kuzyakov et al. 2014).

Specific critical parameters of the pyrolysis process that affect the properties of the biochar are the highest treatment temperature, the rate of temperature increase, and the duration of the heating period (Keiluweit et al. 2010). Molecular structural changes to the feedstocks (plant biomass) with increasing temperatures from 100°C to 700°C are shown in Figure 12.3. As pyrolysis temperature increases, there is a progression in the transformation of the various component fractions of the plant tissue: cellulose at 240 to 350°C, hemicelluloses at 200 to 260°C, and lignin at >280°C. The detailed steps in the chemical transformations of feedstocks are outlined below (Figure 12.3a):

1. Unaltered plant biomass feedstock: chemical structures are preserved at temperatures up to 100°C.
2. Transition: Alterations in basic chemical structure of plant biomass begin to occur at temperatures ranging from 200°C to 300°C. At this stage, the produced biochar contains large quantities of labile C, including volatile organic compounds (VOCs), ketones, and carboxyl C, which are derived from cellulose and lignin and unpyrolyzed feedstock (Keiluweit et al. 2010; Ameloot et al. 2013). Biochar VOC content and biochar-C mineralization rates have been shown to be positively related (Bruun et al. 2011; Mašek et al. 2013).
3. Formation of aromatic-C structure: Biochars formed at temperatures >400°C have decreased aliphatic C and sharply increased aromatic C contents.
4. Formation of aromatic and amorphous biochar: Biochars formed at temperatures ranging from 500°C to 700°C are a composite of turbostratic crystallites surrounded by amorphous aromatic C structures and O-containing compounds.
5. Formation of turbostratic biochar: Biochars formed at temperatures >700°C show increased formation of graphitic crystallites.

Biochar yields decrease with increase in pyrolysis temperatures from 250°C to 300°C without affecting C content. At temperatures >500°C, biochar yields and their C content remain constant (Figure 12.3b).

12.2.3 Molar Ratio of O:C

The molar ratio of oxygen to carbon (O:C) of biochar has been proposed as a potential indicator of biochar-C stability in

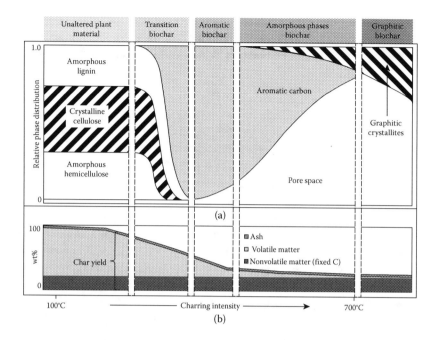

Figure 12.3 Dynamic molecular structure changes in plant biomass and plant biomass–derived biochar. (a) Physical and chemical characteristics of organic phases. Exact temperature ranges for each category are controlled by both charring conditions (i.e., temperature, duration, and atmosphere) and relative contents of plant biomass components (i.e., hemicellulose, cellulose, and lignin). (b) Char composition as inferred from gravimetric analysis. (From Keiluweit, M. et al. 2010. *Environmental Science & Technology* 44 (4): 1247–1253.)

the soil (Spokas 2010). Analysis of biochars produced via various methods, pyrolysis temperatures, feedstock types, and postproduction conditions showed that an O:C molar ratio range of 0 to 0.6 indicates chemical stability of biochar. A low O:C ratio represents a biochar more resistant to decay with a longer half-life in the soil due to the stable structure of its graphitic sheets. For example, biochars with O:C molar ratios ranging from 0.2 to 0.6 can have half-lives >1000 years, whereas O:C molar ratios >0.6 are associated with half-lives <100 years. The positive relationship between biochar O:C molar ratios and their VOC contents indicates that biochars with lower O:C ratios will have a lower labile C content (Spokas 2010).

12.3 Biochar Effects on SOC Storage

12.3.1 Field Studies

The return of crop residues to the soil is a common practice in agroecosystems to promote SOC storage. The SOC sequestration in the long term is primarily dependent on the quality and quantity of the crop residue-C returned. Crop residue decomposition rates are controlled by environmental conditions (soil moisture and temperature), fertilization, and soil tillage practices (Voroney et al. 1989). Early studies of ^{14}C-labeled ryegrass decomposition under field conditions in southeastern England showed a 85–90% loss after 10 years (Jenkinson 1977). Similar results have been reported for wheat straw decomposition in Saskatchewan, Canada (Voroney et al. 1989), and for crops commonly grown in southern Ontario, Canada (Beyaert and Voroney 2011). In these studies, only 40–50% of the added crop residue-C remained in the soil 1 year after application. Given that only a minor fraction of crop residue (~10–15%) becomes stabilized SOC in the long term (decades) (Voroney et al. 1989; Beyaert and Voroney 2011), there is great potential for increasing long-term SOC storage by converting crop residue and other plant biomass to biochar (Sohi 2012). Several recent studies have recommended conversion of various sources of organic C to biochar-C for use as an amendment to benefit soil properties and to increase SOC storage (Lehmann 2007; Luo et al. 2011; Sohi 2012). However, few studies have examined the long-term effects of biochar amendments on SOC storage under field conditions.

A field study conducted by Jones et al. (2012) found that biochar, produced from hardwood at 450°C and amended at a rate 50 Mg·ha^{-1}, increased SOC content by 22.5% over 3 years. Haefele et al. (2011) reported that biochar amendments to a field soil increased SOC contents by 8–93%.

The potential for increasing SOC storage by using biochar amendments varies widely depending on the chemical properties of the soil and the biochar, biochar application rates, soil management practices, and climatic conditions (Table 12.2). Studies have shown that SOC storage was increased by 2–93%

TABLE 12.2
Changes in Soil Organic C (%) Resulting from Biochar Amendments in Agricultural Field Studies

Biochar Feedstock	Production Conditions	Region and Climate	Soil C Content (g·kg⁻¹)	Biochar-C Application Rate (Mg·ha⁻¹)	Ratio of Stabilization of Added Biochar-C[a] (%)	Increased SOC Content (%)	Reference
Rice husks	Husks pile was heated and turned black	Tropical Philippines	14.9	16.4	100	93	Haefele et al. (2011)
			29.6		87	35	
		Thailand	5.5		3	8	
Wheat straw	350–550°C Vertical kiln	Subtropical China	23.2	4.7	81	11	Zhang et al. (2010)
				9.4	93	25	
				18.8	100	55	
Wheat straw	400°C, 1.5 h Slow pyrolysis	Subtropical China	18.3	7.6	84	19	Xie et al. (2013)
			6.4		100	70	
Corn stover	600°C, 0.5 h Slow pyrolysis	Temperate USA	16.0	4.4	100	56	Domene et al. (2014)
Oak and cherry	400°C, 24 h Slow pyrolysis	Oceanic climate UK	18.6	35.4	100	115	Ameloot et al. (2014)

(Continued)

TABLE 12.2 (Continued)
Changes in Soil Organic C (%) Resulting from Biochar Amendments in Agricultural Field Studies

Biochar Feedstock	Production Conditions	Region and Climate	Soil C Content $(g \cdot kg^{-1})$	Biochar-C Application Rate $(Mg \cdot ha^{-1})$	Ratio of Stabilization of Added Biochar-C[a] (%)	Increased SOC Content (%)	Reference
Pruning orchard	500°C Retort kiln	Oceanic climate	11.8	12.8	69	15	Ameloot et al. (2014)
	500°C Retort kiln	Italy	6.6	12.8	25	74	
Beech, hazel, oak, and birch	500°C Charcoal kiln		14.3	16.2	8	2	

SOC, soil organic carbon.

[a] Ratio of stabilization of added biochar-C = (change in soil organic carbon content $[g \cdot kg^{-1}]$/amount of biochar-C $[g \cdot kg^{-1}]$) × 100%; ratio >100% is equivalent to 100% because the increased C content resulting from not only biochar-C but also other C resources addition, such as root biomass and litter.

in tropical and temperate regions in Asia and North America with applications of biochars from rice husks, wheat straw, and corn stover produced at temperatures ranging from 350°C to 600°C and applied at rates ranging from 10 to 40 $Mg \cdot ha^{-1}$. Table 12.2 shows that the net increase in SOC with biochar-C addition has significant variation ranging from 2 to 100%, suggesting that biochar-C stability is affected by climatic conditions. Most studies confirm biochar-C recalcitrance enhances SOC storage under field conditions. It should be noted that compared to the long-term studies of crop residue decomposition by Jenkinson (1977) and Voroney et al. (1989), there are few long-term (decade-long) field studies that have examined the decay of biochar-C or its priming effects on SOC (Lehmann et al. 2011). Thus, more long-term field studies are needed to advance our understanding of the effects biochar-C on SOC and associated benefits to soil quality.

12.3.2 Laboratory Incubation Studies

There have been several recent laboratory studies reporting on biochar's resistance to biodegradation. These studies have investigated how biochar-C mineralization rates are affected by soil type and by biochar production method (Luo et al. 2011; Zimmerman et al. 2011; Kuzyakov et al. 2014; Zhang et al. 2014). Others have studied how the chemical properties of biochar affect its decay rate and its contribution to SOC using ^{14}C- and ^{13}C-labeled biochars, over a range of soil types, incubation temperatures, and water regimes, and for both the short- (months) and long-term (years) laboratory studies (Table 12.3).

The proportions of biochar-C mineralized in the soil over months to years range from 0.3 to 28% (Table 12.3). Kuzyakov et al. (2009) reported that only 0.3–0.95% of biochar-C was assimilated by soil microorganisms over a 3.5-year incubation. Using ^{14}C-labeled biochar, Kuzyakov et al. (2014) found that only ~6% biochar-C was mineralized in 8 years. These authors suggested that the high concentration of benzenepoly-carboxylic acids in the biochar contributed to its stability. There have also been reports of rapid biochar-C mineralization rates during the early period of the incubation, reflecting the decomposition of the labile C components, followed by slower rates of decay

TABLE 12.3
Biochar-C Mineralization at Various Soil Incubation Studies under Laboratory Conditions

Biochar Feedstock	Production Conditions	SOC (%)	Incubation Conditions[a]	Biochar-C Addition $(g \cdot kg^{-1}$ Soil)	Biochar-C Mineralization (%)	Biochar-C Decay Constant $(year^{-1})$	Biochar-C Half-Life[b] (years)	Reference
Maize	Slow pyrolysis at 350°C for 2 h	1.9	20°C/60 days/60% WHC	66.4	0.8	0.0476	15	Hamer et al. (2004)
Rye straw				66.3	0.7	0.0435	16	
Oak wood	Slow pyrolysis at 800°C for 24 h			78.5	0.3	0.0159	44	
Glucose	Hydrothermal pyrolysis at 200°C for 4–24 h	2.5	20–25°C/189 days	11.6	28	0.2500	3	Steinbeiss et al. (2009)
		5.5			8	0.0714	10	
Yeast		2.5		11.5	12	0.1667	4	
		5.5			22	0.0345	20	
Ryegrass	Slow pyrolysis at 400°C for 13 h	1.2	20°C/1095 days/70% WHC	56.0	3.8	0.0005	1400	Kuzyakov et al. (2009)
		0.12			3.9			

(Continued)

TABLE 12.3 (Continued)
Biochar-C Mineralization at Various Soil Incubation Studies under Laboratory Conditions

Biochar Feedstock	Production Conditions	SOC (%)	Incubation Conditions[a]	Biochar-C Addition (g·kg⁻¹ Soil)	Biochar-C Mineralization (%)	Biochar-C Decay Constant (year⁻¹)	Biochar-C Half-Life[b] (years)	Reference
Wood	Slow pyrolysis at 400°C for 0.7 h with activation[c]	0.42	22°C/1825 days/ adjusted to 67% WHC with nutrient solution	55.5	1.8	0.0031	226	Singh et al. (2012)
Tree leaves				53.0	2.5	0.0037	187	
Wood	Slow pyrolysis at 400°C for 0.7 h			55.8	2.0	0.0034	204	
Poultry litter				34.5	6.9	0.0078	89	
Cow manure				14.0	7.3	0.0111	63	
Wood	Slow pyrolysis at 550°C for 0.7 h with activation			63.4	0.5	0.0007	881	
Tree leaves				57.5	1.2	0.0017	397	
Papermill sludge				25.2	9.0	0.0098	71	
Poultry litter				33.0	2.1	0.0025	274	
Cow manure				13.2	2.2	0.0032	217	

(Continued)

TABLE 12.3 (*Continued*)
Biochar-C Mineralization at Various Soil Incubation Studies under Laboratory Conditions

Biochar Feedstock	Production Conditions	SOC (%)	Incubation Conditions[a]	Biochar-C Addition (g·kg⁻¹ Soil)	Biochar-C Mineralization (%)	Biochar-C Decay Constant (year⁻¹)	Biochar-C Half-Life[b] (years)	Reference
Grass	Slow pyrolysis at 250°C for 3 h	0.7	32°C/365 days/ adjusted to 50% WHC with nutrient solution	52.7	1.1	0.0108	65	Zimmerman et al. (2011)
Oak				55.2	1.5	0.0154	45	
Grass		5.5		52.7	1.2	0.0118	59	
Oak				55.2	2.5	0.0256	27	
Oak	Slow pyrolysis at 650°C for 3 h	5.5		78.8	0.8	0.0081	86	
Grass				63.8	1.1	0.0109	64	
Oak		0.7		78.8	0.3	0.0030	228	
Grass				63.8	0.5	0.0045	153	
Eucalyptus	Slow pyrolysis at 450°C for 0.7 h	1.0	20°C/365 days/45% water content	13.5	0.7	0.0039	179	Fang et al. (2014)
		2.5			1.1	0.0047	148	
		4.4			0.9	0.0043	163	
		2.3			0.7	0.0049	151	

(*Continued*)

TABLE 12.3 (Continued)
Biochar-C Mineralization at Various Soil Incubation Studies under Laboratory Conditions

Biochar Feedstock	Production Conditions	SOC (%)	Incubation Conditions[a]	Biochar-C Addition (g·kg⁻¹ Soil)	Biochar-C Mineralization (%)	Biochar-C Decay Constant (year⁻¹)	Biochar-C Half-Life[b] (years)	Reference
		1.0	40°C/365 days/45% water content		2.3	0.0098	71	
		2.5			2.6	0.0161	43	
		4.4			1.5	0.0094	73	
		2.3			2.7	0.0227	30	
	Slow pyrolysis at 550°C for 0.7 h	1.0	20°C/365 days/45% water content		0.4	0.0029	238	
		2.5			0.4	0.0016	422	
		4.4			0.4	0.0017	401	
		2.3			0.3	0.0022	312	
		1.0	40°C/365 days/45% water content		1.2	0.0022	304	
		2.5			1.0	0.0056	123	
		4.4			1.1	0.0052	132	
		2.3			1.0	0.0079	87	

(Continued)

TABLE 12.3 (*Continued*)
Biochar-C Mineralization at Various Soil Incubation Studies under Laboratory Conditions

Biochar Feedstock	Production Conditions	SOC (%)	Incubation Conditions[a]	Biochar-C Addition (g·kg^{-1} Soil)	Biochar-C Mineralization (%)	Biochar-C Decay Constant (year^{-1})	Biochar-C Half-Life[b] (years)	Reference
Ryegrass	Slow pyrolysis at 400°C for 0.7 h	1.2 0.12	20°C/3102 days/70% WHC	56.0	5.3 5.7	0.0025	278 5–8.5	Kuzyakov et al. (2014)

SOC, soil organic carbon; WHC, water-holding capacity.

[a] Incubation conditions included temperature, day, and water content.

[b] Biochar-C half-life: $t_{1/2} = \ln(2)/k$, where k represents decay constant of biochar-C (year^{-1}).

[c] Activation indicates steam activation.

attributed to recalcitrant, dominantly aromatic C constituents (Jones et al. 2011; Zimmerman et al. 2011). The decay rates during the slower period range from 0.0013 to 0.24% biochar-C per day, depending on biochar feedstock and production method. For example, biochars produced from glucose at 200°C and ryegrass at 400°C had half-lives of 4 and 1400 years, respectively (Table 12.3).

The results shown in Table 12.3 suggest that biochars should be produced at >400°C from the perspective of enhancing soil C storage. Other factors that may affect biochar stability in the soil are its oxidation state (Nguyen and Lehmann 2009) and its effect on soil water availability (Jones et al. 2011). Because habitable space, nutrients, and water available to soil microorganisms are central to biochar-C mineralization, its surface area and soil oxidation status and water availability play a significant role in promoting soil microbial activity.

The data in Table 12.3 show that the potential for biochar to contribute to SOC storage depends on (1) biochar production temperature: biochars produced at high temperatures are more recalcitrant than those produced at low temperatures (Ameloot et al. 2013); (2) feedstock type: biochar produced from grasses are mineralized more readily than those produced from woody materials. Biochars produced from grasses have a lower aromatic C content and a higher ash content than those produced from hardwood (Brewer et al. 2011); and (3) soil chemical properties and incubation temperature: biochar-C mineralization rates are greater in soils with high than those with low SOC contents; higher incubation temperatures generally result in greater biochar-C mineralization rates. Therefore, the SOC storage potential of a biochar amendment is determined by its mineralization rate in the soil, a rate that can be predicted from the production method, the feedstock used, the chemical properties of soils, and laboratory incubation conditions. Although biochar-C mineralization during a short-term incubation under optimal conditions is not directly comparable to that under field conditions, an understanding of biochar-C mineralization rates related to the above-mentioned parameters is important for evaluation of biochar-C storage potential in soils and for selecting biochar production parameters when used as a climate change mitigation strategy.

12.3.3 Estimating Biochar-C Storage by Century and RothC Models

Few studies have attempted to predict long-term effects of biochar amendments on the accumulation of SOC. Dil and Oelbermann (2014), using the Century model, predicted that a biochar amendment at an annual rate of 1 and 2 Mg·ha^{-1} increased SOC storage by 17% during 20 years and by up to 58.7% after 150 years.

Changes in SOC with or without biochar amendments were predicted using the RothC model for three study regions (Colombian, West Kenyan, and Iowa, USA) cropped to continuous maize over 100 years. The treatments included the following: (1) annual return of crop residues (CR), (2) annual removal of half of the crop residues (HCR), and (3) annual conversion of half of the crop residues to biochar and returned as an amendment (BC) (Figure 12.4) (Woolf and Lehmann 2012). Compared to CR, HCR reduced SOC by 21–28%, whereas BC increased SOC five times.

12.4 Biochar Contribution to Soil C Storage

12.4.1 Biochar Chemical Property Effects on Decomposition

An understanding of biochar chemical properties and their effects on the decomposition of different biochar-C fractions is essential for evaluating the contribution of biochar-C to SOC. The highly variable response in soil microbial activity in short-term incubation studies has led to reports of increased, decreased, or no effects on soil CO_2 respiration, finding that can be explained by the variable quantity and quality of biochar labile C (Kuzyakov et al. 2009; Zimmerman et al. 2011; Dempster et al. 2012; Luo et al. 2013). The quantity and quality of labile C, determined by the biochar production temperature (Bruun et al. 2011) and the feedstock type, are key factors regulating the variability in biochar decay rates and soil microbial activity (Bruun et al. 2008; Zimmerman et al. 2011). Other factors, such as biochar carbonate content, aromatic C content, pH, and cation exchange capacity, also play a role in affecting

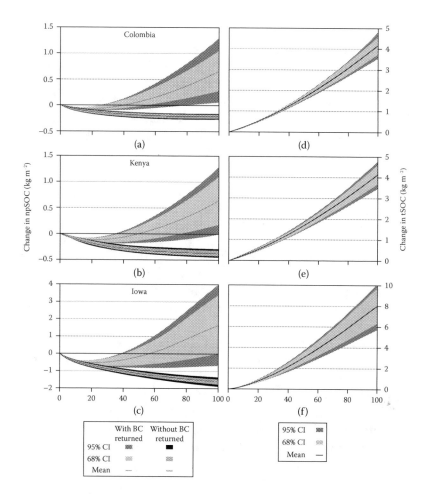

Figure 12.4 Predicted net changes in soil organic C (SOC) over 100 years of annually removing 50% of crop residues with and without returning as biochar to the soil. (a–c) Change in SOC. Narrow zone, no biochar is produced and returned to the soil; wide zone, change in SOC including biochar priming effects on organic C decomposition, and on plant net primary production. (d–f) Changes in SOC, including biochar amendments to the soil. (From Woolf D., and J. Lehmann. 2012. *Biogeochemistry* 111 (1–3): 83–95.)

biochar-C decomposition rates (Lehmann et al. 2011; Luo et al. 2011; Zhang et al. 2015). However, the effects of chemical properties of biochar on the biological processes involved in soil C dynamics are not well understood (Lehmann et al. 2011).

Decomposition rates of biochar-C have been intensely studied for determining their potential to increase SOC (Table 12.2).

These studies concluded that biochar decomposition rates ranged from decades to hundreds of years and were not significantly affected by incubation temperature or water content of amended soil. Thus, the contribution of the labile C component of biochar would have no significant effect on SOC levels, whereas the aromatic C fraction of biochar would contribute to SOC due to its slow decay rate (Zimmerman et al. 2011; Luo et al. 2013; Kuzyakov et al. 2014).

12.4.2 Resistance of Biochar to Microbial Decomposition

The labile components of plant tissue, including carbohydrates, cellulose, and hemicellulose, are converted to aromatic C compounds during pyrolysis in an O_2-limiting environment (Keiluweit et al. 2010). This chemical conversion has been used to explain the increased resistance of biochars to microbial decomposition and to their contribution to SOC storage (Kuzyakov et al. 2014).

It has been suggested that biochar may have positive effects on soil microbial colonization due to its large surface areas and numerous pores providing habitats for soil microorganisms and protection against fungal grazers (Pietikäinen et al. 2000) and increase biochar-C degradation. However, Quilliam et al. (2013) reported no changes in levels of microbial colonization on the internal and external surfaces of biochar buried for 3 years under field conditions. In addition, a lower C-use efficiency by microbial colonization on biochar surfaces compared to surrounding soils suggests that biochars are relatively inert and do not provide additional habitats for soil microbes (Quilliam et al. 2013).

Biochars labeled with ^{14}C have been used to quantify biochar-C incorporation into soil microbial biomass (Kuzyakov et al. 2014). The uptake of biochar-C by the soil microbial biomass after 3.5 years ranged from 0.3 to 0.95%. Microbial biomass-C derived from biochar decomposed slower than their C derived from other SOC fractions after 8.5 years, suggesting that biochar-C is extremely stable and not available as a food source for microorganisms.

Biochar application may affect soil physical properties, i.e., bulk density, porosity, and moisture content (Akhtar et al. 2014), properties that might impact how effectively soil microorganisms are able to degrade biochar. Enhanced physical stabilization of labile C in the soil (Keith et al. 2011) and sorption onto biochar surfaces (LeCroy et al. 2013) can also affect substrate decomposability and C-use efficiency, as otherwise labile C may diffuse into the biochar porous structure and be physically protected from microbial decay (Hamer et al. 2004). Studies have shown that changes to soil physical properties due to biochar amendment do not affect soil microbial activity related to biochar degradation (Jones et al. 2011). Therefore, these results confirm that it is the aromatic-C structure of biochar that defines its resistance to microbial decomposition.

12.4.3 Biochar Effects on Decomposition of Native SOC

Although biochar amendments can promote loss of native SOC, these losses are relatively small compared to the amount of C returned to soils as biochar-C (Jones et al. 2011; Luo et al. 2011). Moreover, the loss of native SOC induced by biochar additions depends on the labile C content of the biochar, which is controlled by biochar production temperature (Keiluweit et al. 2010; Zimmerman et al. 2011). Changes in SOC mineralization can result from interaction between the added biochar and native SOC (Hamer et al. 2004; Kuzyakov et al. 2009; Cross and Sohi 2011; Luo et al. 2011; Zimmerman et al. 2011). Zimmerman et al. (2011) reported that SOC mineralization was enhanced in soils amended with biochars produced at temperatures ranging from 250°C to 400°C, whereas SOC mineralization was reduced in soils amended with biochar produced at higher temperatures (525–650°C). These results suggest that the potential for priming with biochar amendment is determined by its labile C content (Cross and Sohi 2011). Increased SOC mineralization can be explained by microbial assimilation of biochar labile C promoting mineralization of native SOC, i.e., a priming effect (Kuzyakov et al. 2009). Decreased SOC mineralization could be the result of physical protection by biochar particulates and SOC sorption to biochar surfaces (Keith et al. 2011; Zimmerman et al. 2011).

Interactions between biochar-C and additions of readily decomposable C (glucose C and plant residue C) on SOC mineralization have been extensively reported (Keith et al. 2011; Kuzyakov et al. 2014). Hamer et al. (2004) and Kuzyakov et al. (2009) showed that the decomposition rate of a ryegrass biochar increased 2–6 times after glucose addition in a short-term soil incubation study, whereas the decomposition rate of glucose or sugarcane residue decreased. This difference was attributed to an altered soil microbial community structure, either not able to quickly respond to the C addition, or to microorganisms having a decreased C-use efficiency (O'Neill et al. 2009; Quilliam et al. 2013). Another mechanism proposed for these interaction effects is the reduced cometabolism of added C caused by the decreased microbial activity in biochar-amended soils (Kuzyakov et al. 2009; Zimmerman et al. 2011). Although additions of readily decomposable C to soils can affect decomposition of both biochar-C and native SOC in biochar-amended soils, increased SOC storage has been observed in all of these studies.

12.4.4 Biochar Promotes Formation of Soil Humic Substances

Positive relationships between quantities of humic acids and charred fragments derived from C_3- and C_4-plant residues in Japanese volcanic ash soils have been reported, suggesting promotion of the formation of soil humic substances from the charred plant biomass C (Shindo et al. 2005). Klüpfel et al. (2014) have shown that biochars produced at high temperatures (600–700°C) have the ability to accept electrons, whereas those produced at low temperatures (400–500°C) can donate electrons. Thus, the redox activity of biochars may directly contribute to promoting formation of humic substances (Figure 12.5). It has been reported that aromatic C fractions contained in biochar enhance formation of the stable proportion of benzenepoly-carboxylic acids contained in humic substances (Kuzyakov et al. 2014). These findings further emphasize the potential of biochars to benefit long-term C storage, and in particular, provide an explanation for the high SOC content of terra preta soils in Amazon (Lehmann 2007).

Figure 12.5 Changes in the electron exchange capacities, electron accepting capacities (EAC), and electron donating capacities (EDC) of grass and wood biochars during redox cycling that promote the formation of the redox properties of humic substances. (From Klüpfel, L. et al. 2014. *Environmental Science & Technology* 48 (10): 5601–5611.)

12.5 Conclusions

Benefits of biochar amendments to soils and to the environment are increased SOC storage and reduced CO_2 emissions. Thus, production of biochars for use as soil amendments has been considered as a strategy both to restore soil C levels and to mitigate climate change. However, the long-term potential of biochar-C storage in soil raises the fundamental question of how long does biochar-C remain stable in the soil? Recent studies of biochar-C stability suggest the need to measure the labile C contents of biochars. Moreover, selection of appropriate feedstock and control of pyrolysis temperatures play key roles in the production of stable forms of biochar-C. The scientific evidence to date confirms the potential of biochar for use as a soil amendment to increase C storage in the soil. Beyond that, biochar amendments have demonstrated benefits to soil quality and to increasing soil water and nutrient retention. Feedstocks originating from municipal, industrial, commercial, and agricultural sectors as organic wastes or by-products that otherwise would be landfilled or disposed of in some manner are potential sources of organic matter for the production of biochar. In contrast, the removal of residues from agricultural fields to produce biochar may not be economically viable or environmentally sustainable. A life-cycle analysis focusing on the linkage between the benefits of biochar-C storage in soils and the greenhouse gas emissions associated with biochar production should be considered in future research.

References

Akhtar, S. S., G. Li, M. N. Andersen, and F. Liu. 2014. Biochar Enhances Yield and Quality of Tomato Under Reduced Irrigation. *Agricultural Water Management* 138: 37–44.

Ameloot, N., E. R. Graber, F. G. A. Verheijen, and S. D. Neve. 2013. Interactions between Biochar Stability and Soil Organisms: Review and Research Needs. *European Journal of Soil Science* 64 (4): 379–390.

Ameloot, N., S. Sleutel, S. D. C. Case, G. Alberti, N. P. McNamara, C. Zavalloni, B. Vervisch, G. d. Vedove, and S. D. Neve. 2014. C Mineralization and Microbial Activity in Four Biochar Field Experiments Several Years after Incorporation. *Soil Biology and Biochemistry* 78: 195–203.

Beyaert, R., and R. P. Voroney. 2011. Estimation of Decay Constants for Crop Residues Measured Over 15 Years in Conventional and Reduced Tillage Systems in a Coarse-Textured Soil in Southern Ontario. *Canadian Journal of Soil Science* 91 (6): 985–995.

Brewer, C. E., R. Unger, K. Schmidt-Rohr, and R. C. Brown. 2011. Criteria to Select Biochars for Field Studies Based on Biochar Chemical Properties. *Bioenergy Research* 4 (4): 312–323.

Bruun, E. W., H. Hauggaard-Nielsen, N. Ibrahim, H. Egsgaard, P. Ambus, P. A. Jensen, and K. Dam-Johansen. 2011. Influence of Fast Pyrolysis Temperature on Biochar Labile Fraction and Short-Term Carbon Loss in a Loamy Soil. *Biomass and Bioenergy* 35 (3): 1182–1189.

Bruun, S., E. S. Jensen, and L. S. Jensen. 2008. Microbial Mineralization and Assimilation of Black Carbon: Dependency on Degree of Thermal Alteration. *Organic Geochemistry* 39 (7): 839–845.

Congreves, K. A., J. M. Smith, D. D. Németh, D. Hooker, and L. L. Van Eerd. 2014. Soil Organic Carbon and Land Use: Processes and

Potential in Ontario's Long-Term Agro-Ecosystem Research Sites. *Canadian Journal of Soil Science* 94 (3): 317–336.

Cross, A., and S. P. Sohi. 2011. The Priming Potential of Biochar Products in Relation to Labile Carbon Contents and Soil Organic Matter Status. *Soil Biology and Biochemistry* 43 (10): 2127–2134.

Dempster, D. N., D. B. Gleeson, Z. M. Solaiman, D. L. Jones, and D. V. Murphy. 2012. Decreased Soil Microbial Biomass and Nitrogen Mineralisation with Eucalyptus Biochar Addition to a Coarse Textured Soil. *Plant and Soil* 354 (1–2): 311–324.

Dil, M., and M. Oelbermann. 2014. Evaluating the Long-Term Effects of Pre-Conditioned Biochar on Soil Organic Carbon in Two Southern Ontario Soils Using the Century Model. In: M. Oelbermann (ed.) *Sustainable Agroecosystems in Climate Change Mitigation*, 251–270. Wageningen Academic Publishers, Wageningen, The Netherlands.

Domene, X., S. Mattana, K. Hanley, A. Enders, and J. Lehmann. 2014. Medium-Term Effects of Corn Biochar Addition on Soil Biota Activities and Functions in a Temperate Soil Cropped to Corn. *Soil Biology and Biochemistry* 72: 152–162.

Fang, Y., B. Singh, B. P. Singh, and E. Krull. 2014. Biochar Carbon Stability in Four Contrasting Soils. *European Journal of Soil Science* 65 (1): 60–71.

Glaser, B., J. Lehmann, and W. Zech. 2002. Ameliorating Physical and Chemical Properties of Highly Weathered Soils in the Tropics with Charcoal—A Review. *Biology and Fertility of Soils* 35 (4): 219–230.

Haefele, S. M., Y. Konboon, W. Wongboon, S. Amarante, A. A. Maarifat, E. M. Pfeiffer, and C. Knoblauch. 2011. Effects and Fate of Biochar from Rice Residues in Rice-Based Systems. *Field Crops Research* 121 (3): 430–440.

Hamer, U., B. Marschner, S. Brodowski, and W. Amelung. 2004. Interactive Priming of Black Carbon and Glucose Mineralisation. *Organic Geochemistry* 35 (7): 823–830.

Ippolito, J. A., D. A. Laird, and W. J. Busscher. 2012. Environmental Benefits of Biochar. *Journal of Environmental Quality* 41 (4): 967–972.

Jenkinson, D. S. 1977. Studies on the Decomposition of Plant Material in Soil. V. The Effects of Plant Cover and Soil Type on the Loss of Carbon from ^{14}C Labelled Ryegrass Decomposing Under Field Conditions. *Journal of Soil Science* 28 (3): 424–434.

Jones, D. L., D. V. Murphy, M. Khalid, W. Ahmad, G. Edwards-Jones, and T. H. DeLuca. 2011. Short-Term Biochar-Induced Increase in Soil CO_2 Release is both Biotically and Abiotically Mediated. *Soil Biology and Biochemistry* 43 (8): 1723–1731.

Jones, D. L., J. Rousk, G. Edwards-Jones, T. H. DeLuca, and D. V. Murphy. 2012. Biochar-Mediated Changes in Soil Quality and Plant Growth in a Three Year Field Trial. *Soil Biology and Biochemistry* 45: 113–124.

Keiluweit, M., P. S. Nico, M. G. Johnson, and M. Kleber. 2010. Dynamic Molecular Structure of Plant Biomass-Derived Black Carbon (Biochar). *Environmental Science & Technology* 44 (4): 1247–1253.

Keith, A., B. Singh, and B. P. Singh. 2011. Interactive Priming of Biochar and Labile Organic Matter Mineralization in a Smectite-Rich Soil. *Environmental Science & Technology* 45 (22): 9611–9618.

Klüpfel, L., M. Keiluweit, M. Kleber, and M. Sander. 2014. Redox Properties of Plant Biomass-Derived Black Carbon (Biochar). *Environmental Science & Technology* 48 (10): 5601–5611.

Kuzyakov, Y., I. Bogomolova, and B. Glaser. 2014. Biochar Stability in Soil: Decomposition during Eight Years and Transformation as Assessed by Compound-Specific [14]C Analysis. *Soil Biology and Biochemistry* 70: 229–236.

Kuzyakov, Y., I. Subbotina, H. Chen, I. Bogomolova, and X. Xu. 2009. Black Carbon Decomposition and Incorporation into Soil Microbial Biomass Estimated by [14]C Labeling. *Soil Biology and Biochemistry* 41 (2): 210–219.

Lal, R. 2004. Soil Carbon Sequestration Impacts on Global Climate Change and Food Security. *Science* 304 (5677): 1623–1627.

Lal, R. 2009. Challenges and Opportunities in Soil Organic Matter Research. *European Journal of Soil Science* 60 (2): 158–169.

Lal, R. 2010. Managing Soils and Ecosystems for Mitigating Anthropogenic Carbon Emissions and Advancing Global Food Security. *Bioscience* 60 (9): 708–721.

LeCroy, C., C. A. Masiello, J. A. Rudgers, W. C. Hockaday, and J. J. Silberg. 2013. Nitrogen, Biochar, and Mycorrhizae: Alteration of the Symbiosis and Oxidation of the Char Surface. *Soil Biology and Biochemistry* 58: 248–254.

Lehmann, J. 2007. A Handful of Carbon. *Nature* 447 (7141): 143–144.

Lehmann, J., and S. Joseph. 2009. Biochar for Environmental Management: An Introduction. In: Lehmann, J., and Joseph, S. (Eds.), *Biochar for Environmental Management: Science and Technology*, 1–12. Earthscan, London.

Lehmann, J., M. C. Rillig, J. Thies, C. A. Masiello, W. C. Hockaday, and D. Crowley. 2011. Biochar Effects on Soil Biota—A Review. *Soil Biology and Biochemistry* 43 (9): 1812–1836.

Luo, Y., M. Durenkamp, M. De Nobili, Q. Lin, and P. C. Brookes. 2011. Short Term Soil Priming Effects and the Mineralisation of Biochar Following Its Incorporation to Soils of Different pH. *Soil Biology and Biochemistry* 43 (11): 2304–2314.

Luo, Y., M. Durenkamp, M. De Nobili, Q. Lin, B. J. Devonshire, and P. C. Brookes. 2013. Microbial Biomass Growth, Following Incorporation of Biochars Produced at 350°C Or 700°C, in a Silty-Clay Loam Soil of High and Low pH. *Soil Biology and Biochemistry* 57: 513–523.

Mašek, O., P. Brownsort, A. Cross, and S. Sohi. 2013. Influence of Production Conditions on the Yield and Environmental Stability of Biochar. *Fuel* 103: 151–155.

Meyer, S., B. Glaser, and P. Quicker. 2011. Technical, Economical, and Climate-Related Aspects of Biochar Production Technologies: A Literature Review. *Environmental Science & Technology* 45 (22): 9473–9483.

Nguyen, B. T., and J. Lehmann. 2009. Black Carbon Decomposition under Varying Water Regimes. *Organic Geochemistry* 40 (8): 846–853.

Oelbermann, M., and R. Voroney. 2011. An Evaluation of the Century Model to Predict Soil Organic Carbon: Examples from Costa Rica and Canada. *Agroforestry Systems* 82 (1): 37–50.

O'Neill, B., J. Grossman, M. T. Tsai, J. E. Gomes, J. Lehmann, J. Peterson, E. Neves, and J. E. Thies. 2009. Bacterial Community Composition in Brazilian Anthrosols and Adjacent Soils Characterized Using Culturing and Molecular Identification. *Microbial Ecology* 58 (1): 23–35.

Pietikäinen, J., O. Kiikkilä, and H. Fritze. 2000. Charcoal as a Habitat for Microbes and Its Effect on the Microbial Community of the Underlying Humus. *Oikos* 89 (2): 231–242.

Quilliam, R. S., H. C. Glanville, S. C. Wade, and D. L. Jones. 2013. Life in the 'Charosphere'—Does Biochar in Agricultural Soil Provide a Significant Habitat for Microorganisms? *Soil Biology and Biochemistry* 65: 287–293.

Shindo, H., M. Yoshida, A. Yamamoto, H. Honma, and S. Hiradate. 2005. Delta ^{13}C Values of Organic Constituents and Possible Source of Humic Substances in Japanese Volcanic Ash Soils. *Soil Science* 170 (3): 175–182.

Singh, B. P., A. L. Cowie, and R. J. Smernik. 2012. Biochar Carbon Stability in a Clayey Soil as a Function of Feedstock and

Pyrolysis Temperature. *Environmental Science & Technology* 46 (21): 11770–11778.

Sohi, S. P. 2012. Carbon Storage with Benefits. *Science* 338 (6110): 1034–1035.

Sohi, S. P., E. Krull, E. Lopez-Capel, and R. Bol. 2010. A Review of Biochar and Its Use and Function in Soil. In: Donald L. Sparks, editor, *Advances in Agronomy*, 105, 47–82. Academic Press, San Diego, CA.

Spokas, K. A. 2010. Review of the Stability of Biochar in Soils: Predictability of O:C Molar Ratios. *Carbon Management* 1 (2): 289–303.

Steinbeiss, S., G. Gleixner, and M. Antonietti. 2009. Effect of Biochar Amendment on Soil Carbon Balance and Soil Microbial Activity. *Soil Biology and Biochemistry* 41 (6): 1301–1310.

Stockmann, U., M. A. Adams, J. W. Crawford, D. J. Field, N. Henakaarchchi, M. Jenkins, B. Minasny, A. B. McBratney, V. de Remy de Courcelles, and K. Singh. 2013. The Knowns, Known Unknowns and Unknowns of Sequestration of Soil Organic Carbon. *Agriculture, Ecosystems & Environment* 164: 80–99.

Van Zwieten, L., S. Kimber, S. Morris, K. Y. Chan, A. Downie, J. Rust, S. Joseph, and A. Cowie. 2010. Effects of Biochar from Slow Pyrolysis of Papermill Waste on Agronomic Performance and Soil Fertility. *Plant and Soil* 327 (1–2): 235–246.

Voroney, R. P., E. A. Paul, and D. W. Anderson. 1989. Decomposition of Wheat Straw and Stabilization of Microbial Products. *Canadian Journal of Soil Science* 69 (1): 63–77.

Woolf, D., J. E. Amonette, F. A. Street-Perrott, J. Lehmann, and S. Joseph. 2010. Sustainable Biochar to Mitigate Global Climate Change. *Nature Communications* 1: 56.

Woolf, D., and J. Lehmann. 2012. Modelling the Long-Term Response to Positive and Negative Priming of Soil Organic Carbon by Black Carbon. *Biogeochemistry* 111 (1–3): 83–95.

Xie, Z., Y. Xu, G. Liu, Q. Liu, J. Zhu, C. Tu, J. E. Amonette, G. Cadisch, J. W. H. Yong, and S. Hu. 2013. Impact of Biochar Application on Nitrogen Nutrition of Rice, Greenhouse-Gas Emissions and Soil Organic Carbon Dynamics in Two Paddy Soils of China. *Plant and Soil* 370 (1–2): 527–540.

Zhang, A., L. Cui, G. Pan, L. Li, Q. Hussain, X. Zhang, J. Zheng, and D. Crowley. 2010. Effect of Biochar Amendment on Yield and Methane and Nitrous Oxide Emissions from a Rice Paddy from Tai Lake Plain, China. *Agriculture, Ecosystems & Environment* 139 (4): 469–475.

Zhang, H., R. P. Voroney, and G. W. Price. 2014. Effects of Biochar Amendments on Soil Microbial Biomass and Activity. *Journal of Environmental Quality* 43: 2104–2114.

Zhang, H., R. P. Voroney, and G. W. Price. 2015. Effects of Temperature and Processing Conditions on Biochar Chemical Properties and Their Influence on Soil C and N Transformations. *Soil Biology and Biochemistry* 83: 19–28.

Zimmerman, A. R., B. Gao, and M.-Y. Ahn. 2011. Positive and Negative Carbon Mineralization Priming Effects among a Variety of Biochar-Amended Soils. *Soil Biology and Biochemistry* 43 (6): 1169–1179.

13

Biochar Effects on Greenhouse Gas Emissions

Chapter 13
Biochar Effects on Greenhouse Gas Emissions

Pranoy Pal

Chapter Outline

13.1 Introduction

Thermal decomposition of biomass such as crop and forestry wastes and animal manure (hereafter termed *feedstock*) under partially anaerobic conditions and at relatively low temperatures ($<700°C$) produces biochar—a relatively more stable form of C due to the formation of condensed aromatic structures (Lehmann and Joseph 2009). Interactions of biochar with soil minerals could further increase its stability in soil (Brodowski et al. 2006), in addition to contributing to C sequestration over time. The ranges for pH, ash content, C content, surface area, and cation exchange capacity (CEC) of biochar are as follows: 5.2–10.3, 1.1–55.8%, 23.6–87.5%, 0–642 $m^2 \cdot g^{-1}$

(Lehmann et al. 2011) and 10–69 $cmol_c \cdot kg^{-1}$ (Kim et al. 2013), respectively; and these properties vary with feedstock type and pyrolysis conditions.

Similar to any natural phenomena, biochar application also has its own set of pros and cons (Mukherjee and Lal 2014; Saggar et al. 2014). The environmental benefits of biochar application to soils include reduced nutrient leaching, improved soil characteristics and crop productivity, reduced greenhouse gas (GHG) emissions, enhanced bioavailability of plant nutrients, and remediation of contaminated soils. However, the potential risks of biochar application can be (1) unbalanced addition of nutrients, heavy metals, or both and of polycyclic aromatic hydrocarbons (PAHs) and dioxins to soil; (2) negative impact on soil biota; (3) sorption of residual pesticides and subsequent decrease in the efficiency of these products; (4) environmental pollution from dust, erosion, and leaching of biochar particles; (5) effect on soil surface albedo; and (6) soil compaction during application.

Biochar has been used in agriculture by several cultures throughout history to improve soil quality; however, in relation to GHG emissions, Rondon et al. (2005) was the first study to show reduced emissions of the GHGs nitrous oxide (N_2O) and methane (CH_4) due to biochar application in soybean–grass-cultivated soils. Subsequently, an exponential increase in biochar research has produced several books, scientific reports, and policies, including many journal special issues, such as *Global Change Biology Bioenergy* (2013), *Agronomy* (2013), *Carbon Management* (2014), and virtual special issues in *Soil Biology and Biochemistry* (2012), *European Journal of Soil Science* (2014), and *Agriculture, Ecosystems and Environment* (2014). On November 1, 2014, the Web of Science™ database listed 104 studies on N_2O and 90 studies on CH_4 between 2010 and 2014 when a combined search of biochar was performed with these GHGs. This wealth of knowledge has undoubtedly helped provide a mechanistic understanding of the underlying processes of GHG reduction by using biochar; however, with "certain uncertainties." For example, two recent publications on long-term (≥1 year) field trials in a vineyard (Verhoeven and Six 2014) and in a grassland (Angst et al. 2014), with similar climatic conditions and using the same biochar, reported contrasting N_2O emissions, thus highlighting the need to further understand the underlying mechanisms that affect GHG emissions from biochar-amended soils. These differences probably occur because the mechanisms of biochar application are interdependent on several factors. For example, the physicochemical properties of biochar (composition, particle and pore size distribution) are mainly dependent on the pyrolysis conditions and feedstock characteristics that, in turn, determine the suitability of its application under specific conditions, as well as affect its behavior and fate in the environment (Figure 13.1), e.g., biochar produced from lignocellulosic feedstock has high C content, whereas biochar from nutrient-rich feedstock such as poultry litter has characteristics similar to a fertilizer (Cantrell et al. 2012) and is better suited for degraded soils. Disturbance of this balance of soil–biochar benefit shown in Figure 13.1 may cause elevated emissions, not achieve the sought-after benefit, or both. To account for the differences in GHG emissions (or other benefits) caused by biochar application, the International Biochar Initiative, an organization that promotes good industry practices and adheres to environmental and ethical standards, has created a biochar certification program. Certifying a biochar guarantees

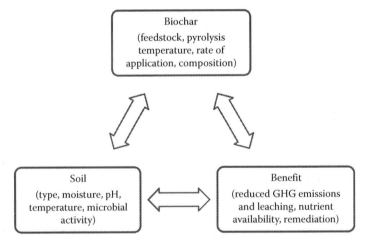

Figure 13.1 Relationships between the type of biochar, the soil to which it is applied, and the sought-after environmental benefit.

the biochar meets the minimum criteria of environmental and ethical safety and still achieves optimum performance. In brief, a biochar would be certified if the following conditions are met: H:C$_{(organic)}$ molar ratio ≤0.7, PAHs range from 6 to 20 mg·kg^{-1} dry weight, polychlorinated biphenyls range from 0.2 to 0.5 mg·kg^{-1} dry weight, dioxins and furans ≤9 ng·kg^{-1} dry weight, negative response to germination inhibition assay, 1.4–3.9 mg·kg^{-1} dry weight Cd content, and 70–500 mg·kg^{-1} dry weight Pb content, along with declarations of total C and N contents, pH, ash, and moisture contents.

This chapter provides a brief introduction to biochar and GHG emissions, summarizes the effects and potential mechanisms of biochar addition on GHGs, and emphasizes that there is still a lack of definitive knowledge on the influence of biochar on biogeochemical cycles before solid recommendations can be made to adopt biochar application as a GHG mitigation option.

13.2 Greenhouse Gases

GHGs in the Earth's atmosphere absorb and re-emit the sun's radiation within the infrared range to cause the "greenhouse effect." The greenhouse effect makes Earth habitable by maintaining surface temperatures, but the imbalance of these gases in the atmosphere due to anthropogenic activities has increased GHG concentrations drastically, resulting in adverse effects such as sea level rise, floods, droughts, adverse seasonal variations, and melting of polar ice caps. Mitigation options are urgently needed to reduce the concentrations of GHGs in the Earth's atmosphere. The main GHGs in the Earth's atmosphere (in the orders of their concentrations) are carbon dioxide (CO_2), CH_4, N_2O, and other fluorinated gases (Figure 13.2). Based on these gases' ability to absorb solar energy and their atmospheric life span, each of these GHGs has a different global warming potential (GWP). The GWPs of CO_2, CH_4, and N_2O over a 100-yr time frame have been calculated to be 1, 25, and 298, respectively. The non-CO_2 GHGs are often expressed as CO_2-equivalents (CO_2-eq).

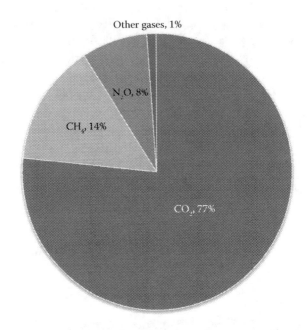

Figure 13.2 Global greenhouse gas emissions. (From IPCC 2007.)

Carbon dioxide is mainly emitted through fossil fuel use (by the industry, transport, and electricity sectors). Although agriculture contributes only 7% to the total CO_2 emissions, agricultural practices contribute 40 and 90% of the global CH_4 and N_2O emissions, respectively. Biochar application to soils produces an initial pulse of CO_2 due to enhanced microbial respiration and decomposition of the labile C fractions, but in the long term, it is expected to sequester C and principally emit negligible quantities of CO_2. Kuzyakov et al. (2014) showed that only 6% of initially added biochar was mineralized to CO_2 during 8.5 years. Because of such lower emissions over time and a smaller agricultural contribution of CO_2, this chapter focuses on biochar effects on the non-CO_2 GHGs.

Soils can act both as a source and as a sink of GHGs. According to the classification of the Intergovernmental Panel on Climate Change (IPCC), anthropogenic activities in the Agriculture, Forestry and Other Land Use (AFOLU) sector are the main contributors to GHG emissions. Emission of GHGs in the AFOLU sector has increased from 9.9 Gt CO_2-eq in 1970 to 12 Gt CO_2-eq in 2010, accounting for ~20–25% of the global GHG emissions in 2010 (IPCC 2014). Between 1970 and 2010, CH_4 emissions increased by 20%, whereas N_2O emissions increased by 45–75% during that period (IPCC 2014).

13.3 Nitrous Oxide

Terrestrial N_2O emissions have increased from 10 to 12 Tg·N_2O-N yr^{-1} between 1900 and 2000 and may reach 16 Tg·N_2O-N yr^{-1} by 2050 (Bouwman et al. 2013). The increase of the atmospheric N_2O concentration is in the order of 0.2–0.3% yr^{-1} due to the growth of emissions caused primarily by anthropogenic activities. It is estimated that 90% of the emissions are derived from agriculture, caused by the intensive use of N fertilizers. In pasture systems, ruminant urine and excreta contribute to a significant amount of the total emissions. Production of N_2O in soils occurs from several distinct and interconnected processes, with major contributions from nitrification, denitrification, and dissimilatory nitrate reduction to ammonium (DNRA), also called nitrate ammonification. To understand the interactions of biochar and N_2O, it is important to understand these production processes.

13.3.1 Biochar Interactions with N_2O Production Pathways

Nitrification is the biological oxidation of ammonium $\left(NH_4^+\right)$ to nitrate $\left(NO_3^-\right)$ by an autotrophic, aerobic group of microbes occurring in two steps. The first step involves the oxidation of ammonia (NH_3) to nitrite $\left(NO_2^-\right)$ by NH_3-oxidizing bacteria, NH_3-oxidizing archaea, and certain fungi; in the second step, oxidation of NO_2^- to NO_3^- by NO_2^--oxidizing bacteria occurs. Nitrification produces H^+ ions that result in a decrease in the soil pH. N_2O is produced during nitrification as a by-product, and this process is dependent on the soil's pH, aeration, temperature (25–35°C), and moisture content (40–65% water-filled pore space [WFPS]). Denitrification is the dissimilatory reduction of NO_3^- to NO_2^- and its sequential reduction via nitric oxide (NO) and N_2O to dinitrogen or nitrogen gas (N_2). Reductase enzymes occur at each step of the process. Denitrifiers are facultative, anaerobic bacteria (and certain fungi) that have the ability to utilize NO_3^- and NO_2^- as electron acceptors when oxygen gas (O_2) is unavailable (Firestone 1982). Denitrification rates and the ratios of its products (e.g., $N_2O:N_2$) are affected by various factors, including oxygen (O_2) diffusion, soil moisture (>60–95% WFPS), temperature, pH, and substrate availability (C and NO_3^-). The optimal pH for both nitrification and denitrification ranges from 7 to 8, whereas depending on the threshold soil moisture contents, either nitrification or denitrification dominates, thereby continuing to affect gaseous diffusion. Specific properties of biochar can affect these processes. Certain toxic compounds in biochar can inhibit the activity of nitrifiers and reduce N_2O emissions via nitrification. The porosity and water-retentive capacity of biochar can alter the soil aeration, gas diffusivities, and moisture content and reduce N_2O emissions via denitrification, whereas certain biochars' relatively higher pH can promote the last step of denitrification to produce N_2 rather than N_2O. Adsorption of NO_3^- in biochar can reduce substrate availability to denitrifiers and lower N_2O production. Biochar might act as a reducing agent itself and also as an electron shuttle, facilitating the electron transfer to microorganisms by acting as an electrical conduit (Cayuela et al. 2013). These attributes of biochar are specifically discussed in the subsequent sections.

Nitrifier-denitrification is the nitrification pathway governed by autotrophic nitrifiers in which NH_3 is oxidized to NO_2^-, followed by the reduction of NO_2^- to NO, N_2O, and N_2 (Ferguson et al. 2007). It is carried out by ammonia-oxidizing bacteria such as *Nitrosomonas* under O_2-depleted conditions. This process was previously thought to be of minor importance in soils, but recent studies have shown that significant N_2O emissions from nitrifier-denitrification can occur, especially under moisture conditions that are suboptimal for heterotrophic NO_3^- denitrification. DNRA is the process where NO_3^- is reduced directly to NH_4^+. Previously, DNRA was only thought to be favored in intensively reduced, C-rich environments. Recent studies show that it is not restricted to reduced environments; it can also occur in rice paddies, in calcareous agricultural soils, and in temperate arable soils where it was shown to be limited by low-molecular-weight C sources (Schmidt et al. 2011). Chemodenitrification refers to any abiotic chemical reaction leading to the formation of NO, N_2O, and N_2, and includes the following: (1) chemical decomposition of hydroxylamine; (2) chemical decomposition of NO_2^-; and (3) the abiotic decomposition of ammonium nitrate in the presence of light, moisture, and reacting surfaces (Rubasinghege et al. 2011). Some fungal species can produce N_2O and N_2 by codenitrification where one N atom from NO_2^- combines with one N atom from a cometabolized compound, other than NO_2^-. Nitrifier-denitrification, chemodenitrification, and codenitrification are equally important N_2O production pathways under specific conditions; however, biochar is not expected to influence these pathways significantly and has not widely been investigated.

It is evident that biochar can influence N_2O production through several interconnected pathways, making it a challenge to pinpoint a specific mechanism for N_2O mitigation. Cayuela et al. (2014), in a meta-analysis, suggested that the basic knowledge of when, how much, and why biochar reduces N_2O emissions under a particular set of conditions is still poorly understood. The biochar characteristics and potential mechanisms that affect N_2O emissions are discussed below.

13.3.2 Biochar Characteristics Affecting N_2O Emissions

Biochar characteristics such as feedstock type, rate of application, and pyrolysis conditions (temperature and duration) directly affect N_2O emissions and also alter soil physical, chemical, and biological properties.

13.3.2.1 Biochemical Composition of Biochar

The inherent biochemical properties of biochar are governed by the type and C:N ratio of the feedstock and its pyrolysis conditions that, in turn, govern the N turnover and extent of N_2O emissions (Cayuela et al. 2014). The importance of feedstock type was highlighted by the study of Zhao et al. (2013) where

12 different biochars were produced using animal manures, waste wood, crop residues, food waste, aquatic plants, and municipal waste as feedstock while varying the pyrolysis conditions. This altered the biochar's total organic C, pH, mineral constituents, and acid functional groups. Yao et al. (2012) tested 13 biochars produced from four different feedstocks (bagasse, bamboo, peanut hull, and Brazilian pepperwood) and found that the biochars produced at higher temperatures (>600°C) had higher NO_3^--adsorbing capabilities but that the sorption characteristics of other nutrients varied. Nelissen et al. (2014) used willow, maize, pine, and wood mixture as feedstock to produce seven different biochars pyrolyzed between 450°C and 650°C and applied to a Luvisol at 20 t·ha^{-1}. Cumulative N_2O emissions decreased by 52–84% across various biochar types, with higher reductions from biochars obtained at higher pyrolysis temperatures. In this study, one of the suggested mechanisms was due to the adsorption of N-containing water onto the biochar, making it unavailable to the microbes. Moreover, the recent meta-analysis of Cayuela et al. (2014) found a positive relation between the C:N ratio of feedstock and reduced N_2O emissions. It was suggested that to achieve a reduction in N_2O emissions, the C:N ratio of the feedstock should be ≥30 so that the resultant biochar can cause C and N immobilization, thereby reducing the emissions.

These altered biochar properties due to feedstock and pyrolysis conditions—especially enhanced organic C, pH, and NO_3^- adsorption—are the key players in governing N_2O emissions. The extent of N_2O emissions is also influenced by the aromaticity and the stability of the oxidizable component of biochar, i.e., initial rapid decomposition of the surface-oriented labile fractions (aliphatic C), followed by the slower decomposition of stable, more recalcitrant fractions (condensed aromatic C). These labile C fractions of the biochar can serve as the substrate to denitrifiers and cause elevated N_2O emissions. Under certain conditions, biochar can potentially stimulate decomposition of native soil organic matter (SOM) by enhancing microbial activity (Wardle et al. 2008). Conversely, biochar may prolong SOM decomposition by enhancing soil aggregation (Liang et al. 2010), thus influencing aeration and ultimately N_2O emissions. Further in situ studies are required because the presence of plant roots, rhizosphere processes (such as exudation), and environmental factors may affect C mineralization and N_2O emissions. It is clear that with an appropriate combination of feedstock and production temperature, it is possible to control the desired biochar properties to different degrees and to provide the opportunity to design biochar for mitigating N_2O emissions in specific soils.

13.3.2.2 Rate of Biochar

Several studies (Bruun et al. 2011; Uzoma et al. 2011; Mukherjee and Lal 2013) emphasized that higher biochar application rates induced relatively quicker and greater effects on soil physical properties such as aeration and macroporosity, and on biological processes, in some cases, causing a toxic or inhibitory effects on soil biota and consequently reducing N_2O emissions. Spokas et al. (2009) applied a biochar produced from fast pyrolysis of sawdust at 550°C to a silt loam soil at a wide application rate ranging from 24 to 720 $t \cdot ha^{-1}$ and found reduced N_2O emissions by up to 74% at rates >100 $t \cdot ha^{-1}$. Bruun et al. (2011) observed that the addition of a wheat straw biochar to a loamy soil at of 3 $t \cdot ha^{-1}$ reduced 47% of the N_2O emissions compared with the treatment with 1 $t \cdot ha^{-1}$ due to biochar-induced changes in water retention and enhanced microbial biomass causing N immobilization. Another field study (Taghizadeh-Toosi et al. 2011) at the application rate of 30 $t \cdot ha^{-1}$ of *Pinus radiata* biochar reduced N_2O emissions from ruminant urine patches by 70% (due to NO_3^- adsorption of urinary-N), but no significant reduction was observed when half the rate was applied. Liu et al. (2014) also found that reduced N_2O emissions were largely dependent on the rate of biochar application. However, Zhang et al. (2012) applied wheat straw biochar at 10, 20, and 40 $t \cdot ha^{-1}$ to a rice paddy and did not observe a consistent trend in the effect of biochar rate on GHG emissions or crop productivity. This exception was possibly due to rice cultivation being adversely different from cropping systems. The merits of higher rates of biochar application are evident, but research should focus on optimizing the threshold rates so that the surplus biochar can be used for other productive purposes.

13.3.2.3 Age of Biochar

It has been speculated that biochar might become more effective over time under particular conditions (Spokas et al. 2009; Liang et al. 2010; Cayuela et al. 2014). Singh et al. (2010) pyrolyzed poultry manure and *Eucalyptus saligna* wood chips at 400 and 550°C, measured the N dynamics over 5 months, and they suggested that biochar could reduce greater N_2O emissions over time due to increased organo-mineral interactions and increased cation sorption capacity through oxidative reactions on biochar surfaces with aging. Aged biochar would have passed the initial rapid decomposition phase of the labile fractions and become more stable over time. However, biochar may also lose its liming properties over time, and oxidation of biochar surfaces may decrease the soil pH in its vicinity. These conditions (pH, sorption capacities) would affect N_2O emissions in different ways; however, longer term (more than 5-year) in situ studies are required to elucidate the effect of age of biochar on N_2O emissions (Cayuela et al. 2014).

13.3.2.4 Biochar Effects on Soil Physical Properties

Application of biochar influences soil physical properties such as porosity (micro-, meso-, and macroporosities), texture, aggregate stability, aeration, and specific surface area. These factors are strongly linked to gas diffusion and soil–water content, which affect soil N_2O emissions. Cayuela et al. (2013), using nine different brush biochars, reported that soil texture was closely related to the ability of biochar to decrease the ratio of $N_2O/(N_2+N_2O)$ at the time of maximum N_2O emissions. Case et al. (2012) amended a sandy-loam soil with hardwood tree biochar at 50–100 $t \cdot ha^{-1}$ and observed increased soil–water-holding capacity that had a direct influence on NO_3^- leaching and indirectly affecting N_2O emissions. Zheng et al. (2012) showed that the particle size of biochar affected N_2O emissions in specific soil types by influencing the soil macroporosity and lowering the chances of denitrification. In terms of soil bulk density, a lower value is the indicator of better soil structure, aggregate stability, and aeration, and studies such as Jones et al. (2010) and Laird et al. (2010) suggested that biochar application at ~2 t·ha^{-1} could be sufficient to decrease the bulk density of amended soils. Jones et al. (2010) applied 4 and 8 t·ha^{-1} of biochar produced from biosolids, spent mushroom compost, greenwaste compost, and greenwaste, and they found that the amendments decreased bulk density and increased mesoporosity, available water-holding capacity, and water retention of the treated soil. Liu et al. (2012) added a pine biochar to silt-loam and sandy-loam soils at 4–16 t·ha^{-1} and found enhanced soil aggregate stability of the silt-loam soil after 11 months of incubation. Moreover, the porous nature of the feedstock is inherited by the biochar, and it may provide ideal conditions for microbial growth and habitat. This can induce N immobilization, thereby lessening the bioavailability of N in soil, and thus decreasing N_2O emissions through limiting substrates for nitrification and denitrification.

13.3.2.5 Biochar Effects on Soil Chemical Properties

The main attributes that directly influence N_2O emissions are liming properties and the presence of functional groups in biochar, C substrate supply to the denitrifiers promoting complete denitrification, and redox properties that facilitate the transfer of electrons to soil denitrifying microorganisms (Singh et al. 2010; Cayuela et al. 2013). Biochar can also reduce the inorganic-N pool available for the nitrifiers, denitrifiers, or both that produce N_2O, as a result of NH_4^+, NO_3^-, or N_2O adsorption, greater plant growth, NH_3 volatilization loss, or immobilization of N. Several studies have shown that biochar produced a liming effect that decreased N_2O resulting from denitrification. Oxygen-containing organic functional groups, mineral deposits such as calcium carbonate, and ample quantities of soluble base cations in biochar can be released rapidly into the soil and increase the soil pH, thereby decreasing the potential for denitrification. For example, van Zwieten et al. (2010) applied biochars derived from poultry litter, paper-mill waste, and biosolids at 10–50 t·ha^{-1} and suggested that increased pH pushed the denitrification process toward completion under flooded conditions, thereby reducing N_2O emissions. Cayuela et al. (2013) suggested that N_2O reduction was mainly governed by the dissolved organic C content of the soil. Biochar, although not being a labile C source itself, sorption of the soil's labile C fractions on the biochar, decreased its bioavailability to the denitrifiers. Zheng et al. (2012) found that N_2O emissions were negatively related to pH in two soils of contrasting pH and also observed an 8% decrease in NO_3^- leaching with coapplication of biochar (derived from oak pellets, pyrolyzed at 550°C) with N fertilizer and attributed it to microbial N immobilization caused by biochar addition. Similarly, Case et al. (2012) found decreased emissions and attributed it to NO_3^- immobilization. Biochars contain acid functional groups (such as carboxyl or hydroxyl groups similar to humic acid amendments), and these functional groups can improve soil characteristics by buffering pH, chelating micronutrients, and helping with soil aggregation. Cayuela et al. (2013) proposed a possible mechanism of biochar facilitating the transfer of electrons to denitrifying microorganisms in soil, thus acting as an electron shuttle. This shuttling, together with its liming effect and high surface area, would promote the reduction of N_2O to N_2. These properties limit the substrates for N_2O production processes, while acting as a slow release N source. For example, Taghizadeh-Toosi et al. (2012) showed that biochar produced from *P. radiata* wood chips adsorbed NH_3 onto the biochar surfaces and is bioavailable to plants. Direct sorption of NH_4^+, NO_3^-, and other organic N species (including N_2O; Cornelissen et al. 2013) can occur on biochar surfaces. Adsorption of N_2O directly onto biochar has not been investigated widely. Cornelissen et al. (2013) investigated this phenomena under anhydrous conditions by using a pinewood biochar produced at 500°C. They found that the Langmuir maximum sorption capacities (Q_{max} [$cm^3 \cdot g^{-1}$]) of N_2O on biochar were 17–73 $cm^3 \cdot g^{-1}$, much higher compared to uncharred wood and metal oxides. This study also showed that biochar can bind N_2O more strongly than both mineral and organic soil materials. However, the mechanism of N_2O adsorption on biochar needs further investigation in situ. The review of Clough et al. (2013) suggested that biochars produced at higher temperatures (>600°C) have higher NO_3^--adsorbing capacities. These indices should be exploited further for the use of biochars as a customized, slow-release fertilizer.

13.3.2.6 Biochar Effects on Soil Biological Properties

Biochar application may alter the soil's microbial community structure, microbial abundance and activity, macrofauna activity, and N cycling enzymes. Van Zwieten et al. (2014) applied biochars produced from poultry litter, wheat chaff, and oil mallee feedstock at 10 $t \cdot ha^{-1}$ to four contrasting soils. They found that among other soils, the lighter soil limited NO_3^- availability to denitrifiers and also increased the *nosZ* gene abundance (responsible for N_2O reductase enzyme activity) by a small yet significant proportion. The reason to why only the lighter soil did this could not be explained. Xu et al. (2014) applied 50 $t \cdot ha^{-1}$ of rice straw biochar pyrolyzed at 500°C to an acidic Acrisol and found positive biochar effects on soil physical and chemical properties and increased crop yield. Moreover, biochar application also stimulated nitrification and denitrification while reducing overall N_2O emissions due to higher *nosZ* gene abundance, indicating a higher $N_2:N_2O$ ratio. Using rice straw and dairy manure biochars applied at 4 $t \cdot ha^{-1}$, Liu et al. (2014) reported reduced N_2O emissions accompanied with lower abundance of ammonia-oxidizing and nitrite-oxidizing bacteria that corresponded with their respective gene copy numbers. In contrast, Anderson et al. (2014) found that adding biochar to the soil potentially increased microbial abundance of N_2O-producing microbes that increased N_2O fluxes and NH_4^+ concentrations. Colonization of arbuscular mycorrhizal fungi, responsible for a symbiotic relationship with a host plant (to help nutrient uptake), in wheat roots was found to increase to 20–40%, 2 years after *Eucalyptus* wood biochar additions of 0.6–6 $t \cdot ha^{-1}$, in comparison to a colonization rate of 5–20% in unamended controls (Solaiman et al. 2010). Harter et al. (2014) performed a microcosm study by applying greenwaste biochar pyrolyzed at 700°C to a sandy-loam soil at 20 and 100 $t \cdot ha^{-1}$ and found reduced N_2O emissions possibly because addition of biochar to a soil changed the denitrifier community composition by promoting the growth and activity of N_2O-reducing bacteria expressing *nosZ* relative to *nirS*- and *nirK*-containing denitrifiers. Biochar application can affect the dominance of either bacterial or fungal communities that are responsible for N_2O production, but the answer to what chemical component, physical component, or combination of biochar induces such changes remains to be explored (Lehmann et al. 2011; Cayuela et al. 2014).

13.4 Methane

The most important sources of CH_4 are natural wetlands (27%); fossil fuel related to natural gas, coal mines, and the coal industry (18%); ruminant animals (18%); rice paddies (11%); biomass burning (10%); landfills (9%); termites (4%); and the ocean and hydrates (3%). In wetlands and rice cultivation systems, CH_4 is produced in flooded anoxic zones through anaerobic decomposition of plant material and SOM; and rice plants acting as conduits from the soil to the atmosphere or via ebullition. Emissions of CH_4 in managed grasslands can originate from enteric fermentation from ruminant animals (~99%), manure and dung deposited on pastoral land (~1%), and smaller contributions from soils and forage plants. The major sink for CH_4 is photochemical oxidation (by hydroxyl radicals) in the atmosphere (~550 $Tg \cdot year^{-1}$), but oxidation of atmospheric CH_4 by aerobic soils also provides a significant additional sink (10–44 $Tg \cdot year^{-1}$) (Lowe 2006).

13.4.1 Production Pathways of CH_4 and Its Interactions with Biochar

The net flux between soils and the atmosphere results from the balance between two microbiological processes: methanogenesis (the source—emission of CH_4 by strict anaerobes of the *Archaea* phylogenetic domain) and methanotrophy (the sink—oxidation and consumption of CH_4 by aerobic proteobacteria). Emission of CH_4 is dependent on soil redox potential, C substrate supply, temperature, diffusion, soil pH, micronutrients availability, competitive inhibition in presence of electron acceptors, and the presence of plant vegetation. Methanogen activity is inhibited in the presence of electron acceptors such as Fe^{3+}, Mn^{2+}, and the oxides of N; NO_3^- among these, causes the strongest inhibition in CH_4 production, followed by NO_2^-, NO, and N_2O (Dalal et al. 2008). Substrates for methanogens are the products of anaerobic decomposition of native SOM, exogenous organic matter introduced to the soil from aboveground processes such as leaf litter and plant biomass, soil microbial biomass, roots, and root exudates.

Based on their physiology, phylogeny, biochemistry, and morphology, methanotrophs have been grouped into types I and II, corresponding to *Gammaproteobecteria* and *Alphaproteobacteria*. Oxidation of CH_4 is an enzyme-controlled reaction carried out by CH_4-assimilating bacteria (methane monooxygenase enzyme) and autotrophic NH_4-oxidizing bacteria (ammonium monooxidase enzyme), both of which require O_2. Methane oxidation is mainly dependent on O_2 availability and soil diffusivity, which is closely related to soil moisture content and soil texture. For example, CH_4 oxidation may reduce under soil compaction and high soil bulk density due to the reduced rate of gas diffusion. In addition, CH_4 consumption decreases with increasing soil–water content, and a wet spell can kick-start methanotrophic activity. Although methanotrophs can survive under extreme acidic and saline conditions, their optimum activity generally occurs in a rather narrow range of 5.0–7.5 pH units. It should be noted that the NH_4^+ produced in soil or added through fertilizers often poses competitive inhibition to CH_4 oxidation, at least over short periods. Elevated CH_4 concentrations, inhibitory chemicals (nitrification inhibitors, urease inhibitors), and pesticides can significantly reduce CH_4 oxidation rates (Dalal et al. 2008).

Studies show that biochar affects these production pathways by hindering CH_4 production and promoting CH_4 oxidation through practices that include enhancing soil O_2 diffusion through soil moisture management, land-use minimized compaction, and soil fertility management. Biochar application supposedly causes enhanced soil aeration (by affecting soil macro- and microporosities) and soil–water content and increased CH_4 diffusion to soil that decreases anoxic conditions, which could decrease CH_4 production, increase CH_4 oxidation, or both. Reduction in CH_4 oxidation may also occur due to suppressed microbial activity caused by toxic or inhibitory compounds present in the biochar (Spokas et al. 2009). Biochar application can tighten the N cycle by limiting the availability of N substrates (including NH_4^+) to microbes, which can reduce its competitive inhibition and ultimately support methanotrophy.

Karhu et al. (2011) observed a 96% increase in CH_4 uptake after application of 9 $t \cdot ha^{-1}$ birch biochar (pyrolyzed at 400°C) to an agricultural soil in Finland, and they suggested that the increased uptake was possibly induced by the elevated soil aeration accompanied with the enhanced CH_4 diffusion from air to soil; however, they did not distinguish the changes caused between the decrease in CH_4 production and the increase in CH_4 oxidation. The effect of biochar on soil aeration and CH_4 oxidation has been reported in several studies (Rondon et al. 2005, 2006; Van Zwieten et al. 2009). Liu et al. (2011) also reported a 51% reduction in CH_4 emissions when a paddy soil was amended with bamboo and rice straw biochar pyrolyzed at 600°C. They could not distinguish between the methanogenic archaeal communities of amended versus unamended soils. The CH_4 emissions were reduced due to the inhibition of methanogenic activity as a result of increased soil pH from biochar and stimulated methanotrophic activity. In a soil applied with pig slurry, Schimmelpfennig et al. (2014) pointed out that biochar application increased CH_4 oxidation via enhanced soil aeration and promoted O_2 and atmospheric CH_4 supply for methanotrophic bacteria (Van Zwieten et al. 2009). Other explanations for an increase in CH_4 oxidation after biochar application are related to the probable adsorption of CH_4 onto biochar surfaces, metals adsorbed on biochar catalyzing CH_4 oxidation, and growth stimulation of methanotrophic populations (Van Zwieten et al. 2009; Feng et al. 2011). Many studies have also found increased CH_4 emissions after biochar application (Spokas et al. 2009; Zhang et al. 2010; Zheng et al. 2012). It is clear that biochar can promote CH_4 oxidation only if the soil under investigation has a relatively lower bulk density (unaffected from agricultural compaction) and has an already established methanotrophic population.

Some studies have found that although suppressing N_2O emissions, biochar application caused an increase in CH_4 emissions (a GHG pollution swap). Mukherjee and Lal (2013) clearly allude to several studies that show a GHG pollution swapping scenario. Zheng et al. (2012) found biochars reduced

N_2O emissions by 3–60% but increased CH_4 emissions by 5–72% after applying oak biochar to a silt-loam soil. Spokas et al. (2009) showed a decrease in CH_4 emissions with biochar addition, whereas Karhu et al. (2011) observed increased CH_4 uptake by 96%, with no changes in N_2O emissions. Similarly, Zhang et al. (2010) found that total soil CH_4 emissions increased by 34–41% in soils amended with biochar (wheat straw feedstock pyrolyzed at 350–550°C) at 40 t·ha^{-1} compared with the treatments without biochar. However, total N_2O emissions decreased by 40–51 and 21–28%, respectively, in biochar-amended soils with or without N fertilization. Pollution swapping does not mitigate GHG emissions; hence, further research is warranted to address these issues.

13.5 Conclusions and Recommendations for Future Research

Biochar can be a powerful GHG mitigation tool when a certified biochar is applied to a specific soil, at a standardized rate, along with fetching additional environmental benefits as is supported by numerous studies and also included in the latest IPCC report (IPCC 2014). At present, we have a comprehensive knowledge of the advantages (Cayuela et al. 2014), disadvantages (Mukherjee and Lal 2014), and relatively stronger confidence in understanding the underlying mechanisms (Lehmann et al. 2011) of GHG emissions and biochar application to soils. Biochar generates significant cobenefits in the form of improved agricultural production systems, resilience, and other ecosystem services. Some suggestions for future research are as follows.

- We now know that the ability of a particular biochar to sequester C and reduce GHGs in soils depends on the characteristics of the receiving soil as well as on the nature of the biochar (Figure 13.1). It has become possible to customize a biochar for desired properties by altering the feedstock type and pyrolysis temperature. With the wealth of literature available, it is possible to estimate the GHG reduction potential of this customized biochar. Hence, research should be focused on optimizing these soil- and crop-specific biochar applications with standardized rates regionally and nationally. This approach will allow biochar to be included in national GHG inventories.

- There is a paucity of data on the timing and method of biochar application to soils. No studies have investigated whether biochars (N-rich) can benefit the crops at different growth stages and act as a slow-release N fertilizer. Moreover, not many studies have investigated the effect of biochar particle size (fine powder, granules, or pellets) and application method (surface versus incorporated) on GHG emissions. These factors require further research followed by optimization.

- Biochar production undoubtedly reduces biomass wastes from agricultural and pastoral systems that are otherwise difficult to dispose. For example, direct incorporation of the straw waste from a crop field causes significant N immobilization and accumulation of organic acids, affecting the yield of subsequent crop and carryover of pathogens, whereas conventional burning causes air pollution. On average, pyrolysis of the straw waste would generate ~8 $t \cdot ha^{-1}$ of biochar (from 15 $t \cdot ha^{-1}$ of straw waste from a typical rice–wheat crop rotation in an intensively managed agricultural system). Most studies have achieved GHG reductions with application rates that vary widely from 10 to 100 $t \cdot ha^{-1}$. Cayuela et al. (2014) concluded that an application rate of 10–20 $t \cdot ha^{-1}$ was sufficient to significantly reduce the N_2O emitted (mean reduction of 27%), whereas high doses (>100 $t \cdot ha^{-1}$) reduced emissions up to 87%. However, reductions in CH_4 emissions have been achieved at much lower application rates. The question posed here is whether intensive agricultural systems have sufficient crop wastes available to be pyrolyzed (e.g., 40–200 $t \cdot ha^{-1}$ of crop wastes) to achieve these biochar rates. If not, are the farmers prepared to pay for the additional costs associated with production and transportation of the extra bio-char, after additional labor costs involved in its application? Sustainable use of biochar is dependent on efficient resource utilization and GHG emissions control during its production. Hence, current research should focus on optimizing a biochar rate that is (1) realistic under field conditions, (2) pyrolyzed from the readily available feedstock on-site, and (3) suitable for the particular regional soil type that will maximize the agronomic benefits along with GHG reductions.

- There is still a paucity of information on how aging will affect the ability of biochar to decrease GHG emissions. In addition, there is a paucity of data from long-term (>5 year) field studies. Research should focus on how the residence times of biochar in soils can be manipulated to design the best possible biochar for a given soil type.

- Mechanisms of biochar with CH_4 emissions and the influence of biotic and biotic factors on these interactions are not fully understood and need further investigation.

- Although some cost–benefit (Gaunt and Cowie 2009) and life-cycle analyses have estimated high profitability with biochar, many soil parameters and crop yields may not be improved with the suggested or assumed biochar application under various soil–biochar combinations. Biochar application can negatively affect soil by reducing crop productivity, increasing GHG emissions, and having adverse effects on soil parameters. Thus, life-cycle analyses of all factors (GHG emissions, soil-C sequestration, crop yield) are required to assess financial implications of biochar amendment. Principal component analyses and cluster analyses of biochars prepared from different feedstock types should be undertaken to identify the salient or harmful aspects of a particular biochar and segregate it with respect to, e.g., phytotoxicity, carcinogenicity, and sorptive abilities.

- Research should focus on novel methods of biochar application. The sorption ability of biochar is well suited for a nutrient carrier that can eventually act as a slow-release fertilizer. However, the manipulation of biochar properties to potentially optimize its use as a fertilizer carrier has not been fully recognized. Three aspects play a key role for evaluating the suitability of the carrier material: (1) the stability of the fertilizer combination during storage, (2) its degradability in the soil, and (3) the carriers' efficiency. To reduce costs associated with biochar application, fertilizer application, or both, research should focus on new methods of coapplication of fertilizer and biochar tested under field conditions. For example, Kim et al. (2014) produced biochar pellets by blending switchgrass biochar, lignin (as a binding agent), and K and P fertilizers and showed that the release rate of the nutrients can be controlled by altering the lignin content and processing temperature of the pellets. Newer methods of applying a mixture of nitrification inhibitors (e.g., dicyandiamide) and urease inhibitors (e.g., Agrotain®) adsorbed onto the biochar should be investigated that may provide additional help in reducing N_2O and NO_3^- leaching, and NH_3 emissions, while also reducing CH_4 emissions by increasing soil aeration. These areas indicate profitable and environmentally responsible future research.

Thus biochar, if used carefully, has the potential to become an ideal GHG mitigation tool along with other environmental benefits, and it provides numerous implications for future research.

References

Anderson CR, Hamonts K, Clough TJ, Condron LM. 2014. Biochar does not affect soil N-transformations or microbial community structure under ruminant urine patches but does alter relative proportions of nitrogen cycling bacteria. *Agriculture, Ecosystems and Environment* 191: 63–72.

Angst T, Six J, Reay DS, Sohi SP. 2014. Impact of pine chip biochar on trace greenhouse gas emissions and soil nutrient dynamics in an annual ryegrass system in California. *Agriculture, Ecosystems and Environment* 191: 17–26.

Bouwman AF, Beusen AHW, Griffioen J, Van Groenigen JW, Hefting MM, Oenema O, Van Puijenbroek PJTM, Seitzinger S, Slomp CP, Stehfest E. 2013. Global trends and uncertainties in terrestrial denitrification and N_2O emissions. *Philosophical Transactions of the Royal Society B* 368: 1621.

Brodowski S, John B, Flessa H, Amelung W. 2006. Aggregate-occluded black carbon in soil. *European Journal of Soil Science* 57: 539–546.

Bruun EW, Müller-Stöver D, Ambus P, Hauggaard-Nielsen H. 2011. Application of biochar to soil and N_2O emissions: Potential effects of blending fast-pyrolysis biochar with anaerobically digested slurry. *European Journal of Soil Science* 62: 581–589.

Cantrell KB, Hunt PG, Uchimiya M, Novak JM, Ro KS. 2012. Impact of pyrolysis temperature and manure source on physicochemical characteristics of biochar. *Bioresource Technology* 107: 419–428.

Case SDC, McNamara NP, Reay DS, Whitaker J. 2012. The effect of biochar addition on N_2O and CO_2 emissions from a sandy loam soil—The role of soil aeration. *Soil Biology and Biochemistry* 51: 125–134.

Cayuela ML, Sanchez-Monedero MA, Roig A, Hanley K, Enders A, Lehmann J. 2013. Biochar and denitrification in soils: When, how much and why does biochar reduce N_2O emissions? *Scientific Reports* 3: 1732.

Cayuela ML, van Zwieten L, Singh BP, Jeffery S, Roig A, Sánchez-Monedero MA. 2014. Biochar's role in mitigating soil nitrous oxide emissions: A review and meta-analysis. *Agriculture, Ecosystems and Environment* 191: 5–16.

Clough TJ, Condron LM, Kammann C, Mueller C. 2013. A review of biochar and soil nitrogen dynamics. *Agronomy Journal* 3: 275–293.

Cornelissen G, Rutherford DW, Arp HPH, Dörsch P, Kelly CN, Rostad CE. 2013. Sorption of pure N_2O to biochars and other organic and inorganic materials under anhydrous conditions. *Environmental Science and Technology* 47: 7704–7712.

Dalal RC, Allen DE, Livesley SJ, Richards G. 2008. Magnitude and biophysical regulators of methane emission and consumption in the Australian agricultural, forest, and submerged landscapes: A review. *Plant and Soil* 309: 43–76.

Feng Y, Xu Y, Yu Y, Xie Z, Lin X. 2011. Mechanisms of biochar decreasing methane emission from Chinese paddy soils. *Soil Biology and Biochemistry* 46: 80–88.

Ferguson SJ, Richardson DJ, Van Spanning RJM. 2007. Biochemistry and molecular biology of nitrification. In: Bothe H, Ferguson SJ, Newton WE, eds. *Biology of the Nitrogen Cycle.* Amsterdam, the Netherlands, Elsevier. pp. 209–222.

Firestone MK. 1982. Biological denitrification. In: Stevenson FJ, ed. *Nitrogen in Agricultural Soils.* Crop Science Society of America, Madison, WI, pp. 289–326.

Gaunt J, Cowie A. 2009. Biochar, greenhouse gas accounting and emissions trading. In: Lehmann J, Joseph S, eds. *Biochar for Environmental Management: Science and Technology.* Earthscan, London, pp. 317–340.

Harter J, Krause H-M, Schuettler S, Ruser R, Fromme M, Scholten T, Kappler A, Behrens S. 2014. Linking N_2O emissions from biochar-amended soil to the structure and function of the N-cycling microbial community. *The ISME Journal* 8(6): 660–674.

IPCC. 2007. *Climate Change 2007: Synthesis Report.* Contribution of Working Groups I, II and III to the Fourth Assessment Report of the Intergovernmental Panel on Climate Change [Core Writing Team, Pachauri, R.K and Reisinger, A. (eds.)]. IPCC, Geneva, Switzerland. pp.35–42.

IPCC. 2014. *Climate Change 2014: Mitigation of Climate Change.* Contribution of Working Group III to the Fifth Assessment Report of the Intergovernmental Panel on Climate Change [Edenhofer, O., R. Pichs-Madruga, Y. Sokona, E. Farahani, S. Kadner, K. Seyboth, A. Adler, I. Baum, S. Brunner, P. Eickemeier, B. Kriemann, J. Savolainen, S. Schlömer, C. von Stechow, T. Zwickel and J.C. Minx (eds.)]. Cambridge University Press, Cambridge, NY.

Jones BEH, Haynes RJ, Phillips IR. 2010. Effect of amendment of bauxite processing sand with organic materials on its chemical, physical and microbial properties. *Journal of Environmental Management* 91: 2281–2288.

Karhu K, Mattila T, Bergström I, Regina K. 2011. Biochar addition to agricultural soil increased CH_4 uptake and water holding capacity—Results from a short-term pilot field study. *Agriculture, Ecosystems and Environment* 140: 309–313.

Kim P, Hensley D, Labbé N. 2014. Nutrient release from switchgrass-derived biochar pellets embedded with fertilisers. *Geoderma* 232–234: 341–351.

Kim P, Johnson AM, Essington ME, Radosevich M, Kwon WT, Lee SH, Rials TG, Labbe N. 2013. Effect of pH on surface characteristics of switchgrass-derived biochars produced by fast pyrolysis. *Chemosphere* 90(10): 2623–2630.

Kuzyakov Y, Bogomolova I, Glaser B. 2014. Biochar stability in soil: Decomposition during eight years and transformation as assessed by compound-specific ^{14}C analysis. *Soil Biology and Biochemistry* 70: 229–236.

Laird DA, Fleming P, Davis DD, Horton R, Wang BQ, Karlen DL. 2010. Impact of biochar amendments on the quality of a typical midwestern agricultural soil. *Geoderma* 158: 443–449.

Lehmann J, Joseph S. 2009. *Biochar for Environmental Management: Science and Technology.* London, UK, Earthscan.

Lehmann J, Rillig MC, Thies J, Masiello CA, Hockaday WC, Crowley D. 2011. Biochar effects on soil biota—A review. *Soil Biology and Biochemistry* 43(9): 1812–1836.

Liang B, Lehmann J, Sohi SP, Thies JE, O'Neill B, Trujillo L, Gaunt J, et al. 2010. Black carbon affects the cycling of non-black carbon in soil. *Organic Geochemistry* 41(2): 206–213.

Liu L, Shen GQ, Sun MX, Cao XD, Shang GF, Chen P. 2014. Effect of biochar on nitrous oxide emission and its potential mechanisms. *Journal of the Air and Waste Management Association* 64(8): 894–902.

Liu X-H, Han F-P, Zhang X-C. 2012. Effect of biochar on soil aggregates in the Loess Plateau: Results from incubation experiments. *International Journal of Agriculture and Biology* 14(6): 975–979.

Liu YX, Yang M, Wu YM, Wang HL, Chen YX, Wu WX. 2011. Reducing CH_4 and CO_2 emissions from waterlogged paddy soil with biochar. *Journal of Soils and Sediments* 11: 930–939.

Lowe DC. 2006. A green source of surprise. *Nature* 439: 148–149.

Mukherjee A, Lal R. 2013. Biochar impacts on soil physical properties and greenhouse gas emissions. *Agronomy* 3: 313–339.

Mukherjee A, Lal R. 2014. The biochar dilemma. *Soil Research* 52: 217–230.

Nelissen V, Saha BK, Ruysschaert G, Boeckx P. 2014. Effect of different biochar and fertiliser types on N_2O and NO emissions. *Soil Biology and Biochemistry* 70: 244–255.

Rondon M, Ramirez JA, Lehmann J. 2005. Charcoal additions reduce net emissions of greenhouse gases to the atmosphere. *Proceedings of the 3rd USDA Symposium on Greenhouse Gases & Carbon Sequestration in Agriculture and Forestry,* 21–24 March 2005; Baltimore, MD, pp. 208.

Rondon M, Molina D, Hurtado M, Ramirez J, Lehmann J, Major J, Amezquita E. 2006. Enhancing the productivity of crops and grasses while reducing greenhouse gas emissions through biochar amendments to unfertile tropical soils. *Proceedings of the 18th World Congress of Soil Science*, Philadelphia, PA, USA, 9–15 July 2006.

Rubasinghege G, Spak SN, Stanier CO, Carmichael GR, Grassian VH. 2011. Abiotic mechanism for the formation of atmospheric nitrous oxide from ammonium nitrate. *Environmental Science and Technology* 45: 2691–2697.

Saggar S, Camps-Arbestain M, Leifeld J. 2014. Editorial: Environmental benefits and risks of biochar application to soil. *Agriculture, Ecosystems and Environment* 191: 1–4.

Schimmelpfennig S, Müller C, Grünhage L, Koch C, Kammann C. 2014. Biochar, hydrochar and uncarbonized feedstock application to permanent grassland - effects on greenhouse gas emissions and plant growth. *Agriculture, Ecosystems and Environment* 191: 39–52.

Schmidt CS, Richardson DJ, Baggs EM. 2011. Constraining the conditions conducive to dissimilatory nitrate reduction to ammonium in temperate arable soils. *Soil Biology and Biochemistry* 43: 1607–1611.

Singh BP, Hatton BJ, Singh B, Cowie AL, Kathuria A. 2010. Influence of biochars on nitrous oxide emission and nitrogen leaching from two contrasting soils. *Journal of Environmental Quality* 39: 1224–1235.

Solaiman ZM, Blackwell P, Abbott LK, Storer P. 2010. Direct and residual effect of biochar application on mycorrhizal colonisation, growth and nutrition of wheat. *Australian Journal of Soil Research* 48: 546–554.

Spokas KA, Koskinen WC, Baker JM, Reicosky DC. 2009. Impacts of woodchip biochar additions on greenhouse gas production and sorption/degradation of two herbicides in a Minnesota soil. *Chemosphere* 77: 574–581.

Taghizadeh-Toosi A, Clough TJ, Condron LM, Sherlock RR, Anderson CR, Craigie RA. 2011. Biochar incorporation into pasture soil suppresses in situ nitrous oxide emissions from ruminant urine patches. *Journal of Environmental Quality* 40: 468–476.

Taghizadeh-Toosi A, Clough TJ, Sherlock RR, Condron LM. 2012. Biochar adsorbed ammonia is bioavailable. *Plant and Soil* 350: 57–69.

Uzoma KC, Inoue M, Andry H, Fujimaki H, Zahoor A, Nishihara E. 2011. Effect of cow manure biochar on maize productivity under sandy soil condition. *Soil Use and Management* 27: 205–212.

Van Zwieten L, Kimber S, Morris S, Downie A, Berger E, Rust J, Scheer C. 2010. Influence of biochars on flux of N_2O and CO_2 from Ferrosol. *Australian Journal of Soil Research* 48: 555–568.

Van Zwieten L, Singh B, Joseph S, Kimber S, Cowie A, Chan KY. 2009. Biochar and emissions of non-CO_2 greenhouse gases from soil. In: Lehmann J, Joseph S, eds. *Biochar for Environmental Management—Science and Technology*. Earthscan. London, pp. 227–249.

Van Zwieten L, Singh BP, Kimber SWL, Murphy DV, Macdonald LM, Rust J, Morris S. 2014. An incubation study investigating the mechanisms that impact N_2O flux from soil following biochar application. *Agriculture, Ecosystems and Environment* 191: 53–62.

Verhoeven EC, Six J. 2014. Biochar does not mitigate fieldscale N_2O emissions in a Northern California vineyard: An assessment across two years. *Agriculture, Ecosystems and Environment* 191: 27–38.

Wardle DA, Nilsson M-C, Zackrisson O. 2008. Fire-derived charcoal causes loss of forest humus. *Science* 320(5876): 629.

Xu H-J, Wang X-H, Li H, Yao H-Y, Su J-Q, Zhu Y-G. 2014. Biochar impacts soil microbial community composition and nitrogen cycling in an acidic soil planted with rape. *Environmental Science and Technology* 48: 9391–9399.

Yao Y, Gao B, Zhang M, Inyang M, Zimmerman AR. 2012. Effect of biochar amendment on sorption and leaching of nitrate, ammonium, and phosphate in a sandy soil. *Chemosphere* 89: 1467–1471.

Zhang A, Bian R, Pan G, Cui L, Hussain Q, Li L, Zheng J, et al. 2012. Effects of biochar amendment on soil quality, crop yield and greenhouse gas emission in a Chinese rice paddy: A field study of 2 consecutive rice growing cycles. *Field Crops Research* 127: 153–160.

Zhang A, Cui L, Pan G, Li L, Hussain Q, Zhang X, Zheng J, Crowley D. 2010. Effect of biochar amendment on yield and methane and nitrous oxide emissions from a rice paddy from Tai Lake plain, China. *Agriculture, Ecosystems and Environment* 139: 469–475.

Zhao L, Cao X, Masek O, Zimmerman A. 2013. Heterogeneity of biochar properties as a function of feedstock sources and production temperatures. *Journal of Hazardous Materials* 256–257: 1–9.

Zheng J, Stewart CE, Cotrufo MF. 2012. Biochar and nitrogen fertiliser alters soil nitrogen dynamics and greenhouse gas fluxes from two temperate soils. *Journal of Environmental Quality* 41(5): 1361–1370.

SECTION V

Future Prospects

14

Emerging Applications of Biochar

Chapter 14

Emerging Applications of Biochar

Daniel C.W. Tsang, Jingzi Beiyuan, and Mei Deng

Chapter Outline

14.1 Motivations

To improve the capacity and selectivity of biochar, various physicochemical methods have been recently proposed, such as acid/alkali modification, oxidation, and chemical grafting (Liu et al. 2012; Wang et al. 2011; Yang and Jiang 2014), which can increase the surface area, tailor the surface functionalities, or both. Doping nanomaterials on the surface of biochar is another innovative way to develop nanocomposite sorbents for the removal of organic and inorganic pollutants, such as dyes, heavy metals, nitrates, and phosphates (Inyang et al. 2014; Song et al. 2014; Zhang et al. 2012, 2013a; Zhou et al. 2013).

Although most of the studies have focused on wastewater treatment and site remediation, biochar has recently demonstrated its efficacy for emerging applications. For example, stormwater in urban areas passes the impervious ground surfaces and carries pathogens, suspended solids, nutrients, heavy metals, and organic chemicals. To improve the effluent quality and harvest the stormwater for nonpotable reuse, at-source bioretention and other means of low-impact development have attracted extensive interest (U.S. EPA 2000), in which biochar can be a promising filter medium considering its extraordinary adsorption capacity (Oregon BEST 2014).

Conversely, in situ remediation of contaminated groundwater by permeable reactive barriers also has been demonstrated to be technically feasible (Erto et al. 2011; Natale et al. 2008; Thiruvenkatachari et al. 2008). With a high surface area and abundant oxygen-containing functional groups, biochar is potentially applicable as the reactive medium for groundwater remediation (Ahmad et al. 2014a; Cao et al. 2009; Chen and Chen 2009; Kumar et al. 2011). Thus, this chapter reviews the latest development of biochar modifications and its emerging applications for stormwater harvesting and groundwater remediation.

14.2 Biochar Modifications for Environmental Management

14.2.1 Physical and Chemical Modifications

Physical activation of biochar involves two stages: pyrolysis and oxidation. The biochar is pyrolyzed in the absence of oxygen, followed by oxidation with steam, carbon dioxide (CO_2), air, or their mixtures. In contrast, chemical activation requires the addition of chemical reagents such as a mineral acids or strong bases and can be performed with a shorter time and lower temperature. The modifications of biochars can result in significant effects on their properties, in particular, increasing the surface area and porosity, enriching the surface sorption sites, and introducing specific surface functional groups. Table 14.1 summarizes the improved effectiveness of biochars modified by different chemical methods.

TABLE 14.1
Chemically Modified Biochars and Their Effectiveness

Raw Material	Target Pollutant	Modification Method	Results	Reference
Mongolian scotch pine tree sawdust	Fluoride	Biochar was soaked in phosphoric acid, radiated by microwave, and then filtered and washed.	885 $mg \cdot kg^{-1}$	Guan et al. (2014)
Acacia saligna, blackbutt wood, Eucalyptus pilularis, Eucalyptus marginata (jarrah)	Dissolved organic matter	Biochars were heated with 1 M phosphoric acid or 0.1 M potassium hydroxide (KOH) at 90°C with a solid-to-solution ratio of 1 g:40 mL.	Not available	Lin et al. (2012)
Rice husk	Tetracycline	Biochar was treated with 10% sulfuric acid (H_2SO_4) or 3 M KOH at 1 g:10 mL solid-to-solution ratio.	58.8 $mg \cdot g^{-1}$ for the alkali-modified biochar	Liu et al. (2012)
Peanut hull	Pb(II), Cu(II), Cd(II), Ni(II)	Peanut hull was added into deionized water at a ratio of 60 g:400 mL and heated to 300°C for 5 h at pressure of 1000 psi. Afterward, 3 g of the biochar was placed into 20 mL of 10% hydrogen peroxide solution for 2 h at room temperature.	22.8 $mg \cdot g^{-1}$ for Pb(II); total sorption site density of 0.11 $mmol \cdot g^{-1}$	Xue et al. (2012)
Sawdust	Cu(II)	Biochar (6 g) was treated with 50 mL of concentrated H_2SO_4 and 50 mL of concentrated nitric acid with mechanical stirring and water ice bath for 2 h.	Five- to eight-fold enhancement in batch and fixed-bed experiments	Yang and Jiang (2014)
Bamboo, sugarcane bagasse, hickory wood, peanut hull	Pb(II), Cu(II), Cd(II)	Biochar (3 g) was mixed with 3 g of chitosan in 180 mL of acetic acid (2%), stirred for 30 min, and then transferred into a 900-mL NaOH (1.2%) solution and kept for 12 h.	Removal increased by 150% for Pb(II), 316% for Cd(II), and 233% for Cu(II)	Zhou et al. (2013)

Lin et al. (2012) and Liu et al. (2012) used phosphoric acid-(H_3PO_4), potassium hydroxide- (KOH), and sulfuric acid (H_2SO_4)-modified biochars for the sorption of organic compounds, in which each gram of biochar was activated by 10–40 mL of concentrated acid or alkali. It has been shown that H_3PO_4 activation resulted in a higher porosity (Hared et al. 2007; Patnukao and Pavasant 2008; Yang et al. 2011), whereas KOH activation increased the amount of O-containing functional groups and enhanced the surface area (Evans et al. 1999; Liu et al. 2012; Rouquerol et al. 1999). In addition, the biochars modified with H_3PO_4, KOH, or potassium carbonate could act as a nutrient source because of the slow release of impregnated potassium or phosphate from the biochar surfaces (Agrafioti et al. 2013). Nevertheless, after acid or alkali modifications, the biochars had to be rinsed with an ample amount of deionized water until the acidic/alkaline pH was neutralized, otherwise the functionality and compatibility of biochars would be compromised.

Alternatively, Zhou et al. (2013) modified biochars with chitosan for heavy metal sorption, thereby exploiting the advantages of the relatively large surface area of biochar and the porous network with a high affinity for chitosan. More recently, amino-modified biochar was found to be more efficient and selective for copper adsorption. The amino modification process involved biochar nitration by electrophilic aromatic substitution in the presence of nitric acid and H_2SO_4, followed by reduction of nitrogen-containing groups by sodium dithionite (Yang and Jiang 2014). This modification chemically bound the amino moiety onto the biochar surface, which possessed high stability constants for complexation, and significantly enhanced copper adsorption.

14.2.2 Nanomaterial Impregnations

Another emerging technology for enhancing the performance of biochar is nanomaterial impregnation. Biochar serves as a good platform for doping and immobilizing nanomaterials that can functionalize the surface and increase the selective sorption of target pollutants (e.g., organic dyes, copper, lead, chromate, arsenate, and phosphate) (Inyang et al. 2014; Yao et al. 2013, 2014; Zhang and Gao 2013; Zhou et al. 2014). Table 14.2 summarizes

TABLE 14.2
Nanomaterials-Impregnated Biochars and Their Effectiveness

Raw Material	Target Pollutant	Nanomaterial	Results	Reference
Straws	Cu(II)	MnO_x at different weight ratios of potassium permanganate (2.5, 10, and 60%)	160 mg·g⁻¹	Song et al. (2014)
Bamboo	Pb(II), Cr(VI), As(V), phosphate, methylene blue	Fine-sized zerovalent iron (ZVI < 850 μm) at different mass ratios of bamboo:chitosan:ZVI	Removal efficiency of 93, 95, 96, 28, and 68% for Pb(II), As(V), phosphate, Cr(VI), and methylene blue, respectively	Zhou et al. (2014)
Cotton wood	As(V), methylene blue, phosphate	Aluminium oxide hydroxide by aluminium chloride pyrolysis with biomass	17.4 mg·g⁻¹ for As(V), 8.5 mg·g⁻¹ for methylene blue, 135 mg·g⁻¹ for phosphate	Zhang and Gao (2013)
Tomato tissues (leaves)	Phosphate	Nanosized magnesium oxide and magnesium hydroxide particles through Mg/Ca bioaccumulation	Not available	Yao et al. (2013)
Cotton wood	Phosphate	Mg/Al-layered double hydroxides	410 mg·g⁻¹	Zhang et al. (2013b)
Cotton wood	Methylene blue	γ-Iron(III) oxide by iron(III) chloride pyrolysis with biomass	3.15 mg·g⁻¹	Zhang et al. (2013a)
Bamboo, bagasse, hickory chips	Methylene blue	Clay (montmorillonite and kaolinite)	11.3 mg·g⁻¹	Yao et al. (2014)
Hickory chips, sugarcane bagasse	Methylene blue	Carbon nanotubes at 0.01–1% by weight	Removal efficiency of 64%	Inyang et al. (2014)

the recent studies of biochar modification with nanomaterials, including manganese, magnesium, aluminium, and iron oxide nanoparticles; layered double hydroxides; nano-zerovalent iron; clay minerals; and carbon nanotubes.

The nanomaterials could be added before or after pyrolysis of biochar. The dip-coating procedure was commonly used for impregnating the nanomaterials, in which biochar was added into suspensions of nanomaterials, stirred using a magnetic stirrer or via ultrasonication, and then dried for characterization and use (Inyang et al. 2014; Song et al. 2014). Another synthesis approach was the self-assembly method, wherein the suspension was allowed to age at a certain temperature without stirring (Chen et al. 2004; Zhang et al. 2013b). Multiple times of rinsing with deionized water are usually needed to wash away the loosely bound nanomaterials for ensuring purity and stability of the modified biochars.

Although these modified biochars feature additional functional groups on the surface (e.g., hydroxyls, carboxyls, amines) that enable high sorption affinity and selectivity, which could be comparable to or better than commercial activated carbons (Song et al. 2014), further investigations are required for understanding their performance in the copresence of a wide spectrum of hydrophilic organic pollutants (e.g., herbicides, pesticides, antibiotics) and natural components (e.g., dissolved organic matter, inorganic colloids) in wastewater and soil solutions.

14.3 Stormwater Harvesting by Bioretention Systems

Stormwater is generated during rainfall and snowmelt events and is increasingly viewed as a valuable resource for sustainable water use in low-impact urban development. However, when flowing through the ground surface, stormwater frequently contains a myriad of pollutants, such as pathogens, heavy metals, organic chemicals, nutrients, and suspended solids. The nature and level of pollutants are largely determined by land use and human activities; for example, urbanization leads to increases in the amount and variety of pollutants (Aryal et al. 2010), often making urban rivers fail to fulfill the water quality standards (Mitchell 2005). The need to implement natural and low-cost approaches for mitigating the adverse impact of stormwater is becoming increasingly recognized.

The U.S. Environmental Protection Agency recommends a bioretention system (also known as bioinfiltration) for the low-impact development practice for stormwater runoff management. A bioretention cell (Figure 14.1) is typically composed of a combination of components (e.g., top vegetation, mulch layer, modified soil layer, aggregate base, subdrain) with various functions of removing pollutants and attenuating stormwater runoff that are designed based on soil type, site condition, and land use (U.S. EPA 2000). Bioretention systems are generally small, aesthetically pleasing, and able to achieve a range of stormwater management objectives through sedimentation, sorption, filtration, biodegradation, and plant uptake (Hatt et al. 2009; Li et al. 2009).

Among the components of bioretention systems, the soil and filter media play the most important roles in contaminant removal. Gravel-based, sand-based, and soil-based filter media are usually deployed with different depths depending on inflow volumes and plant species (Hipp et al. 2006).

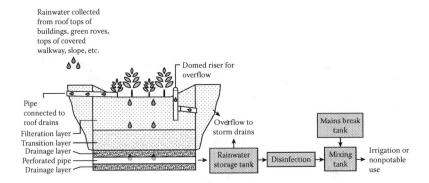

Figure 14.1 Bioretention system for stormwater harvesting. (From China Water Risk. 2012. *Harvesting Hong Kong's Rain*. Retrieved July 9, 2014, from http://chinawaterrisk.org/opinions/harvesting-hong-kongs-rain/.)

Because biochar has a high sorption capacity for heavy metals and organic chemicals, it can potentially be incorporated into the soil or sand filter media for adsorbing pollutants and retaining pathogenic microbes from the stormwater harvested through the bioretention systems. Oregon BEST (2014) reported the successful removal of nearly 100% of the zinc and copper in stormwater by a new filter media that included a mixture of waste biochar produced from lumber mills and hazelnut shells, compost, and oyster shells. In addition to the improvement in the water quality, the waste biochar–based filter media were estimated to cost 60% less than activated carbon filter systems (Oregon BEST 2014).

Research studies have also revealed that adding biochar to the soil or sand filter media can achieve higher effectiveness on the removal of bacteria from stormwater. Mohanty et al. (2014) showed that biochar-amended sand retained up to three orders of magnitude more *Escherichia coli* than original sand filter and that it reduced the detachment and remobilization of *E. coli* during intermittent infiltration of stormwater. In the biochar-amended sandy soil columns, biochar pyrolyzed at high temperature (700°C) resulted in a significant increase in *E. coli* retention compared to biochar pyrolyzed at a low temperature (300°C) (Abit et al. 2012; Bolster and Abit 2012). The removal efficiency of *E. coli* increased with an increasing mass ratio of biochar to sand in the filter media, where five orders of magnitude increase could be achieved compared to biochar-free sandy soil. The effectiveness of biochar incorporation for the removal of *E. coli* from stormwater was attributed to the highly porous structure of biochar, which could augment the overall attachment sites for bacteria in the soil or sand filter media (Mohanty et al. 2014). Furthermore, biochar elevated the ionic strength of the solution and reduced the thickness of the electrical double layer around bacteria and soil particles (Abit et al. 2012; Bolster and Abit 2012). Thus, according to Derjaguin–Landau–Verwey–Overbeek theory, the reduced electrostatic repulsion between bacteria and filter media facilitated the deposition of bacteria (Elimelech et al. 1995).

14.4 Groundwater Remediation by Permeable Reactive Barriers

Permeable reactive barriers (PRBs) are one of the most promising in situ remediation technologies for contaminated groundwater. This technology involves an emplacement of reactive media in the subsurface to intercept a contaminated plume and immobilize or transform the contaminants into environmentally acceptable forms for attaining remediation goals downgradient of the barriers (U.S. EPA 1998). Therefore, PRBs have an outstanding advantage in exploiting the natural hydraulic gradient as a passive and in-place treatment approach that does not require energy or chemical consumption or extraction and transportation (Hashim et al. 2011; Natale et al. 2008; Thiruvenkatachari et al. 2008). It is important to ensure that the hydraulic conductivity of reactive materials used in the PRBs is higher than the surrounding soils, such that contaminated groundwater would pass through the barriers spontaneously. With a prudent and proper design based on hydrogeological and chemical properties, only periodic replacement of the reactive materials is needed upon media aging or clogging by surface passivation (mineral precipitation, microbial growth, or gas accumulation); hence, the operation and maintenance costs are very low and competitive compared to other remediation technologies (Erto et al. 2011, 2013; Thiruvenkatachari et al. 2008). Figure 14.2 illustrates the basic design of PRBs.

The selection of reactive media depends on the targeted pollutants in groundwater. Zerovalent iron has been the most widely used media because of its strong reductive power and low cost as an industrial by-product; however, its nonselective reduction generates unwanted solid precipitation and gas release, reducing the porosity and conductivity as well as shortening the longevity of PRBs (Erto et al. 2011; Natale et al. 2008). Therefore, alternative materials with strong adsorption affinity and capacity have also attracted attention. Activated carbon is considered as a suitable adsorbent for PRBs because of its high surface area ($\sim$1000 $m^2 \cdot g^{-1}$) and excellent adsorption of a wide range of organic and inorganic

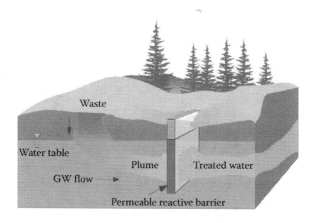

Figure 14.2 Basic design of permeable reactive barriers. (From U.S. EPA. 1998. *Permeable Reactive Barrier Technologies for Contaminant Remediation.* EPA/600/R-98/125, U.S. EPA, Washington, DC.)

pollutants (Erto et al. 2011; Thiruvenkatachari et al. 2008). Natale et al. (2008) explored the possible use of activated carbon in PRBs for removing cadmium from a shallow aquifer. The results of laboratory tests and modeling calculations indicated that the activated carbon PRBs effectively prevented the river pollution. Kalinovich et al. (2011) showcased another successful design and construction of activated carbon PRBs for cleaning up polychlorinated biphenyls contamination at a former military defense site on the summit of Resolution Island, Nunavut. Erto et al. (2011) further suggested that various low-cost materials such as zeolite, fly ash, and other industrial by-products can be used as the media of PRBs.

Biochar, as an effective adsorbent similar to activated carbon, is potentially a good candidate for the application of PRBs. As a by-product of biomass-to-energy pyrolysis and carbon sequestration against climate change, biochar is more economical and environmentally friendly than activated carbon because no extra chemicals or energy is consumed for activation processes (Ahmad et al. 2012; Cao et al. 2009; Lehmann 2007; Zhang et al. 2013c). The aromatic carbon structure and high surface-to-volume ratio of biochar contribute to its strong potential affinity to the adsorption of hydrophobic, nonpolar organic compounds (Ahmad et al. 2014a; Cao et al. 2009; Chen and Chen 2009; Chen et al. 2008; Mohan et al. 2014).

In addition, biochar has an array of surface functional groups (e.g., carboxyls, phenols), high cation exchange capacity, and alkaline pH, all of which are favorable surface characteristics for immobilization of heavy metals, inorganic pollutants, and hydrophilic polar organic compounds (Beesley and Marmiroli 2011; Karami et al. 2011; Kumar et al. 2011; Liu and Zhang 2009; Rajapaksha et al. 2014; Xie et al. 2014; Yao et al. 2012).

For example, the importance of deprotonation of surface oxygen-containing groups on biochar was demonstrated for lead sorption via electrostatic interactions and surface complexation in an acidic pH range (Liu and Zhang 2009). In contrast, in an alkaline pH range, the formation of stable lead hydroxide and lead phosphate precipitates (due to high pH of biochar and its release of soluble phosphorus) played a predominant role in lead sequestration as confirmed by spectroscopic investigations (Ahmad et al. 2014b; Cao et al. 2009). More importantly, biochar has been shown to be useful as a PRB medium for remediation of groundwater contaminated by radioactive uranium (Kumar et al. 2011). However, the effectiveness highly depended on the solution pH; maximum sorption was found at pH 5.9, whereas at alkaline pH uranium formed soluble cationic complexes and the biochar became protonated and positively charged, as indicated by speciation calculation.

Regarding the sorption of nonpolar (e.g., naphthalene, trichloroethylene) and polar (e.g., 1-naphthol) organic pollutants from water, the reduction of polar surface functional groups and the increase of surface area with increasing pyrolysis temperature were shown to be the most critical factors (Ahmad et al. 2012, 2013; Chen and Chen 2009; Qiu et al. 2009). For the hydrophobic organics, the surface area and coverage were important for the π–π interactions with the aromatic carbon structure, whereas for the polar and ionic organics, the oxygen- or hydrogen-containing functional groups on the biochar surface determined the extent of electrostatic attraction and repulsion and intermolecular hydrogen bonding (Ahmad et al. 2014a; Qiu et al. 2009; Xu et al. 2011; Yuan et al. 2011). Therefore, in view of the abovementioned findings, the biochars for PRBs should be tailored (by pyrolysis conditions, chemical modifications, or both) for the specific pollutants of concern in the contaminated groundwater, a research area that warrants further investigation.

14.5 The Way Forward

Motivated by the strong adsorption capacity of biochar, extensive studies have investigated the applications of biochar in environmental remediation of contaminated soils and industrial and municipal wastewaters. In the literature, two operating conditions of biomass pyrolysis are believed to be most influential for controlling the physicochemical properties of biochars: feedstock and pyrolysis temperature. This chapter reviews the latest findings and shows that physical and chemical modifications and nanomaterial impregnations can significantly enhance the sorption capacity and selectivity of biochars, rendering the biochars comparable to or even more versatile than commercial activated carbons. More importantly, cost-effective and environmentally friendly biochars can serve as an excellent candidate for new applications for stormwater harvesting by bioretention systems and groundwater remediation by PRBs. These emerging applications require the biochars to be specifically tailored for sequestering the contaminants of concern in the stormwater or groundwater; therefore, outcome-based modifications and the full potential of tailor-made biochars in wide-ranging environmental technologies are waiting to be explored and applied in the future.

References

Abit, S. M., Bolster, C. H., Cai, P. and Walker, S. L. 2012. Influence of feedstock and pyrolysis temperature of biochar amendments on transport of *Escherichia coli* in saturated and unsaturated soil. *Environmental Science and Technology* 46, 8097–8105.

Agrafioti, E., Bouras, G., Kalderis, D. and Diamadopoulos, E. 2013. Biochar production by sewage sludge pyrolysis. *Journal of Analytical and Applied Pyrolysis* 101, 72–78.

Ahmad, M., Lee, S. S., Dou, X., Mohan, D., Sung, J., Yang, J. E. and Ok, Y. S. 2012. Effects of pyrolysis temperature on soybean stover- and peanut shell-derived biochar properties and TCE adsorption in water. *Bioresource Technology* 118, 536–544.

Ahmad, M., Lee, S. S., Lim, J. E., Lee, S. E., Cho, J. S., Moon, D. H., Hashimoto, Y. and Ok, Y. S. 2014b. Speciation and phytoavailability of lead and antimony in a small arms range soil amended with mussel shell, cow bone and biochar: EXAFS spectroscopy and chemical extractions. *Chemosphere* 95, 433–441.

Ahmad, M., Lee, S. S., Rajapaksha, A. U., Vithanage, M., Zhang, M., Cho, J. S., Lee, S. E. and Ok, Y. S. 2013. Trichloroethylene adsorption by pine needle biochars produced at various pyrolysis temperatures. *Bioresource Technology* 143, 615–622.

Ahmad, M., Rajapaksha, A. U., Lim, J. E., Zhang, M., Bolan, N., Mohan, D., Vithanage, M., Lee, S. S. and Ok, Y. S. 2014a. Biochar as a sorbent for contaminant management in soil and water: A review. *Chemosphere* 99, 19–33.

Aryal, R., Vigneswaran, S., Kandasamy, J. and Naidu, R. 2010. Urbanstormwater quality and treatment. *Korean Journal of Chemical Engineering* 27(5), 1343–1359.

Beesley, L. and Marmiroli, M. 2011. The immobilisation and retention of soluble arsenic, cadmium and zinc by biochar. *Environmental Pollution* 159, 474–480.

Bolster, C. H. and Abit, S. M. 2012. Biochar pyrolyzed at two temperatures affects *Escherichia coli* transport through a sandy soil. *Journal of Environmental Quality* 41, 124–133.

Cao, X., Ma, L., Gao, B. and Harris, W. 2009. Dairy-manure derived biochar effectively sorbs lead and atrazine. *Environmental Science Technology* 43, 3285–3291.

Chen, B. and Chen, Z. 2009. Sorption of naphthalene and 1-naphthol by biochars of orange peels with different pyrolytic temperatures. *Chemosphere* 76, 127–133.

Chen, B., Zhou, D. and Zhu, L. 2008. Transitional adsorption and partition of nonpolar and polar aromatic contaminants by biochars of pine needles with different pyrolytic temperatures. *Environmental Science Technology* 42, 5137–5143.

Chen, W. and Qu, B. J. 2004. LLDPE/ZnAlLDH-exfoliated nanocomposites: Effects of nanolayers on thermal and mechanical properties. *Journal of Materials Chemistry* 14, 1705–1710.

China Water Risk. 2012. *Harvesting Hong Kong's Rain*. Retrieved July 9, 2014, from http://chinawaterrisk.org/opinions/harvesting-hong-kongs-rain/.

Elimelech, M., Gregory, J., Jia, X. and Williams, R. A. 1995. *Particle Deposition and Aggregation: Measurement, Modeling and Simulation*. Butterworth-Heinemann, Woburn, MA.

Erto, A., Lancia, A., Bortone, I., Nardo, A. D., Natale, M. D. and Musmarra, D. 2011. A procedure to design a permeable adsorptive barrier (PAB) for contaminated groundwater remediation. *Journal of Environmental Management* 92, 23–30.

Erto, A., Giraldo, L., Lancia, A. and Moreno-Piraján, J. C. 2013. A comparison between a low-cost sorbent and an activated carbon for the adsorption of heavy metals from water. *Water Air Soil Pollution* 224, 1531–1540.

Evans, M. J. B., Halliop, E. and MacDonald, J. A. F. 1999. The production of chemically activated carbon. *Carbon* 37, 269–274.

Guan, X., Zhou, J., Ma, N., Chen, X., Gao, J. and Zhang, R. 2014. Studies on modified conditions of biochar and the mechanism for fluoride removal. *Desalination and Water Treatment*, pp. 1–8.

Hared, I. A., Dirion, J. L., Salvador, S., Lacroix, M. and Rio, S. 2007. Pyrolysis of wood impregnated with phosphoric acid for the production of activated carbon: Kinetics and porosity development studies. *Journal of Analytical and Applied Pyrolysis* 79, 101–105.

Hashim, M. A., Mukhopadhyay, S., Sahu, J. N. and Sengupta, B. 2011. Remediation technologies for heavy metal contaminated groundwater. *Journal of Environmental Management* 92, 2355–2388.

Hatt, E., Fletcher, D. and Deletic, A. 2009. Hydrologic and pollutant removal performance of stormwater biofiltration systems at the field scale. *Hournal of Hydrology* 365, 310–321.

Hipp, J. A., Oqunseitan, O., Lejano, R. and Smith, C. S. 2006. Optimization of stormwater filtration at the urban/watershed interface. *Environmental Science and Technology* 40(15), 4794–4801.

Inyang, M., Gao, B., Zimmerman, A., Zhang, M. and Chen, H. 2014. Synthesis, characterization, and dye sorption ability of carbon nanotube–biochar nanocomposites. *Chemical Engineering Journal* 236, 39–46.

Kalinovich, I. K., Rutter, A., Rowe, R. K. and Poland, J. S. 2011. Design and application of surface PRBs for PCB remediation in the Canadian Arctic. *Journal of Environmental Management* 101, 124–133.

Karami, N., Clemente, R., Moreno-Jimenez, E., Lepp, N. W. and Beesley, L. 2011. Efficiency of green waste compost and biochar soil amendments for reducing lead and copper mobility and uptake to ryegrass. *Journal of Hazardous Materials* 191, 41–48.

Kumar, S., Loganathan, V. A., Gupta, R. B. and Barnett, M. O. 2011. An assessment of U(VI) removal from groundwater using biochar produced from hydrothermal carbonization. *Journal of Environmental Management* 92, 2504–2512.

Lehmann, J. 2007. A handful of carbon. *Nature* 447(10), 143–144.

Li, H., Sharkey, L. J., Hunt, W. F. and Davis, A. P. 2009. Mitigation of impervious surface hydrology using bioretention in North Carolina and Maryland. *Journal of Hydrologic Engineering* 14, 407–415.

Lin, Y., Munroe, P., Joseph, S., Henderson, R. and Ziolkowski, A. 2012. Water extractable organic carbon in untreated and chemical treated biochars. *Chemosphere* 87, 151–157.

Liu, P., Liu, W. J., Jiang, H., Chen, J. J., Li, W. W. and Yu, H. Q. 2012. Modification of bio-char derived from fast pyrolysis of biomass and its application in removal of tetracycline from aqueous solution. *Bioresource Technology* 121, 235–240.

Liu, Z. and Zhang, F. 2009. Removal of lead from water using biochars prepared from hydrothermal liquefaction of biomass. *Journal of Hazardous Materials* 167, 933–939.

Mitchell, G. 2005. Mapping hazard from urban non-point pollution: A screening model to support sustainable urban drainage planning. *Journal of Environmental Management* 74(1), 1–9.

Mohan, D., Sarswat, A., Ok, Y. S. and Pittman, C. U., Jr. 2014. Organic and inorganic contaminants removal from water with biochar, a renewable, low cost and sustainable adsorbent—A critical review. *Bioresource Technology* 160, 191–202.

Mohanty, S. K., Cantrell, K. B., Nelson, K. L. and Boehm, A. B. 2014. Efficacy of biochar to remove *Escherichia coli* from stormwater under steady and intermittent flow. *Water Research* 61, 288–296.

Natale, F. D., Natale, M. D., Greco, R., Lancia, A., Laudante, C. and Musmarra, D. 2008. Groundwater protection from cadmium contamination by permeable reactive barriers. *Journal of Hazardous Materials* 160, 428–434.

Oregon BEST. 2014. *Using Waste Biochar to Clean Stormwater.* Retrieved July 9, 2014, from http://oregonbest.org/news-events/news/item/news/News/action/detail/story/using-lumber-mill-waste-biochar-to-clean-stormwater/.

Patnukao, P. and Pavasant, P. 2008. Activated carbon from *Eucalyptus camaldulensis* Dehn bark using phosphobic acid activation. *Bioresource Technology* 99, 8540–8543.

Qiu, Y., Zheng, Z., Zhou, Z. and Sheng, G. D. 2009. Effectiveness and mechanisms of dye adsorption on a straw-based biochar. *Bioresource Technology* 100, 5348–5351.

Rajapaksha, A. U., Vithanage, M., Lim, J. E., Ahmad, M., Zhang, M., Lee, S. S. and Ok, Y. S. 2014. Invasive plant-derived biochar inhibits sulfamethazine uptake by lettuce in soil. *Chemosphere* 111, 500–504.

Rouquerol, F., Rouquerol, I. and Sing, K. 1999. Adsorption by Powders and Porous Solids. Academic Press, London, UK.

Song, Z., Lian, F., Yu, Z., Zhu, L., Xing, B. and Qiu, W. 2014. Synthesis and characterization of a novel MnOx-loaded biochar and its adsorption properties for Cu^{2+} in aqueous solution. *Chemical Engineering Journal* 242, 36–42.

Thiruvenkatachari, R., Vigneswaran, S. and Naidu, R. 2008. Permeable reactive barrier for groundwater remediation. *Journal of Industrial and Engineering Chemistry* 14, 145–156.

U.S. EPA. 1998. *Permeable Reactive Barrier Technologies for Contaminant Remediation*. EPA/600/R-98/125, U.S. EPA, Washington, DC.

U.S. EPA. 2000. *Low Impact Development (LID): A Literature Review*. EPA-841-B-00-005, U.S. EPA, Washington, DC.

Wang, C., Feng, C., Gao, Y., Ma, X., Wu, Q. and Wang, Z. 2011. Preparation of a graphene-based magnetic nanocomposite for the removal of an organic dye from aqueous solution. *Chemical Engineering Journal* 173, 92–97.

Xie, T., Reddy, K. R., Wang, C., Yargicoglu, E. and Spokas, K. 2014. Characteristics and applications of biochar for environmental remediation: A review. *Critical Reviews in Environmental Science and Technology* 45, 939–969.

Xu, R. K., Xiao, S. C., Yuan, J. H. and Zhao, A. Z. 2011. Adsorption of methyl violet from aqueous solutions by the biochars derived from crop residues. *Bioresource Technology* 102, 10293–10298.

Xue, Y., Gao, B., Yao, Y., Inyang, M., Zhang, M., Zimmerman, A. R. and Ro, K. S. 2012. Hydrogen peroxide modification enhances the ability of biochar (hydrochar) produced from hydrothermal carbonization of peanut hull to remove aqueous heavy metals: Batch and column tests. *Chemical Engineering Journal* 200–202, 673–680.

Yang, G. X. and Jiang, H. 2014. Amino modification of biochar for enhanced adsorption of copper ions from synthetic wastewater. *Water Research* 48, 396–405.

Yang, R., Liu, G., Xu, X., Li, M., Zhang, J. and Hao, X. 2011. Surface texture, chemistry and adsorption properties of acid blue 9 of hemp (Cannabis sativa L.) bast-based activated carbon fibers prepared by phosphoric acid activation. *Biomass and Bioenergy* 35, 437–445.

Yao, Y., Gao, B., Chen, H., Jiang, L., Inyang, M., Zimmerman, A. R., Cao, X., Yang, L., Xue, Y. and Li, H. 2012. Adsorption of sulfamethoxazole on biochar and its impact on reclaimed water irrigation. *Journal of Hazardous Materials* 209–210, 408–413.

Yao, Y., Gao, B., Chen, J., Zhang, M., Inyang, M., Li, Y., Ashok Alva, A. and Yang, L. 2013. Engineered carbon (biochar) prepared by direct pyrolysis of Mg-accumulated tomato tissues: Characterization and phosphate removal potential. *Bioresource Technology* 138, 8–13.

Yao, Y., Gao, B., Fang, J., Zhang, M., Chen H., Zhou, Y., Creamer, A. E., Sun, Y. and Yang, L. 2014. Characterization and environmental applications of clay–biochar composites. *Chemical Engineering Journal* 242, 136–143.

Yuan, J. H., Xu, R. K. and Zhang, H. 2011. The forms of alkalis in the biochar produced from crop residues at different temperatures. *Bioresource Technology* 102, 3488–3497.

Zhang, M. and Gao, B. 2013. Removal of arsenic, methylene blue, and phosphate by biochar/AlOOH nanocomposite. *Chemical Engineering Journal* 226, 286–292.

Zhang, M., Gao, B., Varnoosfaderani, S., Hebard, A., Yao, Y. and Inyang, M. 2013a. Preparation and characterization of a novel magnetic biochar for arsenic removal. *Bioresource Technology* 130, 457–462.

Zhang, M., Gao, B., Yao, Y. and Inyang, M. 2013b. Phosphate removal ability of biochar/MgAl-LDH ultra-fine composites prepared by liquid-phase deposition. *Chemosphere* 92, 1042–1047.

Zhang, M., Gao, B., Yao, Y., Xue, Y. and Inyang, M. 2012. Synthesis of porous MgO-biochar nanocomposites for removal of phosphate and nitrate from aqueous solutions. *Chemical Engineering Journal* 210, 26–32.

Zhang, W., Mao, S., Chen, H., Huang, L. and Qiu, R. 2013c. Pb(II) and Cr(VI) sorption by biochars pyrolyzed from the municipal wastewater sludge under different heating conditions. *Bioresource Technology* 147, 545–552.

Zhou, Y. M., Gao, B., Zimmerman, A. R., Fang, J., Sun, Y. N. and Gao, X. D. 2013. Sorption of heavy metals on chitosan-modified biochars and its biological effects. *Chemical Engineering Journal* 231, 512–518.

Zhou, Y. M., Gao, B., Zimmerman, A. R., Chen, H., Zhang, M. and Cao, X. 2014. Biochar-supported zerovalent iron for removal of various contaminants from aqueous solutions, *Bioresource Technology* 152, 538–542.

Index

Note: Page numbers ending in "f" refer to figures. Page numbers ending in "t" refer to tables.

Printed in the United States
by Baker & Taylor Publisher Services